Atlas of Stress-Strain Curves

Atlas of Stress-Strain Curves

Edited by

Howard E. Boyer

Senior Technical Editor
American Society for Metals

ASM INTERNATIONAL

Metals Park, Ohio 44073

Editorial and production coordination by
Carnes Publication Services, Inc.
Project manager: Edward C. Huddleston

SEP - 3 1988

Library of Congress Catalog Card Number: 86-72116
ISBN: 0-87170-240-1
SAN 204-7586

Preface

Stress-strain curves are a fundamental part of engineering. From one stress-strain curve many important mechanical properties of a metal or alloy can be extrapolated. Tensile strength, breaking strength, yield strength, proportional elastic limit, and other properties can be derived. Because stress-strain curves are so basic to the engineer, and because there has been no real consolidation of these curves in one book, the *Atlas of Stress-Strain Curves* was developed.

The Atlas covers ferrous as well as nonferrous metals, and consists of more than 550 curves. The book is divided into 23 sections covering all the major metals and alloys, including aluminum, copper, nickel, titanium, magnesium, beryllium, zinc, zirconium, and refractory metals. In addition, there are 12 sections devoted to irons and steels. Throughout these various sections, standard alloy designations are used to classify the metals whenever possible. There is one stress-strain curve per page, and each curve is fully referenced. Many of the curves contain, in addition to expanded and descriptive captions, summarized paragraphs of additional information from their original sources, to make each curve as useful and complete as possible.

The introductory chapter outlines and discusses the engineering tension test and how it is widely used to provide basic design information on the strength of materials. Compression, shear, torsion, and bending tests are also covered. True stress-strain as well as engineering stress-strain curves are explained, and the influence of test-procedure variables, temperature, and strain rate results on material flow properties are examined.

This initial collection of curves is the beginning of an ongoing program by the Reference Publications Division of ASM to consolidate and organize stress-strain data. This program will also tie in with the ASM computer software program, and most of the curves in this publication will eventually be computerized for analysis by computer programs. Possible contributors to this data collection program should contact the Editors, Technical Books, ASM, Metals Park, OH 44073.

Contents

Contents

SECTION 2: Low-Carbon Steels   117

SECTION 12: Superalloys   323

SECTION 13: Cast Irons   359

SECTION 14: Aluminum and Aluminum Alloys   374

SECTION 15: Beryllium Alloys   424

SECTION 16: Copper and Copper Alloys   426

Atlas of
Stress-Strain
Curves

Procedures for Stress-Strain Testing

Stress-strain data that are most valuable to design engineers and metallurgists alike may be obtained by several different procedures. Tension testing is the most frequently used approach, but stress-strain data are also obtained by testing in compression, shear, torsion, and bending.

All these methods of testing are covered in this introductory section.

TENSION TESTING

The engineering tension test is widely used to provide basic design information on the strength of materials and as an acceptance test for the specification of materials. In this test procedure, a specimen is subjected to a continually increasing uniaxial load (force), while simultaneous observations are made of the elongation of the specimen. In this article, emphasis is placed on the interpretation of these observations and on the effect of metallurgical variables on mechanical behavior rather than on the procedures for conducting the tests. Subsequent paragraphs in this section discuss the influence of test procedure variables, temperature, and strain rate on test results and material flow properties.

Engineering Stress-Strain Curve

In the conventional engineering tension test, an engineering stress-strain curve is constructed from the load-elongation measurements made on the test specimen (Fig. 1). The engineering stress, s, used in this stress-strain curve is the average longitudinal stress in the tensile specimen. It is obtained by dividing the load, P, by the original area of the cross section of the specimen, A_0:

$$s = \frac{P}{A_0} \qquad \text{(Eq 1)}$$

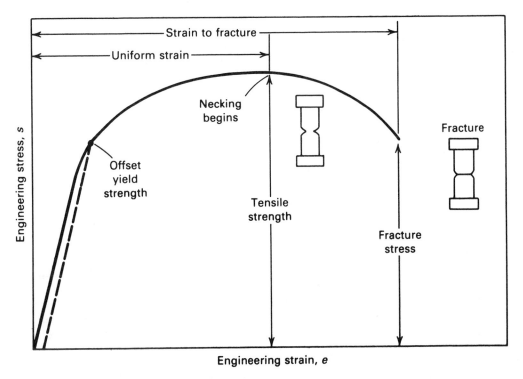

Intersection of the dashed line with the curve determines the offset yield strength. See also Fig. 2 and corresponding text.

Fig. 1 Engineering stress-strain curve

The strain, e, used for the engineering stress-strain curve is the average linear strain, which is obtained by dividing the elongation of the gage length of the specimen, δ, by its original length, L_0:

$$e = \frac{\delta}{L_0} = \frac{\Delta L}{L_0} = \frac{L - L_0}{L_0} \qquad \text{(Eq 2)}$$

Because both the stress and the strain are obtained by dividing the load and elongation by constant factors, the load-elongation curve has the same shape as the engineering stress-strain curve. The two curves frequently are used interchangeably.

The shape and magnitude of the stress-strain curve of a metal depend on its composition, heat treatment, prior history of plastic deformation, and the strain rate, temperature, and state of stress imposed during the testing. The parameters that are used to describe the stress-strain curve of a metal are tensile strength, yield strength or yield point, percent elongation, and reduction in area. The first two are strength parameters; the last two indicate ductility.

The general shape of the engineering stress-strain curve (Fig. 1) requires further explanation. In the elastic region, stress is linearly proportional to strain. When the stress exceeds a value corresponding to the yield strength, the specimen undergoes gross plastic deformation. If the load is subsequently reduced to zero, the specimen will remain permanently deformed. The stress required to produce continued plastic deformation increases with increasing plastic strain; that is, the metal strain hardens. The volume of the specimen (area × length) remains constant during plastic deformation, $AL = A_0 L_0$, and as the specimen elongates, its cross-sectional area decreases uniformly along the gage length.

Initially, the strain hardening more than compensates for this decrease in area, and the engineering stress (proportional to load P) continues to rise with increasing strain. Eventually, a point is reached where the decrease in specimen cross-sectional area is greater than the increase in deformation load arising from strain hardening. This condition will be reached first at some point in the specimen that is slightly weaker than the rest. All further plastic deformation is concentrated in this region, and the specimen begins to neck or thin down locally. Because the cross-sectional area now is decreasing far more rapidly than the deformation load is increased

by strain hardening, the actual load required to deform the specimen falls off and the engineering stress defined in Eq 1 continues to decrease until fracture occurs.

The tensile strength, or ultimate tensile strength, s_u, is the maximum load divided by the original cross-sectional area of the specimen:

$$S_u = \frac{P_{max}}{A_0} \qquad \text{(Eq 3)}$$

The tensile strength is the value most frequently quoted from the results of a tension test. Actually, however, it is a value of little fundamental significance with regard to the strength of a metal. For ductile metals, the tensile strength should be regarded as a measure of the maximum load that a metal can withstand under the very restrictive conditions of uniaxial loading. This value bears little relation to the useful strength of the metal under the more complex conditions of stress that are usually encountered.

For many years, it was customary to base the strength of members on the tensile strength, suitably reduced by a factor of safety. The current trend is to the more rational approach of basing the static design of ductile metals on the yield strength. However, because of the long practice of using the tensile strength to describe the strength of materials, it has become a familiar property, and as such, it is a useful identification of a material in the same sense that the chemical composition serves to identify a metal or alloy. Furthermore, because the tensile strength is easy to determine and is a reproducible property, it is useful for the purposes of specification and for quality control of a product. Extensive empirical correlations between tensile strength and properties such as hardness and fatigue strength are often useful. For brittle materials, the tensile strength is a valid design criterion.

Measures of Yielding. The stress at which plastic deformation or yielding is observed to begin depends on the sensitivity of the strain measurements. With most materials, there is a gradual transition from elastic to plastic behavior, and the point at which plastic deformation begins is difficult to define with precision. In tests of materials under uniaxial loading, three criteria for the initiation of yielding have been used: the elastic limit, the proportional limit, and the yield strength.

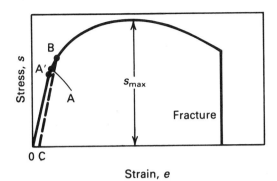

Point A, elastic limit; point A', proportional limit; point B, yield strength; line C-B, offset yield strength; 0, intersection of the stress-strain curve with the strain axis

Fig. 2 Typical tension stress-strain curve for ductile metal indicating yielding criteria

Elastic limit, shown at point A in Fig. 2, is the greatest stress the material can withstand without any measurable permanent strain remaining after the complete release of load. With increasing sensitivity of strain measurement, the value of the elastic limit is decreased until it equals the true elastic limit determined from microstrain measurements. With the sensitivity of strain typically used in engineering studies (10^{-4} in./in.), the elastic limit is greater than the proportional limit. Determination of the elastic limit requires a tedious incremental loading-unloading test procedure. For this reason, it is often replaced by the proportional limit.

Proportional limit, shown at point A' in Fig. 2, is the highest stress at which stress is directly proportional to strain. It is obtained by observing the deviation from the straight-line portion of the stress-strain curve.

The yield strength, shown at point B in Fig. 2, is the stress required to produce a small specified amount of plastic deformation. The usual definition of this property is the offset yield strength determined by the stress corresponding to the intersection of the stress-strain curve and a line parallel to the elastic part of the curve offset by a specified strain (see Fig. 1 and 2). In the United States, the offset is usually specified as a strain of 0.2% or 0.1% ($e = 0.002$ or 0.001):

$$s_0 = \frac{P_{(\text{strain offset} = 0.002)}}{A_0} \qquad \text{(Eq 4)}$$

Offset yield strength determination requires a specimen that has been loaded to its 0.2% offset

yield strength and unloaded so that it is 0.2% longer than before the test. The offset yield strength is often referred to in Great Britain as the proof stress, where offset values are either 0.1% or 0.5%. The yield strength obtained by an offset method is commonly used for design and specification purposes, because it avoids the practical difficulties of measuring the elastic limit or proportional limit.

Some materials, for example, soft copper or gray cast iron, have essentially no linear portion to their stress-strain curve. For these materials, the offset method cannot be used, and the usual practice is to define the yield strength as the stress to produce some total strain, for example, $e = 0.005$.

Many metals, particularly annealed low-carbon steel, show a localized, heterogeneous type of transition from elastic to plastic deformation that produces a yield point in the stress-strain curve. Rather than having a flow curve with a gradual transition from elastic to plastic behavior, such as in Fig. 1 and 2, metals with a yield point produce a flow curve or a load-elongation diagram similar to Fig. 3. The load increases steadily with elastic strain, drops suddenly, fluctuates about some approximately constant value of load, and then rises with further strain.

The load at which the sudden drop occurs is called the upper yield point. The constant load is called the lower yield point, and the elonga-

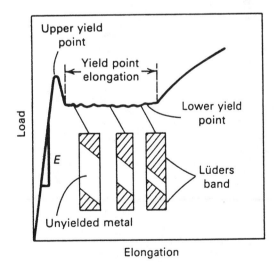

The slope of the initial linear portion of the stress-strain curve, designated by E, is the modulus of elasticity.

Fig. 3 Typical yield point behavior of low-carbon steel

tion that occurs at constant load is called the yield-point elongation. The deformation occurring throughout the yield-point elongation is heterogeneous. At the upper yield point, a discrete band of deformed metal, often readily visible, appears at a stress concentration such as a fillet. Coincident with the formation of the band, the load drops to the lower yield point. The band then propagates along the length of the specimen, causing the yield-point elongation.

In typical cases, several bands form at several points of stress concentration. These bands are generally at approximately 45° to the tensile axis. They are usually called Lüders bands, Hartmann lines, or stretcher strains, and this type of deformation is sometimes referred to as the Piobert effect. When several Lüders bands are formed, the flow curve during the yield-point elongation is irregular, each jog corresponding to the formation of a new Lüders band. After the Lüders bands have propagated to cover the entire length of the specimen test section, the flow will increase with strain in the typical manner. This marks the end of the yield-point elongation. Lüders bands formed on a rimmed 1008 steel are shown in Fig. 4.

Fig. 4 Rimmed 1008 steel with Lüders bands on the surface as a result of stretching the sheet just beyond the yield point during forming

Measures of Ductility. Currently, ductility is considered a qualitative, subjective property of a material. In general, measurements of ductility are of interest in three respects:

- To indicate the extent to which a metal can be deformed without fracture in metalworking operations such as rolling and extrusion.
- To indicate to the designer the ability of the metal to flow plastically before fracture. A high ductility indicates that the material is "forgiving" and likely to deform locally without fracture should the designer err in the stress calculation or the prediction of severe loads.
- To serve as an indicator of changes in impurity level or processing conditions. Ductility measurements may be specified to assess material quality, even though no direct relationship exists between the ductility measurement and performance in service.

The conventional measures of ductility that are obtained from the tension test are the engineering strain at fracture e_f (usually called the elongation) and the reduction in area at fracture q. Elongation and reduction in area are usually expressed as a percentage. Both of these properties are obtained after fracture by putting the specimen back together and taking measurements of the final length, L_f, and final specimen cross section, A_f:

$$e_f = \frac{L_f - L_0}{L_0} \qquad \text{(Eq 5)}$$

$$q = \frac{A_0 - A_f}{A_0} \qquad \text{(Eq 6)}$$

Because an appreciable fraction of the plastic deformation will be concentrated in the necked region of the tension specimen, the value of e_f will depend on the gage length L_0 over which the measurement was taken (see the section of this article on ductility measurement in tension testing). The smaller the gage length, the greater the contribution to the overall elongation from the necked region and the higher the value of e_f. Therefore, when reporting values of percentage elongation, the gage length, L_0, should always be given.

Reduction in area does not suffer from this difficulty. These values can be converted into an equivalent zero-gage-length elongation, e_0.

Table 1   Typical values of modulus of elasticity at different temperatures

Material	Room temperature	Modulus of elasticity, GPa (10^6 psi), at:			
		250 °C (400 °F)	425 °C (800 °F)	540 °C (1000 °F)	650 °C (1200 °F)
Carbon steel	207 (30.0)	186 (27.0)	155 (22.5)	134 (19.5)	124 (18.0)
Austenitic stainless steel	193 (28.0)	176 (25.5)	159 (23.0)	155 (22.5)	145 (21.0)
Titanium alloys	114 (16.5)	96.5 (14.0)	74 (10.7)	70 (10.0)	...
Aluminum alloys	72 (10.5)	65.5 (9.5)	54 (7.8)	...	...

From the constancy of volume relationship for plastic deformation, $AL = A_0L_0$:

$$\frac{L}{L_0} = \frac{A_0}{A} = \frac{1}{1 - q}$$

$$e_0 = \frac{L - L_0}{L_0} = \frac{A_0}{A} - 1 = \frac{1}{1 - q} - 1$$

$$= \frac{q}{1 - q} \qquad \text{(Eq 7)}$$

This represents the elongation based on a very short gage length near the fracture.

Another way to avoid the complications resulting from necking is to base the percentage elongation on the uniform strain out to the point at which necking begins. The uniform elongation, e_u, correlates well with stretch-forming operations. Because the engineering stress-strain curve often is quite flat in the vicinity of necking, it may be difficult to establish the strain at maximum load without ambiguity.

Modulus of Elasticity. The slope of the initial linear portion of the stress-strain curve is the modulus of elasticity, or Young's modulus, as shown in Fig. 3. The modulus of elasticity, E, is a measure of the stiffness of the material. The greater the modulus, the smaller the elastic strain resulting from the application of a given stress. Because the modulus of elasticity is needed for computing deflections of beams and other members, it is an important design value.

The modulus of elasticity is determined by the binding forces between atoms. Because these forces cannot be changed without changing the basic nature of the material, the modulus of elasticity is one of the most structure-insensitive of the mechanical properties. Generally, it is only slightly affected by alloying additions, heat treatment, or cold work. However, increasing the temperature decreases the modulus of elasticity. At elevated temperatures, the modulus is often measured by a dynamic method. Typical values of modulus of elasticity for common engineering metals at different temperatures are given in Table 1.

Resilience. The ability of a material to absorb energy when deformed elastically and to return it when unloaded is called resilience. This property usually is measured by the modulus of resilience, which is the strain energy per unit volume, U_0, required to stress the material from zero stress to the yield stress, σ_0. The strain energy per unit volume for uniaxial tension is:

$$U_0 = \frac{1}{2} \sigma_x e_x \qquad \text{(Eq 8)}$$

From the above definition, the modulus of resilience, U_R, is:

$$U_R = \frac{1}{2} s_0 e_0 = \frac{1}{2} s_0 \frac{s_0}{E} = \frac{s_0^2}{2E} \qquad \text{(Eq 9)}$$

This equation indicates that the ideal material for resisting energy loads in applications where the material must not undergo permanent distortion, such as mechanical springs, is one having a high yield stress and a low modulus of elasticity.

For various grades of steel, the modulus of resilience ranges from 100 to 4500 kJ/m^3 (14.5 to 650 lbf-in./in.^{3}), with the higher values representing steels with higher carbon or alloy contents. The cross-hatched regions in Fig. 5 indicate the modulus of resilience for two steels. Because of its higher yield strength, the high-carbon spring steel has the greater resilience.

The toughness of a material is its ability to absorb energy in the plastic range. The ability to withstand occasional stresses above the yield stress without fracturing is particularly desirable in parts such as freight-car couplings, gears, chains, and crane hooks. Toughness is a commonly used concept that is difficult to define precisely. Toughness may be considered to be the total area under the stress-strain curve. This area, which is referred to as the modulus of toughness, U_T, is an indication of the amount of work per unit volume that can be done on the material without causing it to rupture.

Figure 5 shows the stress-strain curves for high- and low-toughness materials. The high-carbon spring steel has a higher yield strength and tensile strength than the medium-carbon structural steel. However, the structural steel is more ductile and has a greater total elongation. The total area under the stress-strain curve is greater for the structural steel; therefore, it is a tougher material. This illustrates that toughness is a parameter that comprises both strength and ductility.

Several mathematical approximations for the area under the stress-strain curve have been suggested. For ductile metals that have a stress-strain curve like that of the structural steel, the area under the curve can be approximated by:

$$U_T \approx s_u e_f \qquad \text{(Eq 10)}$$

or

$$U_T \approx \frac{s_0 + s_u}{2} e_f \qquad \text{(Eq 11)}$$

For brittle materials, the stress-strain curve is sometimes assumed to be a parabola, and the area under the curve is given by:

$$U_T \approx \frac{2}{3} s_u e_f \qquad \text{(Eq 12)}$$

True Stress/True Strain Curve

The engineering stress-strain curve does not give a true indication of the deformation characteristics of a metal, because it is based entirely on the original dimensions of the specimen, and these dimensions change continuously during the test. Also, ductile metal that is pulled in tension becomes unstable and necks down during the course of the test. Because the cross-sectional area of the specimen is decreasing rapidly at this stage in the test, the load required to continue deformation falls off.

The average stress based on the original area likewise decreases, and this produces the fall-off in the engineering stress-strain curve beyond the point of maximum load. Actually, the metal continues to strain harden to fracture, so that the stress required to produce further deformation

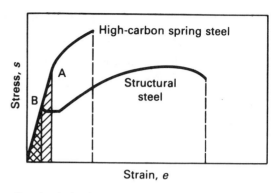

Cross-hatched regions in this curve represent the modulus of resilience, U_R, of the two materials. The U_R is determined by measuring the area under the stress-strain curve up to the elastic limit of the material. Point A represents the elastic limit of the spring steel; point B that of the structural steel.

Fig. 5　Comparison of stress-strain curves for high- and low-toughness steels

should also increase. If the true stress, based on the actual cross-sectional area of the specimen, is used, the stress-strain curve increases continuously to fracture. If the strain measurement is also based on instantaneous measurement, the curve that is obtained is known as true stress/true strain curve. This is also known as a flow curve, because it represents the basic plastic-flow characteristics of the material.

Any point on the flow curve can be considered the yield stress for a metal strained in tension by the amount shown on the curve. Thus, if the load is removed at this point and then reapplied, the material will behave elastically throughout the entire range of reloading. The true stress, σ, is expressed in terms of engineering stress, s, by:

$$\sigma = \frac{P}{A_0}(e + 1) = s(e + 1) \qquad \text{(Eq 13)}$$

The derivation of Eq 13 assumes both constancy of volume and a homogeneous distribution of strain along the gage length of the tension specimen. Thus, Eq 13 should be used only until the onset of necking. Beyond the maximum load, the true stress should be determined from actual measurements of load and cross-sectional area.

$$\sigma = \frac{P}{A} \qquad \text{(Eq 14)}$$

The true strain, ϵ, may be determined from the engineering or conventional strain, e, by:

$$\epsilon = \ln(e + 1) = \ln \frac{L}{L_0} \qquad \text{(Eq 15)}$$

This equation is applicable only to the onset of necking for the reasons discussed above. Beyond maximum load, the true strain should be based on actual area or diameter, D, measurements:

$$\epsilon = \ln \frac{A_0}{A}$$

$$= \ln \frac{\left(\dfrac{\pi}{4}\right) D_0^2}{\left(\dfrac{\pi}{4}\right) D^2} = 2 \ln \frac{D_0}{D} \qquad \text{(Eq 16)}$$

Figure 6 compares the true stress/true strain curve with its corresponding engineering stress-strain curve. Note that, because of the relatively large plastic strains, the elastic region has been compressed into the y axis. In agreement with Eq 13 and 15, the true stress/true strain curve is always to the left of the engineering curve until the maximum load is reached.

However, beyond maximum load, the high, localized strains in the necked region that are used in Eq 16 far exceed the engineering strain calculated from Eq 2. Frequently, the flow curve is linear from maximum load to fracture, while in other cases its slope continuously decreases to fracture. The formation of a necked region or mild notch introduces triaxial stresses that make

it difficult to determine accurately the longitudinal tensile stress from the onset of necking until fracture occurs. This concept is discussed in greater detail in the section of this article on instability in tension. The following parameters are usually determined from the true stress/true strain curve.

The true stress at maximum load corresponds to the true tensile strength. For most materials, necking begins at maximum load at a value of strain where the true stress equals the slope of the flow curve. Let σ_u and ϵ_u denote the true stress and true strain at maximum load when the cross-sectional area of the specimen is A_u. The ultimate tensile strength can be defined as:

$$s_u = \frac{P_{\text{max}}}{A_0} \qquad \text{(Eq 17)}$$

and

$$\sigma_u = \frac{P_{\text{max}}}{A_u} \qquad \text{(Eq 18)}$$

Eliminating P_{max} yields:

$$\sigma_u = s_u \frac{A_0}{A_u} \qquad \text{(Eq 19)}$$

and

$$\sigma_u = s_u e^{\epsilon_u} \qquad \text{(Eq 20)}$$

The true fracture stress is the load at fracture divided by the cross-sectional area at fracture. This stress should be corrected for the triaxial state of stress existing in the tensile specimen at fracture. Because the data required for this correction frequently are not available, true fracture stress values are frequently in error.

The true fracture strain, ϵ_f, is the true strain based on the original area, A_0, and the area after fracture, A_f:

$$\epsilon_f = \ln \frac{A_0}{A_f} \qquad \text{(Eq 21)}$$

This parameter represents the maximum true strain that the material can withstand before fracture and is analogous to the total strain to fracture of the engineering stress-strain curve. Because Eq 15 is not valid beyond the onset of necking, it is not possible to calculate ϵ_f from

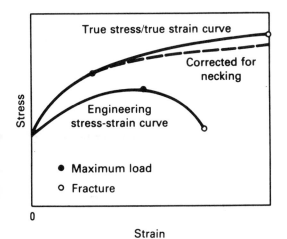

Fig. 6 Comparison of engineering and true stress/true strain curves

measured values of e_f. However, for cylindrical tensile specimens, the reduction in area, q, is related to the true fracture strain by:

$$\epsilon_f = \ln \frac{1}{1 - q} \qquad \text{(Eq 22)}$$

The true uniform strain, ϵ_u, is the true strain based only on the strain up to maximum load. It may be calculated from either the specimen cross-sectional area, A_u, or the gage length, L_u, at maximum load. Equation 15 may be used to convert conventional uniform strain to true uniform strain. The uniform strain frequently is useful in estimating the formability of metals from the results of a tension test:

$$\epsilon_u = \ln \frac{A_0}{A_u} \qquad \text{(Eq 23)}$$

The true local necking strain, ϵ_n, is the strain required to deform the specimen from maximum load to fracture:

$$\epsilon_n = \ln \frac{A_u}{A_f} \qquad \text{(Eq 24)}$$

The flow curve of many metals in the region of uniform plastic deformation can be expressed by the simple power curve relation:

$$\sigma = K\epsilon^n \qquad \text{(Eq 25)}$$

where n is the strain-hardening exponent, and K is the strength coefficient. A log-log plot of true stress and true strain up to maximum load will result in a straight line if Eq 25 is satisfied by the data (Fig 7).

The linear slope of this line is n, and K is the true stress at $\epsilon = 1.0$ (corresponds to $q = 0.63$). As shown in Fig. 8, the strain-hardening exponent may have values from $n = 0$ (perfectly plastic solid) to $n = 1$ (elastic solid). For most metals, n has values between 0.10 and 0.50 (see Table 2).

The rate of strain hardening $d\sigma/d\epsilon$ is not identical to the strain-hardening exponent. From the definition of n:

$$n = \frac{d(\log \sigma)}{d(\log \epsilon)} = \frac{d(\ln \sigma)}{d(\ln \epsilon)} = \frac{\epsilon}{\sigma} \frac{d\sigma}{d\epsilon}$$

or

$$\frac{d\sigma}{d\epsilon} = n \frac{\sigma}{\epsilon} \qquad \text{(Eq 26)}$$

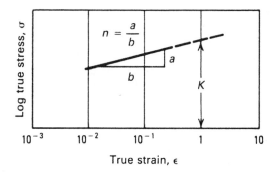

n is the strain-hardening exponent; *K* is the strength coefficient.

Fig. 7 Log-log plot of true stress/true strain curve

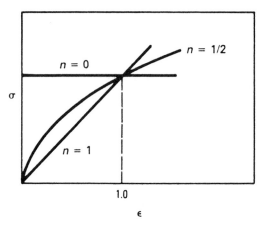

Fig. 8 Various forms of power curve $\sigma = K\epsilon^n$

Deviations from Eq 25 frequently are observed, often at low strains (10^{-3}) or high strains ($\epsilon \approx 1.0$). One common type of deviation is for a log-log plot of Eq 25 to result in two straight lines with different slopes. Sometimes data that do not plot according to Eq 25 will yield a straight line according to the relationship:

$$\sigma = K(\epsilon_0 + \epsilon)^n \qquad \text{(Eq 27)}$$

ϵ_0 can be considered to be the amount of strain hardening that the material received prior to the tension test. Another common variation on Eq 25 is the Ludwik equation:

$$\sigma = \sigma_0 + K\epsilon^n \qquad \text{(Eq 28)}$$

where σ_0 is the yield stress, and K and n are the same constants as in Eq 25. This equation may be more satisfying than Eq 25, because the latter implies that at zero true strain the stress is zero. It has been shown that σ_0 can be obtained from the intercept of the strain-hardening por-

Table 2 Values for *n* and *K* for metals at room temperature

Metal	Condition	*n*	*K* MPa	*K* ksi
0.05% carbon steel	Annealed	0.26	530	77
SAE 4340 steel	Annealed	0.15	641	93
0.6% carbon steel	Quenched and tempered at 540 °C (1000 °F)	0.10	1572	228
0.6% carbon steel	Quenched and tempered at 705 °C (1300 °F)	0.19	1227	178
Copper	Annealed	0.54	320	46.4
70/30 brass	Annealed	0.49	896	130

tion of the stress-strain curve and the elastic modulus line by:

$$\sigma_0 = \left(\frac{K}{E^n}\right)^{1/1-n} \qquad \text{(Eq 29)}$$

The true stress/true strain curve of metals such as austenitic stainless steel, which deviate markedly from Eq 25 at low strains, can be expressed by:

$$\sigma = K\epsilon^n + e^{K_1}e^{n_1\epsilon} \qquad \text{(Eq 30)}$$

where e^{K_1} is approximately equal to the proportional limit, and n_1 is the slope of the deviation of stress from Eq 25 plotted against ϵ. Other expressions for the flow curve are available. The true strain term in Eq 25 to 28 properly should be the plastic strain, $\epsilon_p = \epsilon_{total} - \epsilon_E = \epsilon_{total} - \sigma/E$, where ϵ_E represents elastic strain.

Instability in Tension

Necking generally begins at maximum load during the tensile deformation of a ductile metal. An exception to this is the behavior of cold rolled zirconium tested at 200 to 370 °C (390 to 700 °F), where necking occurs at a strain of twice the strain at maximum load. An ideal plastic material in which no strain hardening occurs would become unstable in tension and begin to neck as soon as yielding occurred. However, an actual metal undergoes strain hardening, which tends to increase the load-carrying capacity of the specimen as deformation increases.

This effect is opposed by the gradual decrease in the cross-sectional area of the specimen as it elongates. Necking or localized deformation begins at maximum load, where the increase in stress due to decrease in the cross-sectional area of the specimen becomes greater than the increase in the load-carrying ability of the metal due to strain hardening. This condition of instability leading to localized deformation is defined by the condition $dP = 0$:

$$P = \sigma A \qquad \text{(Eq 31)}$$

$$dP = \sigma dA + A d\sigma = 0 \qquad \text{(Eq 32)}$$

From the constancy-of-volume relationship:

$$\frac{dL}{L} = -\frac{dA}{A} = d\epsilon \qquad \text{(Eq 33)}$$

and from the instability condition:

$$-\frac{dA}{A} = \frac{d\sigma}{\sigma} \qquad \text{(Eq 34)}$$

so that at a point of tensile instability:

$$\frac{d\sigma}{d\epsilon} = \sigma \qquad \text{(Eq 35)}$$

Therefore, the point of necking at maximum load can be obtained from the true stress/true strain curve by finding the point on the curve having a subtangent of unity (Fig. 9a), or the point where the rate of strain hardening equals the stress (Fig. 9b). The necking criterion can be expressed more explicitly if engineering strain is used. Starting with Eq 35:

$$\frac{d\sigma}{d\epsilon} = \frac{d\sigma}{de}\frac{de}{d\epsilon} = \frac{d\sigma}{de}\frac{\dfrac{dL}{L_0}}{\dfrac{dL}{L}} = \frac{d\sigma}{de}\frac{L}{L_0}$$

$$= \frac{d\sigma}{de}(1 + e) = \sigma$$

$$\frac{d\sigma}{de} = \frac{\sigma}{1 + e} \qquad \text{(Eq 36)}$$

Equation 36 permits an interesting geometrical construction for the determination of the point

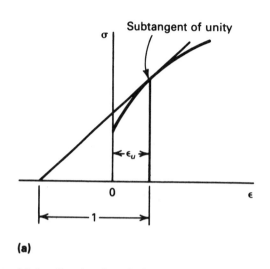

(a)

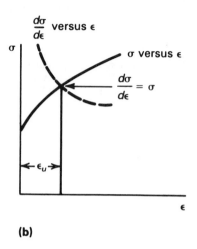

(b)

The point of necking at maximum load can be obtained from the true stress/true strain curve by finding (a) the point on the curve having a subtangent of unity or (b) the point where $d\sigma/d\epsilon = \sigma$.

Fig. 9   Graphical interpretation of necking criterion

of maximum load. In Fig. 10, the stress-strain curve is plotted in terms of true stress against conventional linear strain. Let point A represent a negative strain of 1.0. A line drawn from point A, which is tangent to the stress-strain curve, will establish the point of maximum load, because according to Eq 36, the slope at this point is $\sigma/(1 + e)$.

By substituting the necking criterion given in Eq 35 into Eq 26, a simple relationship for the strain at which necking occurs is obtained. This strain is the true uniform strain, e_u:

$$\epsilon_u = n \qquad \text{(Eq 37)}$$

Although Eq 26 is based on the assumption that the flow curve is given by Eq 25, it has been shown that $\epsilon_u = n$ does not depend on this power law behavior.

SHEAR TESTING

Shear data are primarily used in the design of mechanically fastened components, shear webs, torsion members, and other structural components subject to the application of parallel, opposing loads. Although industry-approved standards do exist for the shear testing of fasteners, adhesive joints, and composites, standards do not currently exist for the development of shear data on mill products such as forgings, extrusions, sheet, and plate.

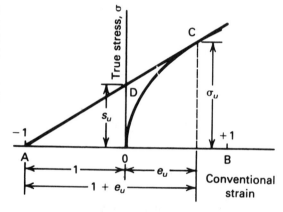

Fig. 10   Considére's construction for the determination of the point of maximum load

Double-Shear Tests

Shear testing for the development of engineering data on mill products is most commonly performed using the double-shear test. An advantage of the double-shear test is that it uses easily machined solid round bars as test specimens. As such, the test can be used on mill products with thicknesses of 4.8 mm (3/16 in.) or more. For sheet less than 1.8 mm (0.071 in.) thick, the blanking-shear test can be used.

The action of the shearing blades and the test procedures used in double-shear testing of mill products are similar to those used in the testing of fasteners, except that additional controls on

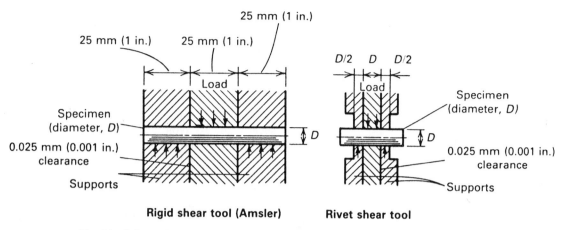

Rigid shear tool (Amsler)   **Rivet shear tool**

Fig. 11 Schematic representation of rigid Amsler and rivet double-shear tooling

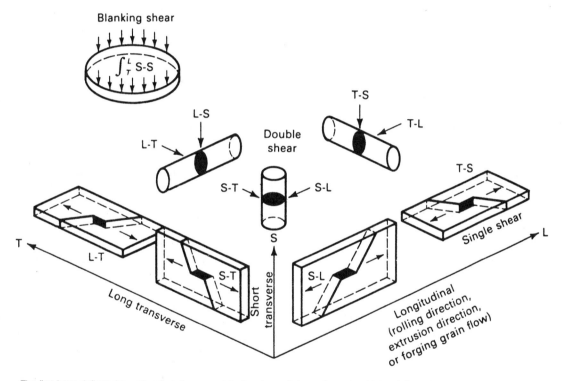

The first letter defines the grain orientation normal to the plane of shear; the second letter defines the direction of loading. L, longitudinal; T, long transverse; S, short transverse.

Fig. 12 Planes of shear and loading directions for shear specimens

fixture rigidity are necessary if the data are to be used for the development of design allowables.

Effect of Test Fixtures. Double-shear tests have been performed with a rigid Amsler tool as well as with a double-shear tool used primarily for testing of rivets. These setups are shown schematically in Fig. 11.

Effect of Specimen Orientation. As shown in Fig. 12, six different combinations of specimen grain orientations (normal to the shear plane) and loading directions are possible in double-shear testing. Test results will vary, depending on the direction of loading and plane of shear. Because of these influences, the specimen ends must be marked (scribed) with the

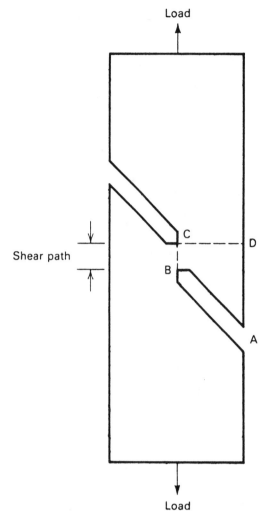

Dashed line between B and C represents failure along the shear path. Tensile failure depicted by dahsed line CD. See text for the relative dimensions of ABCD and the specimen thickness needed to ensure shear failure.

Fig. 13   Single-shear slotted-sheet specimen

correct orientation prior to insertion into the test fixture.

Single-Shear Tests

The single-shear test is used to obtain ultimate shear strength values for sheet and thin plate. The test is conducted on standard tension testing equipment with machined (slotted) sheet specimens of the design shown in Fig. 13. To ensure that failure occurs along the shear path of the specimen, the geometric dimensions of the specimen are critical. In a study of the fail-

ure mechanisms of single-shear sheet specimens, three possible modes of failure were examined, as discussed below.

Single-Shear Failure Modes. Ideally, failure should occur in shear along the plane normal to the surface of the specimen and along the axis of loading. This is depicted in Fig. 13, with the dashed line between B and C representing the shear path. To avoid undesirable tensile failure (indicated by the dashed line CD in Fig. 13), the dimension CD must be sufficiently greater than the dimension CB so that the critical shear stress is developed before the critical tension stress. Depending on the material being tested, this ratio (CD/CB) can vary from a minimum of 2 to 14. To achieve this ratio, the dimension CB cannot be reduced to the extent that the specimen becomes fragile and easily damaged upon application of load.

AXIAL COMPRESSION TESTING

Compression loads are applied to many engineering structures that vary in dimension from massive suspension bridge piers to the thin sheet of aircraft wings. In addition, metalforming processes involve large compressive deformations. Analyses of structural behavior or metalforming require knowledge of compression stress-strain properties. This is best obtained by using the procedures described in ASTM Standard E 9. This section discusses factors that contribute to obtaining valid test data, including the effects of barreling and buckling, a definition of compressive failure, and compression test procedures or techniques that differ from those of tension testing.

Several assumptions are inherent in stress-strain testing. In any test used to obtain uniaxial compression stress-strain properties, the measured quantities are generally load and strain, if a strain gage is used, or displacement, if a compressometer is used. Strain or displacement is measured at the surface of the specimen, and longitudinal and transverse strains are assumed to be uniform in each cross section along the entire gage length. The measured surface strain is assumed to be the same as the internal strain. The stress is always an inferred quantity that is calculated by dividing the applied load by the cross-sectional area of the test piece.

Additionally, it is assumed that the cross-sectional area is constant over the gage length and that the stress is uniaxial and uniform in each cross section along the gage length. Errors in stress or strain occur if the assumptions of uniformity do not exist in a test. Buckling and barreling cause nonuniform stress and strain distributions, and elimination of these phenomena in compression tests can lead to more accurate stress-strain data.

Specimen Buckling

When a specimen buckles during a compression test, the stress data calculated from the applied load will be erroneous. However, the risk of specimen buckling can be reduced by careful attention to alignment of the loading train and by careful manufacture of the specimen according to the specifications of flatness, parallelism, and perpendicularity. Even with well-made specimens tested in a carefully aligned loading train, buckling may still occur. Conditions that typically induce buckling are discussed below.

Alignment. The loading train, including the loading faces, must maintain initial alignment throughout the entire loading process. Alignment, parallelism, and perpendicularity tests should be conducted at maximum load conditions of the testing apparatus.

Specimen Tolerances. The tolerances for specimen end-flatness, end-parallelism, and end-perpendicularity should be considered as upper limit values. This is also true for concentricity of outer surfaces in cylindrical specimens and uniformity of dimensions in rectangular sheet specimens. If tolerances are reduced from these values, the risk of premature buckling is also reduced.

Inelastic Buckling. For the most slender specimen recommended, the calculated elastic buckling stresses are higher than can be achieved in a test. Assume a specimen has a length-to-diameter ratio of $L/D = 10$. An approximate calculation using the elastic Euler equation for a steel specimen with flat ends on a flat surface (assumed value of end-fixity coefficient is 3.5) yields a buckling stress in excess of 4100 MPa (600 ksi); the comparable value for an aluminum specimen would be 1380 MPa (200 ksi). These values, however, are not realistic.

Buckling stress in the above example should not be calculated by an elastic formula but by an inelastic buckling relation. In terms of inelastic buckling it has been concluded that the following relation appropriately calculates inelastic buckling stresses:

$$S_{cr} = C\pi^2 \left[\frac{E_t}{\left(\dfrac{L}{\rho}\right)^2} \right] \qquad \text{(Eq 1)}$$

where S_{cr} is the buckling stress, MPa (ksi); C is the end-fixity coefficient; E_t is the tangent modulus of the stress-strain curve, MPa (ksi); L is the specimen length, mm (in.); and ρ is the radius of gyration of specimen cross section, mm (in.). Equation 1 reduces to the Euler equation if E, the modulus of elasticity, is substituted for E_t.

Rearranging Eq 1 to combine the stress-related factors results in:

$$\left(\frac{1}{C\pi^2}\right)\left(\frac{L}{\rho}\right)^2 = \frac{E_t}{S_{cr}} \qquad \text{(Eq 2)}$$

Note that the value of the right side of Eq 2 decreases as stress increases in a stress-strain curve. In a material whose response is elastic-pure plastic, the right side of Eq 2 vanishes, because E_t becomes zero, and buckling will always occur at the yield stress. When the material exhibits strain hardening, calculations using Eq 2 will yield the appropriate specimen dimensions to resist buckling for given values of stress.

Side Slip. One form of buckling of cylindrical specimens that may result from misalignment of the loading train under load or from loose tolerances on specimen dimensions is illustrated in Fig. 14. The ends of the specimen undergo side slip, resulting in a sigmoidal central axis. This form of buckling could be described by Eq 1 and 2, provided an appropriate value of the end-fixity coefficient can be assigned.

Thin Sheet Specimens. In testing thin sheet in a compression jig, approximately 2% of the specimen length protrudes from the jig. Buckling of this unsupported length can occur if there is misalignment of the loading train such that it does not remain coaxial with the specimen throughout the test. A typical compression jig and contact-point compressometer are shown in Fig. 15(a) and 15(b).

Barreling of Cylindrical Specimens

When a cylindrical specimen is compressed, Poisson expansion occurs. If this expansion is restrained by friction at the loading faces of the

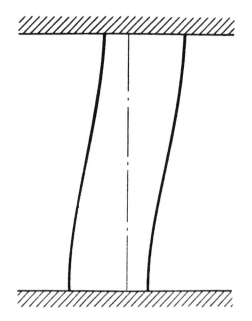

The original position of the specimen center line is indicated by the dashed line.

Fig. 14   Schematic diagram of side-slip buckling

Fig. 15(a)   Sheet compression jig suitable for room-temperature or elevated-temperature testing

Contact points fit in pre-drilled shallow holes in edge of specimen.

Fig. 15(b)   Contact-point compressometer installed on specimen removed from jig

specimen, nonuniform states of stress and strain in the entire specimen result. The specimen acquires a barreled shape, as shown in Fig. 16. The effect on the stress and strain distributions is of consequence only when the deformations are on the order of 10% or more.

Friction at the Loading Surfaces. Friction on the loading face causes roll-over. As shown in Fig. 16, points originally on the sides of the specimen are ultimately located on the specimen end face. A computer code has been used to study the details of a compression test when friction is present. Figure 17 shows the computer results after a 50% height reduction in a steel specimen (loading axis is horizontal). In the computer simulation, a friction coefficient of 0.3 was assumed. The grid in the undeformed state consisted of orthogonal sets of equally spaced lines. Points *A* were originally on the circumference of the undeformed face (see also Fig. 16).

TORSION TESTING

Torsion tests can be carried out on most materials to determine mechanical properties such as modulus of elasticity in shear, yield shear strength, ultimate shear strength, modulus of rupture in shear, and ductility. The torsion test can also be used on full-size parts (shafts, axles,

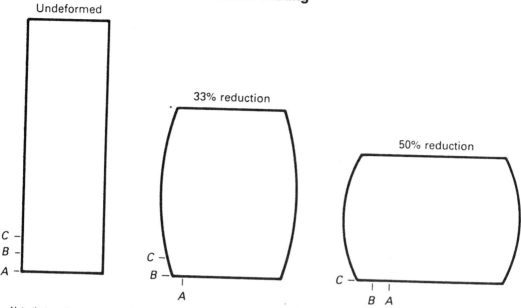

Note that as the deformation increases, points A, B, and C originally on the specimen sides move to the loading face.

Fig. 16   Barreling during a test when the friction coefficient is 1.00 at the specimen loading face

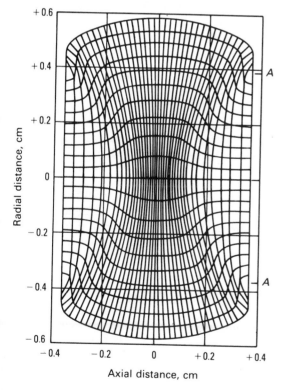

The loading axis is horizontal. This specimen had a length-to-diameter ratio (*L/D*) of 2. The original grid was square. The points marked *A* were on the circumference of the undeformed specimen.

Fig. 17   Computer-code simulation for 50% compression of a steel specimen

and twist drills) and structures (beams and frames) to determine their response to torsional loading. In torsion testing, unlike tension testing and compression testing, large strains can be applied before plastic instability occurs, and complications due to friction between the test specimen and dies do not arise.

Torsion tests are most frequently carried out on prismatic bars of circular cross section by applying a torsional moment about the longitudinal axis. The shear stress versus shear strain curve can be determined from simultaneous measurements of the torque and angle of twist of the test specimen over a predetermined gage length.

The following section discusses the torsional deformation of prismatic bars of circular cross section.

Torsion Testing of Prismatic Bars of Circular Cross Section

In torsion testing of prismatic bars of circular cross section it is assumed that:

- Bar material is homogeneous and isotropic.
- Twist per unit length along the bar is constant.
- Sections that are originally plane to the torsional axis remain plane after deformation.
- Initially straight radii remain straight after deformation.

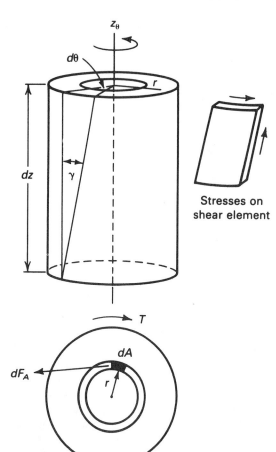

Fig. 18 **Torsion of a solid circular prismatic section**

Figure 18 shows the torsional deformation of a long, straight, isotropic prismatic bar of circular section. Assuming the above-mentioned constraints, the displacements are given by:

$$u_r = 0$$

$$u_z = 0$$

$$u_\theta = rz \left(\frac{d\theta}{dz} \right) \qquad \text{(Eq 1)}$$

where $d\theta/dz$ is the angle of twist per unit length (θ/L), and L is the gage length of the test specimen. The strains are given by:

$$\epsilon_{zz} = 0$$

$$\epsilon_{rr} = 0$$

$$\epsilon_{\theta\theta} = 0$$

$$\gamma_{zr} = 0$$

$$\gamma_{r\theta} = 0$$

$$\gamma_{z\theta} = \frac{rd\theta}{dz} \qquad \text{(Eq 2)}$$

For an isotropic material that obeys Hooke's law, the corresponding stress state is given by:

$$\sigma_{zz} = 0$$

$$\sigma_{rr} = 0$$

$$\sigma_{\theta\theta} = 0$$

$$\tau_{zr} = 0$$

$$\tau_{r\theta} = 0$$

$$\tau_{z\theta} = G\gamma_{z\theta} \qquad \text{(Eq 3)}$$

where G is the shear modulus that is related to Young's modulus (E) and Poisson's ratio (v) by:

$$G = \frac{E}{2(1 + v)} \qquad \text{(Eq 4)}$$

The stress distribution across the prismatic bar of circular cross section is given by:

$$\tau_{z\theta} = \frac{Gr\theta}{L} \qquad \text{(Eq 5)}$$

The maximum shear stress can be calculated from knowledge of the torsional loading and bar geometry. When the surface shear stress ($\tau_{z\theta max}$) reaches the yield shear stress (k) of the test material, plastic deformation (flow) occurs. The deformation zone begins at the surface of the bar and advances inward as an annulus surrounding an elastic core. The stress distributions are shown schematically in Fig. 19 for a non-work-hardening and a work-hardening material.

BENDING STRENGTH TESTS

Bending strength tests offer a means of determining the modulus of elasticity in bending and the bending strength of flat metallic materials in the form of strip, sheet, or plate. Three standard bending load-deflection tests are discussed herein: the cantilever beam bend test, the three-point bend test, and the four-point bend test. The variables that can affect the results of these test procedures are also examined.

The cantilever beam bend test measures the

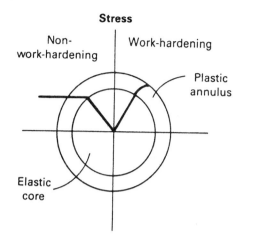

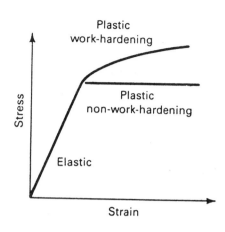

Fig. 19 Section through prismatic bar of circular section

angle of deflection and corresponding bending moment of metallic strip or sheet subjected to continuous loading in bending. A bending stress-strain curve analogous to the stress-strain curve in tension is developed from the data. The modulus of elasticity is determined at stresses below the proportional limit. The bending yield strength is determined by the offset method similar to that used for stress-strain curves in tension or compression.

In the three-point bend test, the test specimen is supported near each end and is loaded at one point equidistant from each support. The modulus of elasticity in bending is obtained by the measurements of load and deflection at stresses below the proportional limit. The bending proof strength is determined by increasing the load in steps and unloading until a specified permanent set is obtained.

The four-point bend test consists of a simple beam resting on two supports and loaded at two points equally spaced from each support. The modulus of elasticity in bending and the bending proof strength in this method are obtained by a procedure similar to the three-point method of bend testing.

Apparatus. A cantilever beam bend test apparatus is shown in Fig. 20. This tester consists of a specimen holder, a pendulum weighing system, a moment scale, and an angular deflection scale.

The specimen holder is a vise, V, to which an angular deflection indicator, I_1, is attached. The specimen holder is rotated counterclockwise about point 0 at a nominal rate of 60° of arc per minute.

The pendulum weighing system is composed of a set of detachable weights, an angular deflection scale with a moment pointer indicator I_2, a 6.35-mm (0.25-in.) diam loading pin that transmits the bending force of the pendulum system to the free end of the cantilever specimen, and a weight to counterbalance the loading pin. The pendulum weighing system pivots about the center of rotation designated by point 0 in Fig. 20; therefore, the span length remains constant at 50.8 mm (2.0 in.).

The specimen is then unloaded. With the specimen in contact with the loading pin, the permanent set in degrees is recorded by reading the angular deflection scale at zero load.

Stress-Strain Relationships. To construct a stress-strain curve in a cantilever beam bend test, the bending moment-deflection data must be normalized with respect to specimen geometry. Also, a relationship must be established between the applied bending moment and the maximum bending stress occurring in the outer fibers at the clamped end of the beam. The angle of deflection must be replaced by the deflection of the loaded end of the beam to estimate the outer fiber bending strain at the clamped end. The following standard formulas for bending stress and deflection in the simple cantilever beam are used to derive these relationships.

The deflection, D, of the loaded end of a cantilever beam is given by:

$$D = \frac{4PL^3}{E_b bh^3} \qquad \text{(Eq 1)}$$

The maximum bending stress, σ_b, in the outer

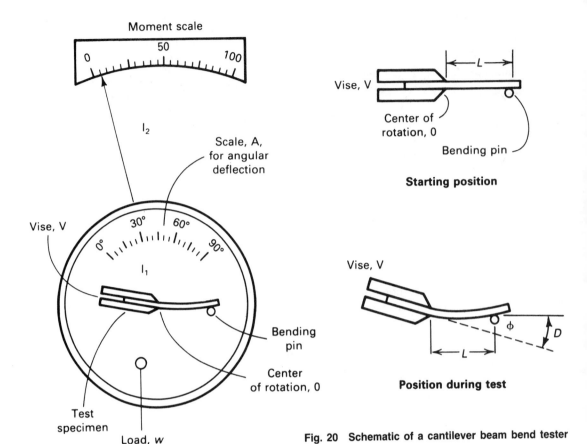

Fig. 20 Schematic of a cantilever beam bend tester

fibers at the clamped end is given by:

$$\sigma_b = \frac{6PL}{bh^2} \qquad \text{(Eq 2)}$$

where P is applied load, N (lbf); E_b is modulus of elasticity in bending, Pa (psi); b is specimen width, mm (in.); h is specimen thickness, mm (in.); and L is span length, mm (in.).

In the cantilever beam test, the maximum bending moment is related to the applied load, P, by:

$$M = PL \qquad \text{(Eq 3)}$$

The maximum attainable bending moment in the pendulum system is expressed in terms of the moment scale reading, f, and the bending moment, M, measured at an angular deflection of ϕ as follows:

$$f = \frac{M}{M_m} \times 100 \qquad \text{(Eq 4)}$$

where a full-scale reading of 100 corresponds to the maximum bending moment of the pendu-

lum. As shown in Fig. 20, the beam deflection, D, is approximated by the length of an arc having radius L and an included angle of ϕ radians. Using this approximation:

$$D = \phi L \qquad \text{(Eq 5)}$$

Combining Eq 1 through 5:

$$E_b = \frac{fM_mL}{25\phi bh^3} \qquad \text{(Eq 6)}$$

Combining Eq 2 through 4:

$$\sigma_b = \frac{3M_mf}{50bh^2} \qquad \text{(Eq 7)}$$

The bending strain in the outer fibers at the clamped end corresponding to the stress given by Eq 7 is:

$$\epsilon_b = \frac{\sigma_b}{E_b} = \frac{3}{2}\frac{\phi h}{L} \qquad \text{(Eq 8)}$$

Stress-strain curves in bending are similar to those in tension or compression.

The modulus of elasticity in bending, E_b, is determined by the slope of a straight line extending from the maximum deflection data point (max) to the permanent set point (ps) obtained from the unloading curve at zero load:

$$E_b = \frac{\dfrac{M_m f}{25bh^2}}{\left(\dfrac{\phi h}{L}\right)_{max} - \left(\dfrac{\phi b}{L}\right)_{ps}} \quad \text{(Eq 9)}$$

To construct the bending stress-strain curve, a straight line having slope E_b is drawn so that it passes through the origin. The actual data points for elastic loading may be slightly displaced from this line due to shape deficiencies in the specimen that would prevent uniform distribution of load over the full width.

The nonlinear portion of the bending stress-strain curve is constructed by drawing a curve through the remaining data points and connecting it with the modulus of elasticity line. To determine the bending yield strength from the stress-strain curve, the "offset method," analogous to that used for tensile or compressive stress-strain curves, is used. The offset yield strengths in bending should be determined for strains of 0.01, 0.05, and 0.10%, provided the maximum allowable deflection angle of 30° is not exceeded.

Example Calculations. Bending moment-deflection data were obtained for spring-tempered C77000 copper alloy strip in the longitudinal orientation using the test procedure described above. The permanent set angle was measured after the load was removed. The bending moment and deflection data are shown in Table 3.

Using Eq 7 and 8, the data in Table 3 were normalized with regard to specimen geometry and plotted on the stress-strain curve shown in Fig. 21. In this diagram, the ordinate is equal to the bending stress multiplied by a constant that is equal to $50bh^2/3M_m$. Similarly, the abscissa is equal to the bending strain multiplied by a constant that is equal to $2L/3h$.

The modulus of elasticity in bending is obtained from Fig. 21 by drawing a straight line from the maximum deflection data point (max) to the permanent set point (ps). The slope of the unloading line is the modulus of elasticity in bending. To construct the linear portion of the bending stress-strain curve, the elastic modulus line is translated along the abscissa until it passes

Table 3 Bending moment-deflection data for spring-tempered C77000 copper alloy strip in the longitudinal orientation

Obtained by cantilever beam bend testing

Angle of deflection, degrees	Bending moment, N·m (lbf·in.)	
	Loading	Unloading
0	0	NA(a)
2	0.2 (2.0)	NA(a)
4	0.5 (4.4)	NA(a)
4.5(b)	No data	0
6	0.8 (6.8)	0.2 (2.0)
8	1.0 (9.2)	0.5 (4.4)
10	1.3 (11.6)	0.8 (6.8)
12	1.6 (14.0)	0.9 (8.8)
15	1.9 (16.8)	1.4 (12.4)
18	2.2 (19.6)	1.8 (16.0)
21	2.5 (22.0)	2.2 (19.6)
24	2.7 (24.0)	2.6 (23.2)
25	2.8 (24.8)	2.75 (24.4)

Note: Span length, L, is 50.8 mm (2.0 in.); specimen width, b, is 25.4 mm (1.0 in.); specimen thickness, h, is 0.805 mm (0.0317 in.); maximum attainable bending moment, M_m, is 4.5 N·m (40 lbf·in.).
(a) Not applicable. (b) Permanent set angle measured during unloading at zero load is 4.5°. Source: Olin Corp.

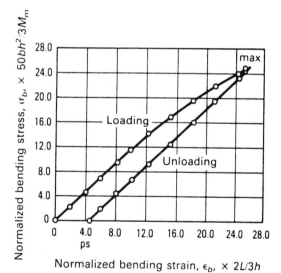

max, maximum deflection; ps, permanent set at zero load.
Source: Olin Corp.

Fig. 21 Cantilever bending normalized stress-strain curve for spring-tempered C77000 copper alloy strip in the longitudinal orientation

through the origin. The nonlinear portion is constructed by drawing a curve through the remaining data points and connecting this curve with the elastic modulus line.

To determine the yield strengths in bending for strains of 0.01, 0.05, and 0.10% from Fig. 21, the offset method can be used.

EFFECT OF STRAIN RATE ON FLOW PROPERTIES

Strain rate, or the rate at which a specimen is deformed, is an important consideration in the production, fabrication, and testing of materials. This rate can have an important influence on the mechanical properties, particularly the flow stress, of a material. During a conventional (pseudostatic) tension test, ASTM prescribes an upper limit for the deformation rate; however, during actual fabrication processes, economics prescribes that the deformation be as rapid as possible. Under these different conditions, the mechanical response of a given material may vary. For most materials, strength properties tend to increase at higher rates of deformation.

Figure 22 illustrates true yield stress at various strains for a low-carbon steel at room temperature. Between strain rates of 10^{-6} s^{-1} and 10^{-3} s^{-1} (a 1000-fold increase), yield stress increases only by 10%. Above 1 s^{-1}, however, an equivalent rate increase doubles the yield stress. For the data in Fig. 22, at every level of strain the flow stress increases with increasing strain rate. However, a decrease in strain-hardening rate is exhibited at the higher deformation rates.

The results of the combined effects of strain rate and temperature at 200, 400, and 600 °C (390, 750, and 1110 °F) are shown in Fig. 23. At the highest temperature (600 °C, or 1110 °F), yield strength increases with increasing strain rate, as do room-temperature results, but strain hard-

ening increases (rather than decreases) with increasing strain rate. At the intermediate temperatures shown in Fig. 23(a) and (b), however, regions of negative strain rate sensitivity are visible; that is, under certain conditions of strain, strain rate, and temperature, the flow stress of carbon steels can decrease with an increase in strain rate. This is opposite to the usual strain rate effect.

Another class of common metals is the structural aluminums, which are less strain-rate-sensitive than steels. Figure 24 shows data obtained for 1060-O aluminum. Between strain rates of 10^{-3} s^{-1} and 10^{3} s^{-1} (a million-fold increase), the stress at 2% plastic strain increases by less than 20%.

EFFECT OF TEMPERATURE ON STRESS-STRAIN BEHAVIOR

Elevated/low temperature tension tests are conducted with basically the same specimens and procedures as room-temperature tension tests. However, the specimens must be heated or cooled in an appropriate environmental chamber (Fig. 25). Also the test fixtures must be sufficiently strong and corrosion resistant, and the strain-measuring system must be usable at the test temperature.

Once determined, load-elongation data can be transformed into engineering stresses and strains using standard formulas. True stresses and strains, at least up to the onset of necking, can also be calculated. The onset of necking is difficult to identify, particularly at elevated temperatures.

General Characteristics

Elevated/low temperature stress-strain diagrams are similar in appearance to those determined at room temperature (Fig. 26). Relative to ambient temperature, materials become stronger, but less ductile, as temperature decreases. Conversely, materials become weaker with increasing temperature (Fig. 27). Although simple, stable alloys exhibit increased ductility with rising temperatures, the temperature-ductility behavior for most engineering materials (Fig. 27b) varies greatly. Such discontinuities in ductility with increasing temperature usually can be traced to metallurgical instabilities—carbide precipitation, for example—that affect the failure mode.

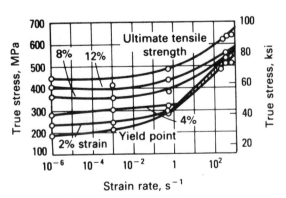

Fig. 22 True yield stresses at various strains versus strain rate for a low-carbon steel at room temperature

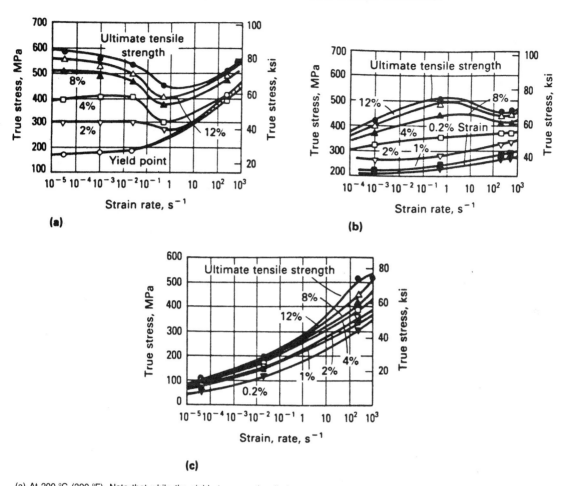

(a) At 200 °C (390 °F). Note that while the yield stress continually increases with an increase in strain rate, the ultimate tensile strength and some of the strengths at various strains decrease with increases in strain rate. This is attributed to dynamic strain aging. (b) At 400 °C (750 °F). The decrease in tensile strength with increase in strain rate is again attributed to dynamic strain aging. This temperature and the temperature in (a) are in the "blue brittle" region for steels. (c) At 600 °C (1110 °F). The abrupt change in the slope of the yield versus strain rate present at room temperature (Fig. 3) is not present at this temperature.

Fig. 23 True stresses at various strains versus strain rate for a low-carbon steel

Because of the relatively high strain rates—usually 8.33×10^{-5} s^{-1} (0.5%/min) and 8.33×10^{-4} s^{-1} (5%/min)—involved in tensile testing, deformation occurs by slip (glide of dislocations along definite crystallographic planes). Thus, changes in strength and ductility with temperature generally can be related to the effect of temperature on slip. At low temperatures (less than 0.3 homologous temperature), the number of slip systems is restricted, and recovery processes are not possible. Therefore, strain-hardening mechanisms, such as dislocation intersections and pileups, lead to the increasingly higher forces required for continued deformation. This continues until the local stresses at pileups exceed the fracture stress, and failure occurs.

At higher temperatures (between 0.3 and 0.5 homologous temperature), thermally activated processes such as multiple slip and cross slip allow the high local stresses to be relaxed, and strength is decreased. For sufficiently high temperatures in excess of half the homologous temperature, diffusion processes become important, and mechanisms such as recovery, dislocation climb, recrystallization, and grain growth can reduce the dislocation density, prevent pileups, and further reduce strength.

Deformation under tensile conditions is governed to some extent by crystal structure. Face-centered cubic materials generally exhibit a gradual change in strength and ductility as temperature decreases. Such changes for type 304 stainless steel are illustrated in Fig. 26. Some

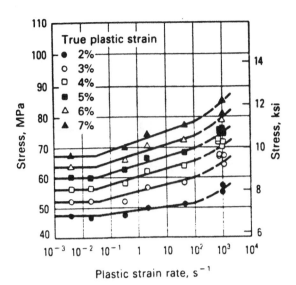

Fig. 25 **Tension testing apparatus with environmental chamber for testing up to 540 °C (1000 °F) (Courtesy of Instron Corp.)**

Fig. 24 **Analysis of the uniaxial stress/ strain/strain rate data for aluminum 1060-O**

body-centered cubic alloys, however, exhibit an abrupt change at the ductile-to-brittle transition temperature (approximately 200 °C, or 390 °F, for tungsten in Fig. 27), below which there is little plastic flow. In close-packed hexagonal and body-centered cubic materials, mechanical twinning can also occur during testing. However, twinning by itself contributes little to the overall elongation; its primary role is to reorient

previously unfavorable slip systems to positions in which they can be activated.

Other factors can affect tensile behavior; however, the specific effects cannot be predicted easily. For example, re-solutioning, precipitation, and aging (diffusion-controlled particle growth) can occur in two-phase alloys during heating prior to testing and during the actual testing. These processes can produce a wide variety of responses in mechanical behavior de-

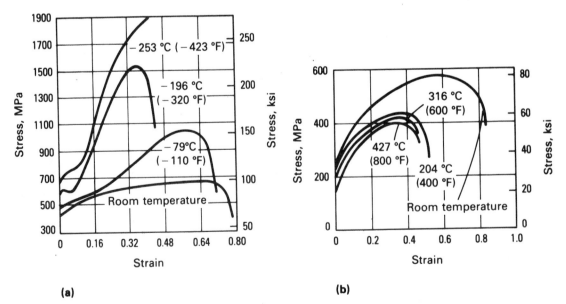

(a) At low temperatures. (b) At elevated temperatures.

Fig. 26 **Stress-strain diagrams for type 304 stainless steel**

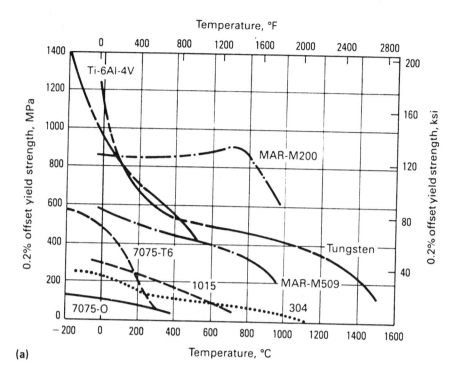

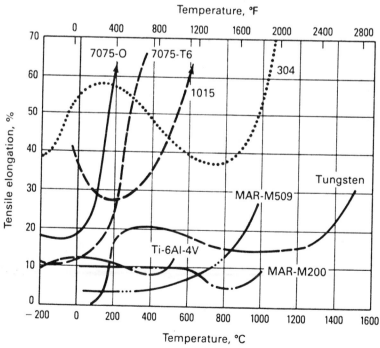

(a) 0.2% offset yield strength. (b) Tensile elongation. Materials tested include aluminum alloy 7075 in two heat-treated conditions; Ti-6Al-4V; AISI 1015 low-carbon steel; type 304 stainless steel; cobalt-base alloy MAR-M509; directionally solidified nickel-base alloy MAR-M200; and pure tungsten.

Fig. 27 Effect of temperature on strength and ductility of various materials

pending on the material. Diffusion processes also are involved in yield point and strain-aging phenomena. Under certain combinations of strain rate and temperature, interstitial atoms can be dragged along with dislocations, or dislocations can alternately break away and be re-pinned, producing serrations in the stress-strain curves.

There are exceptions to the above generalizations, particularly at elevated temperatures. For example, at sufficiently high temperatures, the grain boundaries in polycrystalline materials are weaker than the grain interiors, and intergranular fracture occurs at relatively low elongation. In complex alloys, hot shortness, in which a liq-

uid phase forms at grain boundaries, or grain boundary precipitation can lead to low strength and/or ductility.

Experimental Methods

Specific procedures and methods for conducting room-temperature and elevated-temperature tension tests have been standardized by ASTM in test methods E 8 and E 21, respectively. Although no standard has been adopted for low-temperature testing, the general requirements of ASTM E 8 and E 21 must also be satisfied at lower temperatures. Assuming that a test machine with appropriate load-measuring and

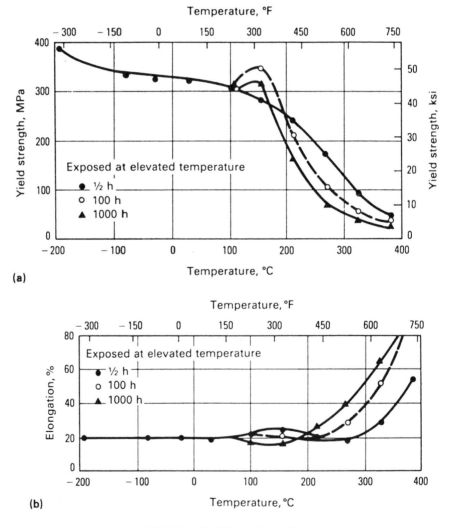

(a) Yield strength. (b) Percent elongation.

Fig. 28 Effect of exposure time and temperature on the tensile properties of naturally aged aluminum alloy 2024-T4

speed-control capabilities is used for testing, the validity of a test rests on proper concern for strain measurement, temperature control, and materials behavior.

Strain Measurement. The simplest method for strain measurement is based on crosshead displacement in the test apparatus. However, the crosshead extension includes not only the strain in the gage section, but also the deformation in the rest of the sample, load train, and testing machine. This method is suitable only for measurement of large plastic strains in which the other factors can be overlooked. For accurate determination of strain, an extensometer or strain gage must be used. These are attached directly to the specimen and thus exclude any contributions from the testing apparatus. However, errors can still arise from bending strains and strain gradients due to nonaxial loading, specimen nonuniformities, and inherent extensometer inaccuracies.

Specimen strain frequently is magnified to increase the accuracy of measurement. Mechanical extensometers that magnify by using lever arms and optical extensometers that magnify by using mirrors and lenses have been employed. Electronic magnification is widely used; one common device is the linear variable differential transformer (LVDT), in which the displacement of the specimen is converted to an electrical signal by motion of a ferrite core within a coil. Electrical resistance gages that make use of the change in resistance of a material with strain are also available. These gages are generally thin and small in size and can be cemented directly on the specimen. They are useful in both elastic and plastic regimes.

Material Behavior. Because alloys undergoing elevated tensile testing will, in effect, be subject to annealing prior to loading, changes in microstructure can occur and can produce a material that is not characteristic of the original stock. Thus, very slow heating or prolonged holds at temperature should be avoided. Figure 28 illustrates the influence of hold time on the yield strength and ductility of a precipitation-strengthened aluminum alloy. Holding at 150 °C (300 °F) changes the amount and distribution of the reinforcing phases in such a manner that strengthening initially occurs. This is subsequently followed by weakening. Clearly, the exposed material is not the same as one that is tested rapidly.

Test environment can also affect the measured properties. Generally, the atmosphere should reflect the intended or proposed use of the material. Although the environment can rarely be a complete simulation of operating conditions, it should produce the same basic effects and should not introduce foreign attack mechanisms. For example, it would be appropriate to test oxidation-resistant alloys at elevated temperature in air; however, such conditions cannot be used for refractory metals that undergo catastrophic oxidation.

1-1. General Stress-Strain Curves

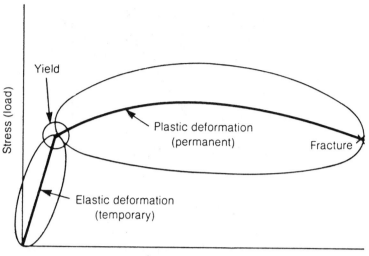

General stress-strain curve showing elastic and plastic portions of a typical curve. Area marked "Yield" is the area of transition from elastic to plastic deformation. Yield strength, yield point, elastic limit, and proportional limit are all in this area.

The terms "elastic deformation" and "plastic deformation" are used widely in failure analysis. Elastic deformation refers to the springiness of a metal, or its ability to return to its original size and shape after being loaded and unloaded. As shown above, elastic deformation occurs on the straight-line portion of the typical stress-strain curve. This is the condition in which most metal parts are used during their period of service. If higher loads are applied to the part, however, the range of elasticity, or elastic deformation, is exceeded and the metal is now permanently deformed; that is, it is in the plastic deformation range, as shown in the graph above.

Source: Donald J. Wulpi, Understanding How Components Fail, American Society for Metals, Metals Park OH, 1985, p 40

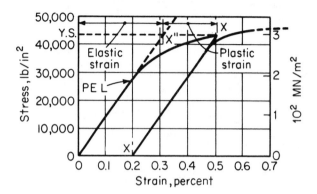

For most engineering materials, the curve will have an initial linear elastic region in which deformation is reversible and time-independent. The slope in this region is **Young's modulus,** E. The **proportional elastic limit** (PEL) is the point at which the curve starts to deviate from a straight line. The **elastic limit** (frequently indistinguishable from PEL) is the point on the curve beyond which plastic deformation is present after release of the load. If the stress is increased further, the stress-strain curve departs more and more from the straight line. Unloading the specimen at point X, the portion XX′ is linear and is essentially parallel to the original line OX″. The horizontal distance OX′ is called the **permanent set** corresponding to the stress at X. This is the basis for the construction of the arbitrary **yield strength.** To determine the yield strength, a straight line XX′ is drawn parallel to the initial elastic line OX″ but displaced from it by an arbitrary value of permanent strain. The permanent strain commonly used is 0.20 percent of the original gage length. The intersection of this line with the curve determines the stress value called the yield strength. In reporting the yield strength, the amount of permanent set should be specified. The arbitrary yield strength is used especially for those materials not exhibiting a natural yield point, such as nonferrous metals, but it is not limited to these. Plastic behavior is somewhat time-dependent, particularly at high temperatures. Also at high temperatures, a small amount of time-dependent reversible strain may be detectable, indicative of **anelastic** behavior.

Source: Marks' Standard Handbook for Mechanical Engineers, Eighth Edition, Theodore Baumeister, Eugene A. Avallone and Theodore Baumeister III, Eds., McGraw-Hill Book Company, New York, 1978, p 5-2

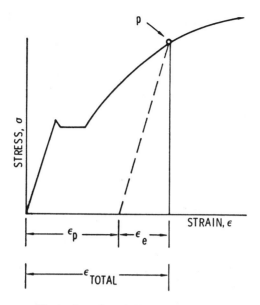

Illustration of total strain components.

Again, note that this relationship is valid *only up to the point of necking* in the specimens during a monotonic tension test.

The total true strain in a tension test may be separated conveniently into two components, as illustrated above.

1. The linear elastic, or that portion of strain that is recovered upon unloading, ϵ_e.
2. The nonlinear plastic strain, which cannot be recovered on unloading, ϵ_p.

Mathematically, this concept is expressed by the equation

$$\epsilon = \epsilon_e + \epsilon_p$$

at any point, P, on the true stress-strain curve.

For most metals, a logarithmic plot of true stress versus true plastic strain is a straight line. It may be expressed by the equation

$$\sigma = K(\epsilon_p)^n$$

or

$$\epsilon_p = \left(\frac{\sigma}{K}\right)^{1/n}$$

where K is the strength coefficient (intercept at $\epsilon_p = 1$) and n is the strain-hardening exponent (slope).

Source: Fatigue and Microstructure, papers presented at the 1978 ASM Materials Science Seminar, 14–15 October 1978, St. Louis MO, sponsored by the Materials Science Division of the American Society for Metals, Metals Park OH, 1979, p 390

1-4. Stress-Strain Curve of a Metal With a Sharp Yield Point

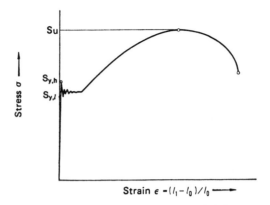

Upper and lower yield points and strain aging.

A peculiarity of various alloys, particularly of those with interstitial solid solution, is the occurrence of a sharp yield point during the tensile test (above). Plastic deformation takes place if the stress exceeds the upper yield point $S_{y,h}$, from which it drops to the lower yield point $S_{y,l}$. The stress rises again only when the strain is further increased by up to several percentage points (above). Usually the upper yield point is determined by tensile tests. Yield-point values that are given in material standards refer to the upper yield point.

After reaching the upper yield point, deformation in the tensile test specimen does not occur homogeneously, but rather in the form of narrow bands that spread over the specimen. These are called *Lüders' lines*. They can be seen clearly on the polished surfaces of the test specimen. During the sheet-metal-forming processes the inhomogeneous deformation within the yield-point zone can lead to rough surfaces, especially in deep-drawing components that thus become unsuitable for usage. Examples of metals with sharp yield points are steels with low carbon content and aluminum-magnesium alloys.

Source: Handbook of Metal Forming, Kurt Lange, Ed., McGraw-Hill Book Co., New York, 1985, p 3.14

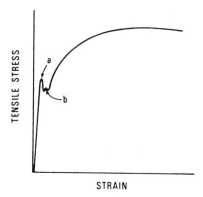

Yield points in stress-strain curve: (a) upper yield stress; (b) lower yield stress.

As a specimen is stretched, the stress suddenly decreases. This phenomenon is referred to as a yield point, load drop, or sometimes sharp yield point to emphasize the suddenness of the change of slope of the stress-strain curve (see above). The maximum value of the stress at the yield point is called the upper yield stress, and the minimum value of the stress immediately following the yield point is the lower yield stress. Surface markings are associated with the yield-point phenomenon. A band of deformed metal is formed at the upper yield point at a stress concentration on the specimen. The load drop to the lower yield point is coincident with the band formation. These bands, called Lüders bands, gradually propagate along the length of the specimen as the specimen is extended.

Source: Materials at Low Temperatures, Richard P. Reed and Alan F. Clark, Eds., American Society for Metals, Metals Park OH, 1983, p 251

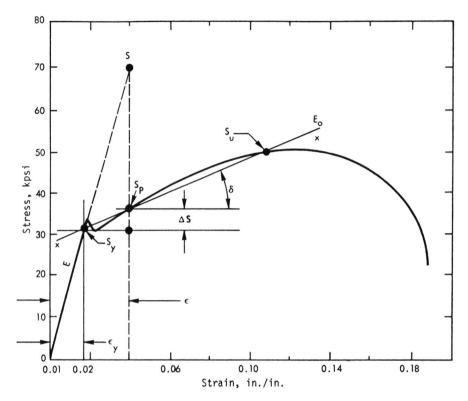

Typical stress-strain curve for a low-strength ductile material.

Let it be assumed that the stress-strain diagram has been established experimentally and that the curve obtained can be approximated by the two straight lines from zero to the yield point and from the yield region to the highest point on the curve, as shown by the line x – x. Let the corresponding elastic moduli be E and E_0, respectively. In the study of plasticity of materials, a bilinear approximation of this kind is well known and indicates that the material in question conforms to the law of linear strain hardening.

In terms of the calculated stress S, shown as 70,000 psi in the above figure, Hooke's law gives

$$\epsilon = \frac{S}{E}$$

The corresponding strain at yield stress S_y is

$$\epsilon_y = \frac{S_y}{E}$$

Source: Alexander Blake, Practical Stress Analysis in Engineering Design, Marcel Dekker, Inc., New York, 1982, p 443

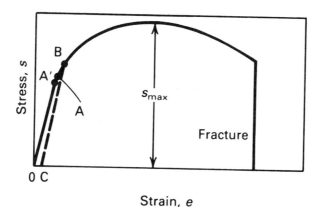

Typical tension stress-strain curve for ductile metal indicating yielding criteria. Point A, elastic limit; point A′, proportional limit; point B, yield strength; line C-B, offset yield strength; 0, intersection of the stress-strain curve with the strain axis.

Elastic limit, shown at Point A, is the greatest stress the material can withstand without any measurable permanent strain remaining after the complete release of load. With increasing sensitivity of strain measurement, the value of the elastic limit is decreased until it equals the true elastic limit determined from microstrain measurements. With the sensitivity of strain typically used in engineering studies (10^{-4} in./in.), the elastic limit is greater than the proportional limit. Determination of the elastic limit requires a tedious incremental loading-unloading test procedure. For this reason, it is often replaced by the proportional limit.

Proportional limit, shown at point A′, is the highest stress at which stress is directly proportional to strain. It is obtained by observing the deviation from the straight-line portion of the stress-strain curve.

The yield strength, shown at point B, is the stress required to produce a small specified amount of plastic deformation. The usual definition of this property is the offset yield strength determined by the stress corresponding to the intersection of the stress-strain curve and a line parallel to the elastic part of the curve offset by a specified strain. In the United States, the offset is usually specified as a strain of 0.2% or 0.1% ($e = 0.002$ or 0.001):

$$s_0 = \frac{P_{(\text{strain offset} = 0.002)}}{A_0}$$

Source: Metals Handbook, Ninth Edition, Volume 8, Mechanical Testing, American Society for Metals, Metals Park OH, 1985, p 21

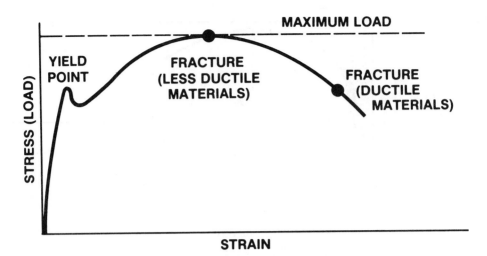

Stress-strain diagram showing fracture points for ductile and less-ductile materials.

Source: Hamilton B. Bowman, Handbook of Precision Sheet, Strip and Foil, American Society for Metals, Metals Park OH, 1980, p 65

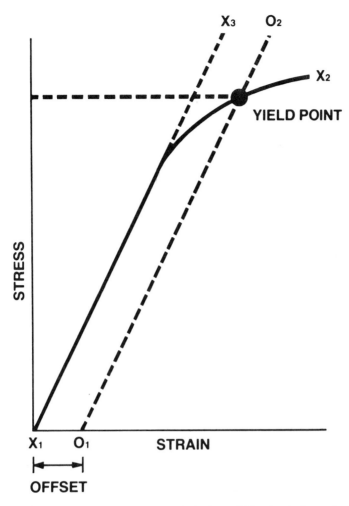

Stress-strain diagram showing a material whose yield point is determined by the offset method.

In this curve, the early portion of a stress-strain diagram extending up to about 1% elongation is shown as $X_1 - X_2$. A distance of 0.2% of the strain within the gage length being measured is marked off on the strain axis of the diagram and the offset line $O_1 - O_2$ is drawn parallel to $X_1 - X_3$. The stress corresponding to the point at which $O_1 - O_2$ intersects the stress-strain curve ($X_1 - X_2$) represents the 0.2%-offset yield strength. The offset method is based on the fact that if the load is released at this point, the specimen will recover along $O_1 - O_2$ until, at zero load, the permanent set $X_1 - O_1$ remains. The yield strength is calculated by dividing the load corresponding to a plastic yield strain of 0.2% by the original cross section of the specimen gage section.

Source: Hamilton B. Bowman, Handbook of Precision Sheet, Strip and Foil, American Society for Metals, Metals Park OH, 1980, p 64

1-10. Typical Stress-Strain Curve for Low-Strength, Easily Deformed Materials

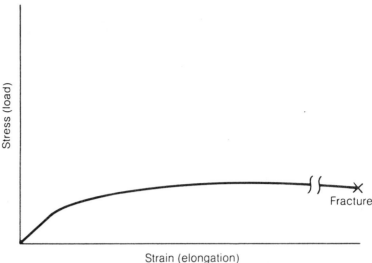

Examples of metals having very low elastic ranges but very great capacity for plastic deformation are aluminum foil, as used for household wrapping, and solder wire. Both are very readily deformed; indeed, aluminum foil must have tremendous ductility (plastic deformation) for it to be useful. If aluminum foil were too strong, it would not wrap properly and could be considered "defective." The graph above shows the type of stress-strain curve desired for this low-strength, easily deformable, highly ductile metal.

A very important feature of the stress-strain curve must be pointed out: the straight-line, or elastic, part of the stress-strain curve of a given metal has a constant slope. That is, it cannot be changed by changing the microstructure or heat treatment. This slope, called the modulus of elasticity, measures the stiffness of the metal in the elastic range; changing the hardness or strength does not change the stiffness of the metal.

Source: Donald J. Wulpi, Understanding How Components Fail, American Society for Metals, Metals Park OH, 1985, p 41

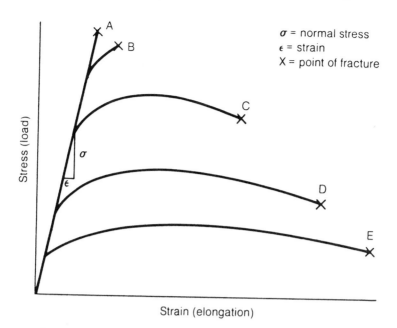

Stress-strain curves for steels of different strength levels ranging from a very hard, strong, brittle steel (A) to a relatively soft, ductile steel (E).

It must be considered that the elastic deflection under load of a given part is a function of the section of the part rather than of the composition, heat treatment, or hardness of the particular steel that may be used. In other words, the modulus of elasticity of all the commercial steels, both plain carbon and alloy types, is the same so far as practical designing is concerned. Consequently if a part deflects excessively within the elastic range, the remedy lies in the field of design, not in the field of metallurgy. Either a heavier section must be used, the points of support must be increased, or some similar change made, since under the same conditions of loading, all steels deflect the same amount within the elastic limit.

The same point may be made with a diagram, as in the graph above, which shows stress-strain curves for steels of different strength levels, which all branch from the same straight line (elastic portion). A very hard, brittle metal, A, such as very strong steel, goes straight up the elastic line with no deviation, then fractures; a file behaves in this manner. A slightly less strong steel, such as B, has slight plastic deformation (ductility). Steel C is of intermediate strength, as is D. Steel E is of the relatively low-strength, high-ductility type desired for deep drawing and severe forming, somewhat similar to the aluminum foil referred to on the previous page. But note that the straight-line (elastic) portion of the curve is identical for all.

Source: Donald J. Wulpi, Understanding How Components Fail, American Society for Metals, Metals Park OH, 1985, p 42

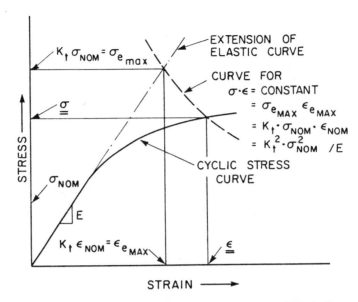

Schematic stress strain curves showing application of Neuber's rule:

$$\sigma \cdot \epsilon = K_t^2 \cdot \sigma_{\text{Nom}}^2/E \qquad \text{(Eq 1)}$$

If an analytic expression is fitted to the actual stress-strain curve of interest, Eq 1 may be solved to find the notch tip stress and strain. A graphical solution to this equation is illustrated in the figure above. A cyclic stress strain curve is indicated since the results will be used later to estimate the cyclic plastic strain range. The linear portion of the stress-strain curve is extended beyond its valid range. The point corresponding to the maximum elastic stress and strain is found from the nominal stress and the stress concentration factor. At this point the product $\sigma_{e\,max} \times \epsilon_{e\,max}$ is computed. It is equal to $K_t^2 \cdot \sigma_{nom}^2/E$. If any value of σ is now chosen, Eq 1 enables the calculation of the corresponding ϵ. A set of these points forms a hyperbola. The actual notch tip stress and strain is given where the hyperbola intersects the real stress-strain curve.

Source: Steel Castings Handbook, Fifth Edition, Peter F. Wieser, Ed., Steel Founders Society of America, Rocky River OH, 1980, p 4-14

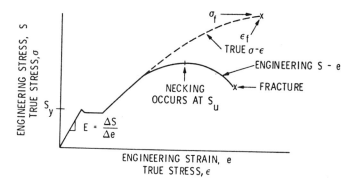

Engineering and true stress-strain curves.

The engineering stress-strain behavior of materials is usually determined from a monotonic tension test on smooth specimens with a cylindrical gage section. For such a test specimen:

$$S = \text{engineering stress} = P/A_o$$

$$e = \text{engineering strain} = \frac{\ell - \ell_o}{\ell_o} = \frac{\Delta \ell}{\ell_o}$$

where P is the applied load, ℓ_o is the original length, d_o is the original diameter, A_o is the original area, ℓ is the instantaneous length, d is the instantaneous diameter, and A is the instantaneous area.

However, because of changes in cross-sectional area during deformation, the *true* stress, which is larger than the engineering stress in tension (conversely, less in compression), is defined by:

$$\sigma = \text{true stress} = P/A$$

Similarly, in tension *true* strain is smaller than engineering strain (up to necking). Ludwik (circa 1909) defined *true* or *natural* strain, based on the instantaneous gage length, ℓ, as:

$$\epsilon = \text{true strain} = \int_{\ell_o}^{\ell} \left(\frac{d\ell}{\ell} \right) = \ell n \frac{\ell}{\ell_o}$$

The use of true stress and true strain merely changes the appearance of the monotonic tension stress-strain curve, as illustrated for a typical low-carbon steel in the figure above.

The engineering stress and strain may be related to true stress and strain as follows:

For strain:

$$\ell = \ell_o + \Delta \ell$$

Source: Fatigue and Microstructure, papers presented at the 1978 ASM Materials Science Seminar, 14–15 October 1978, St. Louis MO, sponsored by the Materials Science Division of the American Society for Metals, Metals Park OH, 1979, p 388

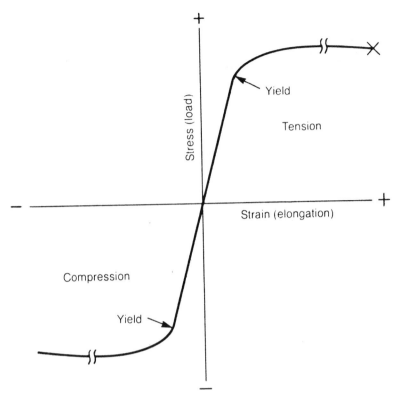

Complete engineering stress-strain curve showing the normally considered tensile region (upper right) and the oft-neglected compression region (lower left).

Normally, only the tensile part of the curve is shown. However, the straight-line portion also extends into the compression region, as shown above. In metals that have yield strengths, the compressive yield strength is usually considered to be approximately equivalent to the tensile yield strength. With ductile metals in compression, there is no definite end point. Consequently, the end point must be an arbitrarily selected value depending upon the degree of distortion that is regarded as indicating complete failure of the material. Certain metals fail in compression by a shattering type of fracture; these are normally the more brittle materials that do not deform plastically. Gray cast iron, which is relatively weak in tension because of the mass of internal graphite flakes, has a compressive strength that is several times its tensile strength.

Source: Donald J. Wulpi, Understanding How Components Fail, American Society for Metals, Metals Park OH, 1985, p 45

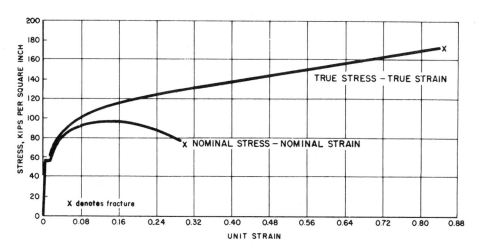

Comparison of the true stress–true strain curve with the engineering (nominal) stress–engineering (nominal) strain diagram.

It is possible to monitor the cross-sectional area of the sample during the test, so that the load can be divided by the actual minimum area in order to obtain the true stress. This is usually plotted against the true strain, which is given by $\ln(l/l_o)$. The figure above compares the two types of curves. Note that the true stress–true strain curve contains no maximum.

Source: Charlie R. Brooks, Heat Treatment, Structure and Properties of Nonferrous Alloys, American Society for Metals, Metals Park OH, 1982, p 3

1-16. Stress-Strain Behavior of a Brittle Material

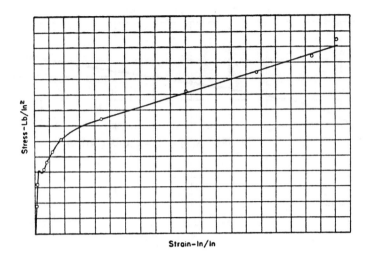

Stress-strain diagram of a brittle material.

Source: E. Paul Degarmo, Materials and Processes in Manufacturing, The Macmillan Co., New York, 1957, p 17

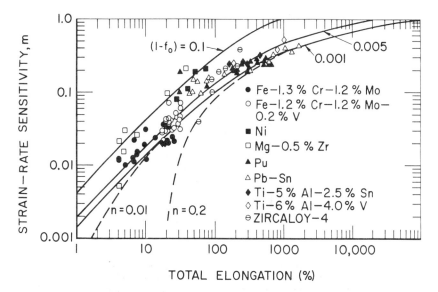

Relationship between total elongation and strain-rate sensitivity for round bar tensile tests on a number of engineering materials. Data are those of Woodford. Solid lines are predictions of Eq 1; broken lines are from Eq 2.

The correlation of expressions such as Eq 1 and Eq 2 below, with experimental data employing various values of f_o (e.g., 0.995, 0.999) and various n's has demonstrated the success of the above formulation in an approximate sense (figure above). From volume constancy, the nominal (engineering) strain outside the neck, e_f, which is a good measure of the total elongation, can be deduced from the above to be

$$e_f = (1 - f_o)^{-m} - 1 \qquad \text{(Eq 1)}$$

for a non-strain-hardening material. This relation can be used to estimate the total elongation of materials with varying initial imperfection sizes and m values. A smiliar expression can be derived and evaluated when the material shows power-law strain hardening:

$$e_f = \exp[n - m\ln(1 - f_o)] - 1 \qquad \text{(Eq 2)}$$

Source: S. L. Semiatin and J. J. Jonas, Formability and Workability of Metals, American Society for Metals, Metals Park OH, 1984, p 160

1-18. Comparison of Stress-Strain Curves: Ferrous vs Nonferrous Metals

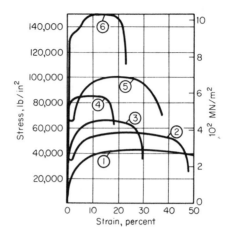

Comparative stress-strain diagrams. (1) Soft brass; (2) low-carbon steel; (3) hard bronze; (4) cold-rolled steel; (5) medium-carbon steel, annealed; (6) medium-carbon steel, heat treated.

The engineering tensile stress-strain curve is obtained by **static loading** of a standard specimen, that is, by applying the load slowly enough that all parts of the specimen are in equilibrium at any instant. The curve is usually obtained by controlling the loading rate in the tensile machine. ASTM Standard E8 requires a loading rate not exceeding 100,000 lb/in^2 (70 kg_f/mm^2)/min. An alternate method of obtaining the curve is to specify the strain rate as the independent variable, in which case the loading rate is continuously adjusted to maintain the required strain rate. A strain rate of 0.05 in/in/min is commonly used. It is measured usually by an extensometer attached to the gage length of the specimen. The figure above shows several stress-strain curves.

Source: Marks' Standard Handbook for Mechanical Engineers, Eighth Edition, Theodore Baumeister, Eugene A. Avallone, and Theodore Baumeister III, Eds., McGraw-Hill Book Co., New York, 1978, p 5-2

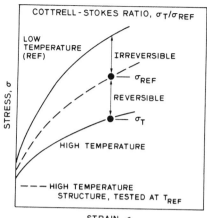

Schematic illustration of the Cottrell-Stokes experiment. The solid curves represent stress-strain curves at temperatures T and T_REF. The dashed curve is obtained by deforming the sample at T and measuring the flow stress at T_REF.

Fundamentals of Thermally Activated Flow. To understand the mechanisms of time-dependent flow at both high and low temperatures it is useful to draw a distinction between the kinetics of flow for a fixed structure and the kinetics of structure evolution. This distinction was first made by Cottrell and Stokes in their now-classic experiments on the reversible and irreversible contributions to the flow stress. The Cottrell-Stokes experiment is illustrated schematically above. The two solid curves represent the stress-strain curves at two different temperatures. The lower temperature is called the reference temperature. Cottrell and Stokes were the first to point out that the flow-stress difference at a particular strain is caused by two fundamentally different effects. The "reversible" flow-stress difference is the difference expected because of the different testing temperatures (the strain rates are the same in this illustration), assuming the defect structures to be the same. Of course, since the deformation experiments are conducted at different temperatures, the structures would not be the same. The sample deformed at the lower temperature (and at a higher stress) is expected to have a "harder" structure. The "irreversible" flow-stress difference arises from the differences in the structures, assuming the testing temperatures to be the same.

Source: Flow and Fracture at Elevated Temperatures, Rishi Raj, Ed., American Society for Metals, Metals Park OH, 1985, p 24

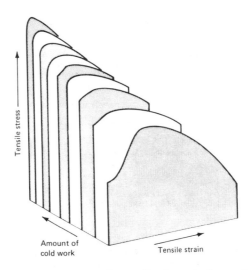

Effect of cold work on the tensile stress-strain curve for low-carbon steel bars.

A major difference between machined and cold drawn round bars is the improvement in tensile and yield strengths that results from the cold work of drawing. Cold work also changes the shape of the stress-strain diagram, as shown above. Within the range of commercial drafts, cold work markedly affects certain mechanical properties.

Source: Metals Handbook, Ninth Edition, Volume 1, Properties and Selection: Irons and Steels, American Society for Metals, Metals Park OH, 1978, p 221

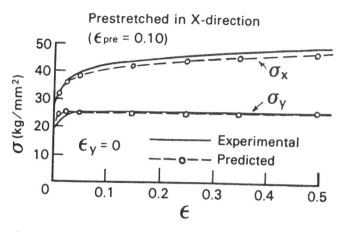

Stress-strain curve in plane strain compression in X-direction for steel (S10C) prestretched in X-direction.

The stress state under a given strain state is obtained from the yield locus by applying the normality rule for the strain vector. The calculated compressive stress-strain relations for plane strain are compared with experiments shown above. It is seen that the formulation described above is satisfactory.

Source: Mechanics of Sheet Metal Forming, Material Behavior and Deformation Analysis, Donald P. Koistinen and Neng-Ming Wang, Eds., proceedings of a symposium sponsored by General Motors Research Laboratories, 17–18 October 1977, Warren MI, Plenum Press, New York, 1978, p 107

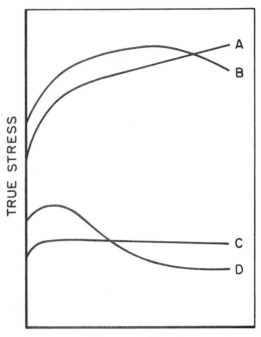

Typical flow curves for metals deformed at cold working temperatures (A–low strain rate, B–high strain rate) and at hot working
temperatures (C, D). Strain hardening persists to large strains for
curve A. The flow stress maximum and flow softening in curve B
arise from deformation heating. The steady-state flow stress exhibited by curve C is typical of metals that dynamically recover.
The flow stress maximum and flow softening in curve D may result from a number of metallurgical processes.

Source: S. L. Semiatin and J. J. Jonas, Formability and Workability of Metals, American Society for Metals, Metals
Park OH, 1984, p 5

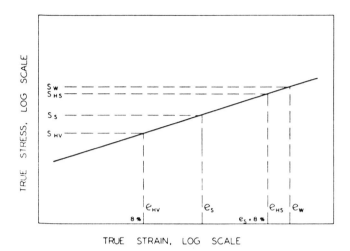

Sketch of the common true stress–true strain relationship in double logarithmic coordinates.

The close correlation between the bulk hardness, the hardness of the abraded surface, and the wear resistance can be explained qualitatively in the following way.

For many metals the relation between true stress and true strain is approximately represented by the equation

$$S = C \cdot e^n$$

where S is the true stress; C is a constant, the "strength coefficient"; e is the true strain; and n is a constant, the "strain-hardening exponent." Thus, the true stress–true strain relationship will fit a straight line in double logarithmic coordinates, as sketched in the figure above. Tabor has shown that the Vickers hardness number for metals is approximately three times the true yield stress at a true strain of 8%. These values are shown above as S_{HV} and e_{HV}. The heavily deformed surface material will be located at a point given by the stress S_S and the strain e_S. When the hardness of the material is measured a point given by S_{HS} and e_{HS} is reached under the indenter, where e_{HS} is e_S + 8%. The strain obtained during metal removal by abrasion is not known. It will probably be in the vicinity of e_{HS}, towards somewhat higher strains. This point is sketched above as S_W, e_W. Its exact location is of limited importance in this connection.

Source: Source Book on Wear Control Technology, David A. Rigney and W. A. Glaeser, Eds., American Society for Metals, Metals Park OH, 1978, p 123

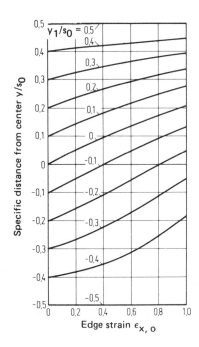

Shift of individual fibers in cross section with increasing boundary strain.

Source: Handbook of Metal Forming, Kurt Lange, Ed., McGraw-Hill Book Co., New York, 1985, p 19.11

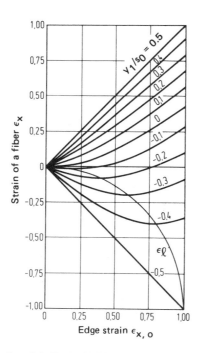

Strain of individual fibers with increasing boundary strain.

Source: Handbook of Metal Forming, Kurt Lange, Ed., McGraw-Hill Book Co., New York, 1985, p 19.11

1-26. Axial Strain in Sheet Metal

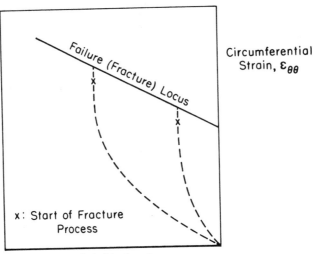

Relation between end of flow localization and start of the fracture process during upsetting of specimens that fail by surface fracture.

The variation in the localization strain, or limit strain as it is often called, predicted by these models follows a trend very similar to the measured failure loci of metals. The major drawback of the analysis lies in the need to set an initial value for f_o. A similar deficiency occurs in the analysis of necking in sheet stretching. However, the analyses do offer considerable insight into the effects of material properties such as n (and m in modified models similar to that just presented) and stress ratio on the limit strains.

Source: S. L. Semiatin and J. J. Jonas, Formability and Workability of Metals, American Society for Metals, Metals Park OH, 1984, p 50

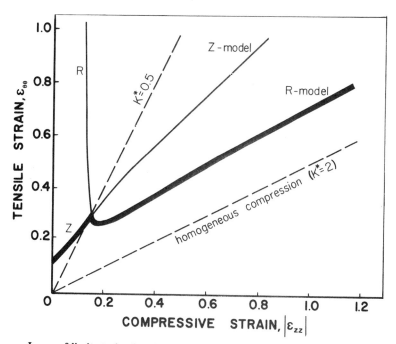

Locus of limit strains for the R and Z models. The lower parts of each model are used to construct the actual failure locus (heavy line). Calculations were based on $f_o = 0.99$, $n = 0.25$, and a prestrain of 0.10, resulting in a flow curve of the form $\bar{\sigma} = K(\bar{\epsilon} + 0.10)^{0.25}$.

Source: S. L. Semiatin and J. J. Jonas, Formability and Workability of Metals, American Society for Metals, Metals Park OH, 1984, p 51

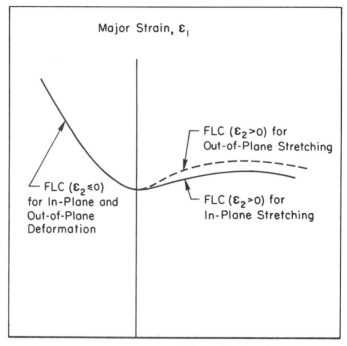

Typical forming limit diagrams, or loci of pincipal surface strains at failure, for sheet metals deformed in-plane and out-of-plane (as in punch forming).

The failure data from both the in-plane and punch-stretching tests are most conveniently presented above in terms of ϵ_1 (denoted the *major* strain) versus ϵ_2 (denoted the *minor* strain) diagrams. Since Keeler, who studied strain states for which $\epsilon_2 \geq 0$, and Goodwin, who studied strain states for which $\epsilon_2 < 0$, were among the first to present data in this way, these are often called Keeler-Goodwin diagrams, especially when the data are for low-carbon steels. Other names, such as forming limit curves (FLC's) and forming limit diagrams (FLD's) are also widely used when referring to these plots.

Source: S. L. Semiatin and J. J. Jonas, Formability and Workability of Metals, American Society for Metals, Metals Park OH, 1984, p 39

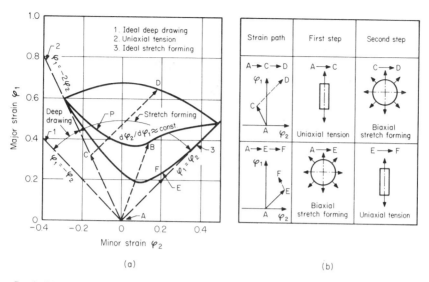

Strain history and forming-limit diagram structure. (a) Influence of strain history on location of forming-limit curve. (b) Schematic presentation of various strain paths.

Source: Handbook of Metal Forming, Kurt Lange, Ed., McGraw-Hill Book Co., New York, 1985, p 18.14

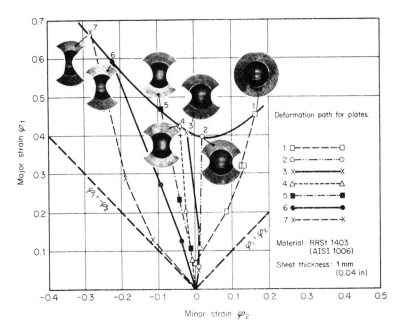

Forming-limit diagram obtained from drawing tests with specially shaped blank.

Improvements on these testing methods have been made by using circular-shaped strips with material removed on the side as shown above. The strips are drawn until cracks occur. With details from these tests, the forming-limit diagram can be obtained for strain paths ranging from biaxial tension (stretch forming) to equal tension and compression (deep drawing). The diagram must be determined for each particular sheet material. The forming-limit diagram is also influenced by the diameter of the circular grids, the relative position of the strip with respect to the rolling direction, the lubricant, the sheet thickness, the deformation (strain) history, and the strain rate.

Source: Handbook of Metal Forming, Kurt Lange, Ed., McGraw-Hill Book Co., New York, 1985, p 18.12

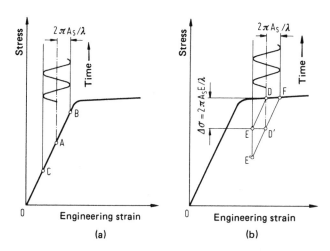

Stress-strain diagrams with superimposed vibrations. (a) In range of elastic recovery. (b) In range of plastic deformation.

This figure shows both cases by means of an idealized stress-strain diagram in which an additional time coordinate is used to show the superposition of a vibrating stress having the strain amplitude $\hat{\epsilon}_S = (2\pi/\lambda) A_S$.

The test specimen is considered to be strained within the elastic range only up to point A, which lies quite below the yield limit (see A, above). The specimen deforms up to point B because of the additional vibrating stress acting on it during the first half-period of the imparted vibration, whereby point B still remains within the elastic range of the stress-strain diagram. In this manner the curve of imparted stress sets forth periodically within the limiting points B and C. However, at the center of the vibration period the magnitude of the static stress remains the same as at point A so that externally no reduction in total stress is registered.

In the second case, however (curve b, above), the specimen is subjected to plastic deformation and elongated up to point D, which lies beyond the elastic range. The superimposed vibration at this stage reduces the stress on the specimen up to point E, from which the plastic deformation of the specimen proceeds beyond point D up to point F. During the following periods of vibration the stress varies between points E' and F, this deformation being of purely elastic nature only. Because of this phenomenon, the static stress is reduced from D to D' in the center of the period of vibration, the amount of stress reduction depending on the strain amplitude of the imparted vibration as well as on the modulus of elasticity of the test material.

Source: Handbook of Metal Forming, Kurt Lange, Ed., McGraw-Hill Book Co., New York, 1985, p 29.4

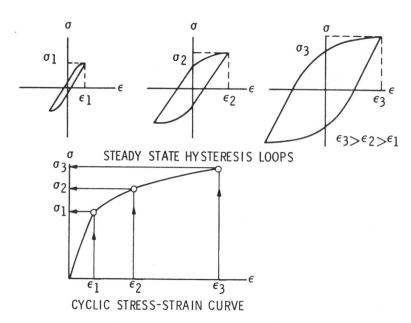

Construction of cyclic stress-strain curve by joining tips of stabilized hysteresis loops.

In the particular example shown above, three companion specimens were tested to failure at three different controlled strain amplitudes. The steady-state stress response, measured at approximately 50% of the life to failure, is thereby obtained. These stress values are then plotted at the appropriate strain levels to obtain the cyclic stress-strain curve.

Source: Fatigue and Microstructure, papers presented at the 1978 ASM Materials Science Seminar, 14–15 October 1978, St. Louis MO, sponsored by the Materials Science Division of the American Society for Metals, Metals Park OH, 1979, p 398

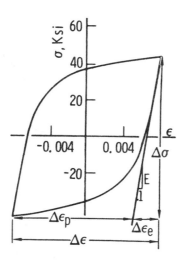

Steady-state stress-strain hysteresis loop.

The cyclic behavior of metals is best described in terms of a stress-strain hysteresis loop, as illustrated above.

For completely reversed, strain-controlled conditions with zero mean strain, the total width of the loop is $\Delta\epsilon$, or total strain range:

$$\Delta\epsilon = 2\epsilon_a \ (\epsilon_a = \text{strain amplitude})$$

The total height of the loop is $\Delta\sigma$, or the total stress range:

$$\Delta\sigma = 2\sigma_a \ (\sigma_a = \text{stress amplitude})$$

The difference between the total and elastic strain amplitudes is the plastic-strain amplitude.

Since:

$$\frac{\Delta\epsilon}{2} = \frac{\Delta\epsilon_e}{2} + \frac{\Delta\epsilon_p}{2}$$

then:

$$\frac{\Delta\epsilon_p}{2} = \frac{\Delta\epsilon}{2} - \frac{\Delta\epsilon_e}{2} = \frac{\Delta\epsilon}{2} - \frac{\Delta\sigma}{2E}$$

Changes in stress response of a metal occur rapidly during the first several percent of the total reversals to failure. The metal, under controlled strain amplitude, will eventually attain a steady-state stress response. To construct a *cyclic* stress-strain curve, one connects the tips of the stabilized hysteresis loops from comparison specimen tests at several controlled strain amplitudes.

Source: Fatigue and Microstructure, papers presented at the 1978 ASM Materials Science Seminar, 14–15 October 1978, St. Louis MO, sponsored by the Materials Science Division of the American Society for Metals, Metals Park OH, 1979, p 397

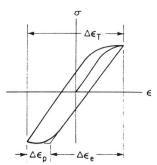

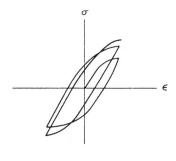

A. TOTAL, PLASTIC AND ELASTIC
   STRAIN RANGES

B. CYCLIC HARDENING

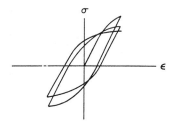

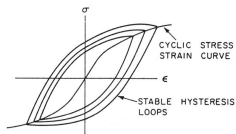

C. CYCLIC SOFTENING

D. STABLE HYSTERESIS LOOPS AND
   CYCLIC STRESS STRAIN CURVE

Cyclic stress-strain behavior.

Fatigue failure in less than about 10^4 cycles typically requires significant cyclic plastic strains. This regime is termed low-cycle fatigue and here the fatigue life is usually correlated with the cyclic plastic strain amplitude. Plots of stress versus strain for cyclic loading between set strain limits are shown above. Results for a single cycle are illustrated in A. The material first yields in tension, then in compression, and again in tension to complete the cycle and generate a hysteresis loop. The total strain range, $\Delta\epsilon_T$, is the sum of the plastic strain range, $\Delta\epsilon_p$, and the elastic strain range, $\Delta\epsilon_e$. The cyclic plastic strain amplitude is $\Delta\epsilon_p/2$. As cycling continues the material may cyclically strain harden or cyclically strain soften as illustrated in B and C, respectively. As a rule of thumb, high-strength alloys cyclically strain soften and low-strength alloys cyclically strain harden. Eventually a stable hysteresis loop is obtained. If stable hysteresis loops are obtained at a number of strain ranges, the tips of these loops may be connected to form a cyclic stress-strain curve. This curve describes the material flow behavior under cyclic plastic straining.

Source: Steel Castings Handbook, Fifth Edition, Peter F. Wieser, Ed., Steel Founders Society of America, Rocky River OH, 1980, p 4-12

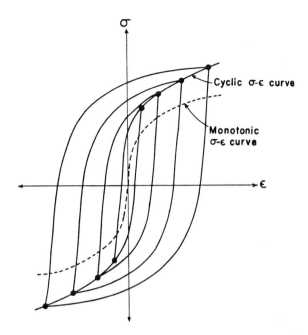

Schematic representation of monotonic and cyclic stress-strain curves for a material that cyclic hardens. Points represent tips of stable loops for a particular plastic-strain amplitude.

The cyclic stress-strain curve may be determined in several ways, but it is usually obtained by connecting the tips of stable hysteresis loops from constant-strain-amplitude fatigue tests of companion samples at different strain amplitudes. A schematic illustration of this method is shown above. Under conditions where no saturation is reached, which is the case for many engineering materials, one normally uses either the stress amplitude obtained at half-life, or the maximum stress amplitude for the case of hardening, or the minimum stress amplitude for the case of softening. However, it should be recognized that the cyclic stress-strain curve derived in this manner is only an approximation of the actual cyclic plastic behavior.

Source: Fatigue and Microstructure, papers presented at the 1978 Materials Science Seminar, 14–15 October 1978, St. Louis MO, sponsored by the Materials Science Division of the American Society for Metals, Metals Park OH, 1979, p 209

1-36. Effect of Metal Composition on Monotonic and Cyclic Stress-Strain Behavior

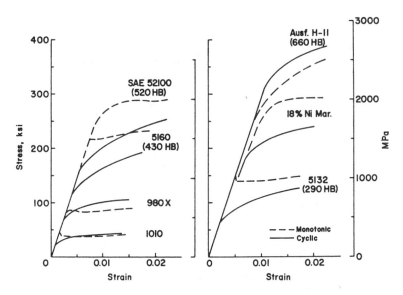

Monotonic and cyclic stress-strain behaivor of representative alloys.

Because material flow properties may be altered by reversed deformation, it is necessary to use a *cyclic* stress-strain relation in fatigue studies. Monotonic and cyclic stress-strain curves for a series of representative steels are compared above. Ferritic-pearlitic steels, such as hot-rolled low-carbon SAE 1010 and microalloyed SAE 980 X, exhibit some cyclic softening at low strains followed by cyclic hardening at higher strains. Quenched-and-tempered martensitic steels SAE 5132, 5160, and 52100 all show significant cyclic softening. Such behavior is typical of a broad range of low- and intermediate-hardness martensitic alloys. Similar softening trends are observed for the precipitation-strengthened 18% Ni maraging steel. Ausformed H-11 steel, on the other hand, is seen to cyclically harden above its already high monotonic level.

Source: Fatigue and Microstructure, papers presented at the 1978 ASM Materials Science Seminar, 14–15 October 1978, St. Louis MO, sponsored by the Materials Science Division of the American Society for Metals, Metals Park OH, 1979, p 442

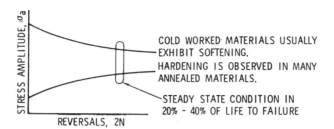

COLD WORKED MATERIALS USUALLY EXHIBIT SOFTENING.

HARDENING IS OBSERVED IN MANY ANNEALED MATERIALS.

STEADY STATE CONDITION IN 20% - 40% OF LIFE TO FAILURE

Steady-state stress response for strain-controlled cycling.

By plotting the stress amplitude versus reversals from controlled-strain test results, one can observe cyclic strain hardening and softening, as illustrated here.

Thus, through cyclic hardening and softening, some intermediate strength level is attained that represents a steady-state condition (in which the stress required to enforce the controlled strain does not vary significantly). Some metals are cyclically stable, in which case their monotonic stress-strain behavior adequately describes their cyclic response. The steady-state condition is usually achieved in about 20 to 40% of the total fatigue life in either hardening or softening materials.

Source: Fatigue and Microstructure, papers presented at the 1978 ASM Materials Science Seminar, 14–15 October 1978, St. Louis MO, sponsored by the Materials Science Division of the American Society for Metals, Metals Park OH, 1979, p 396

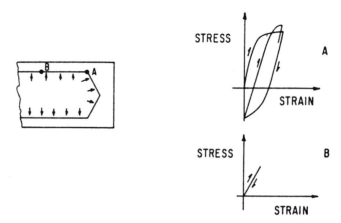

Cyclic effects in pressure vessels.

Many potential fatigue concerns arise in start-stop operation from low-cycle fatigue, where the source of cyclic strain is of mechanical rather than thermal origin and arises because of geometric effects such as notches. A schematic view of how this comes about is shown in the pressure vessel above. Because of repeated loading and unloading, cyclic stresses are induced in the structure. Two points, A and B, are considered. Region B responds entirely elastically to the cyclic pressure, while region A, because of the strain concentration, develops a plastic zone. The accompanying stress-strain behavior of the two regions shows the cyclic elastic strain produced at B and the development of a hysteresis loop of stress versus strain at A, indicating cyclic plastic strain. In the absence of any well-defined plastic zone, failure at B would be a case of high-cycle fatigue, while crack initiation at A is a low-cycle-fatigue problem. Obviously, region A is the more critical one from design and failure-analysis considerations.

Source: Fatigue and Microstructure, papers presented at the 1978 Materials Science Seminar, 14–15 October 1978, St Louis MO, sponsored by the Materials Science Division of the American Society for Metals, Metals Park OH, 1979, p 6

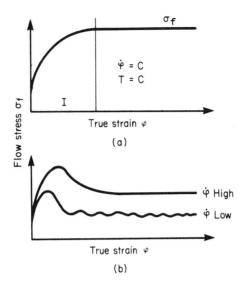

Dynamic recovery and recrystallization processes affect the course of the flow curves in a typical manner. Hot flow curves with constant or slightly falling yield stress are typical for dynamic recovery (see a). On the contrary, the flow curves during dynamic recrystallization (after initial hardening) show a sudden decrease in yield stress which then adopts a constant or slightly falling course (see b).

Source: Handbook of Metal Forming, Kurt Lange, Ed., McGraw-Hill Book Co., New York, 1985, p 3.21

1-40. Load-Strain Plot

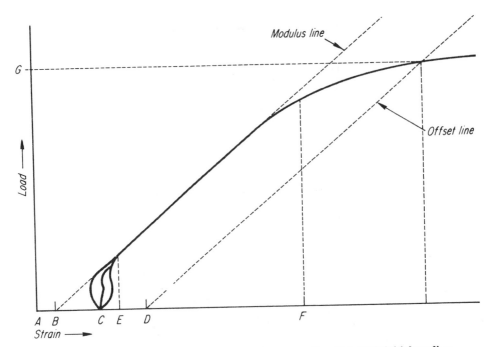

Load-strain plot shows zero error sources and compensation. Point C is initial reading (tare) for strain. Value is subtracted from all subsequent readings to correct measurement zero error. Point E is strain threshold for start of least-squares fit. Fit operation continues between points E and F. Point B is intercept point, where zero strain would have been if sample and extensometer were perfect. Point G is load for offset yield.

Source: Metal Progress, August 1980, American Society for Metals, Metals Park OH, p 46-5

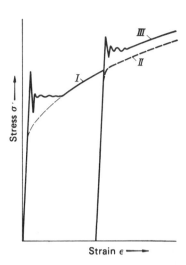

This figure illustrates the course of stepped tensile tests. Here curve II indicates the course during faster reloading and curve III during loading after aging has taken place. Apart from the increase in yield point mentioned above, the aging of steels also causes embrittlement. The elongation values are reduced and the ductile-brittle transition temperature of notch toughness is raised. In the case of components under load, this effect must be considered carefully.

The rate of diffusion rises with increasing temperature. In Cottrell's view there should be a temperature range where the rate of diffusion of carbon and nitrogen atoms matches the rate of the dislocation movement. In this temperature range we encounter a higher yield stress since the dragged atoms hinder dislocation movement considerably. In fact, steels with low carbon content show an increase in yield stress of between about 400 and 750 K. Because of a simultaneous drop of the elongation values, this process is also called *blue brittleness*.

Source: Handbook of Metal Forming, Kurt Lange, Ed., McGraw-Hill Book Co., New York, 1985, p 3.15

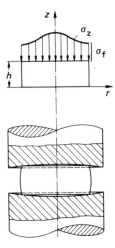

**Normal stress distribution and tool-
workpiece deformation in coining.**

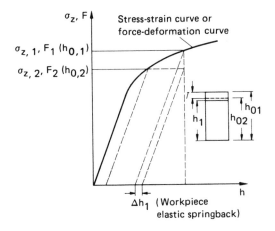

**Schematic representation of variations
of elastic springback in coining.**

Source: Handbook of Metal Forming, Kurt Lange, Ed., McGraw-Hill Book Co., New York, 1985, p 9.13

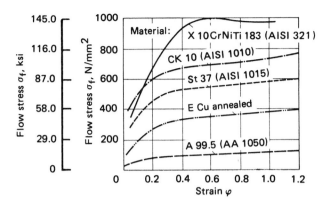

Flow curves of some metals at room temperature.

From the viewpoint of metal physics, the room temperature is of no importance with respect to the elementary processes of plastic deformation. However, for practical reasons, cold and hot forming are defined as deformation at room temperature and deformation at elevated temperatures, respectively. The figure above shows the flow curves of some metals which at room temperature are not affected by thermally activated processes. In such cases the flow stress increases monotonically with growing strain, the slope $d\sigma_f/d\varphi$ of the curve being highest at low strains.

Source: Handbook of Metal Forming, Kurt Lange, Ed., McGraw-Hill Book Co., New York, 1985, p 4.3

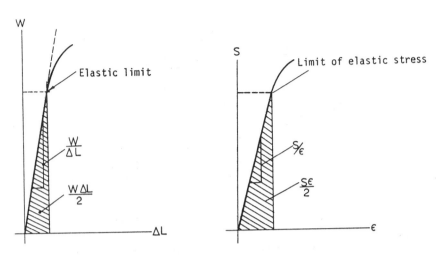

(a) Load-deflection diagram. (b) Stress-strain diagram.

The original load-deflection diagram at left can now be replotted as shown at right, where the slope of the curve is also constant below the elastic limit of the material. Here the proportionality between the stress and strain determines the fundamental quantity in the mechanics of materials known as the *modulus of elasticity:*

$$E = \frac{S}{\epsilon}$$

Since S is expressed, for instance, in pounds per square inch, and strain ϵ is nondimensional by definition, the dimensions of the modulus of elasticity must be the same as those of the stress.

Source: Alexander Blake, Practical Stress Analysis in Engineering Design, Marcel Dekker, Inc., New York, 1982, p 6–7

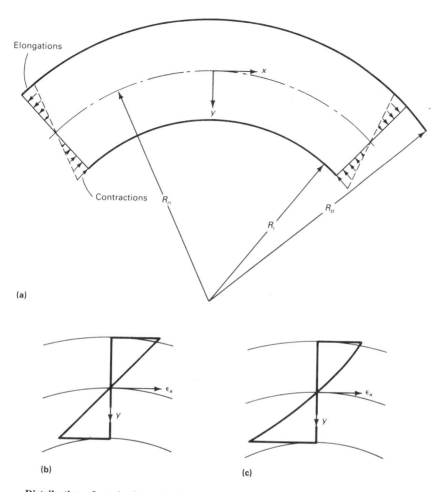

(a)

(b) (c)

Distribution of strain determined by the simple-beam theory. (a) Linear distribution for fiber elongations and contractions. (b) Distribution of engineering strain. (c) Distribution of true strain. R_n = radius of neutral axis; R_i = inner radius; R_o = outer radius

Elastic analysis of bending deformation can be performed by simple-beam theory, elasticity solutions, and numerical methods such as the finite-difference and Rayleigh-Ritz methods. Generally, numerical methods are suitable for bending of specimens that are subjected to complex loading patterns and that have irregular and/or varying cross-sectional areas. Elasticity solutions are useful when accuracies better than ~5% are desired. Simple-beam theory is used in most testing applications in which plates, strips, bars, and rods are bent in three- or four-point bending modes. The basic assumptions of the simple-beam theory for pure elastic bending (shear force = 0) are: (1) all sections that are initially plane and perpendicular to the axis of the beam remain plane and perpendicular to it after bending; (2) all longitudinal elements (fibers) bend into concentric circular arcs (hence, cylindrical bending); and (3) a one-dimensional stress state is assumed, and the same stress-strain relationship is used for tension and compression.

The first assumption implies a linear distribution for fiber elongations and contractions, as shown in (a) above. The resulting strain distributions (b) and (c) above are given by:

Engineering bending strain,

$$\epsilon_x = -\frac{y}{R_n}$$

True bending strain,

$$\epsilon_x = \ln\left(1 - \frac{y}{R_n}\right)$$

where R_n is the radius of curvature of the neutral axis. For $\epsilon_x \leq 0.1$, the difference between these two strain definitions is $\leq 5\%$. Therefore, for most elastic bendings, the engineering strain definition is sufficiently accurate and more convenient.

Source: Metals Handbook, Ninth Edition, Volume 8, Mechanical Testing, American Society for Metals, Metals Park OH, 1985, p 119

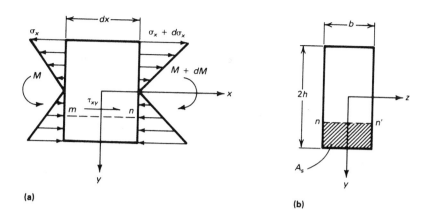

(a)

(b)

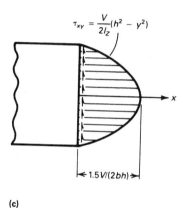

$$\tau_{xy} = \frac{V}{2I_z}(h^2 - y^2)$$

1.5 V/(2bh)

(c)

Distribution of shear stress, τ_{xy}, for a rectangular specimen

In a majority of testing applications such as three-point bending, roll bending, and press-brake forming, the applied bending moment varies along the length of the specimen. Because shear force $V = (dM/dx)$, such variations in the bending moment imply a nonzero shear force. Therefore, condition 1 for cylindrical bending is not met. The resulting shear stress, τ_{xy}, which is determined from the equilibrium considerations at a typical section m-n in (a) above, is:

$$\tau_{xy} = \frac{VQ}{I_z b}$$

where

$$Q = \int_{A_s} y\,dA$$

is the first moment of the shaded area (A_s) with respect to the neutral axis (b) above. The first moment of the unshaded area with respect to the neutral axis gives the same Q. The distribution of τ_{xy} for a rectangular cross section is shown in (c) above.

Source: Metals Handbook, Ninth Edition, Volume 8, Mechanical Testing, American Society for Metals, Metals Park OH, 1985, p 121

1-47. Stress Distribution in a Beam

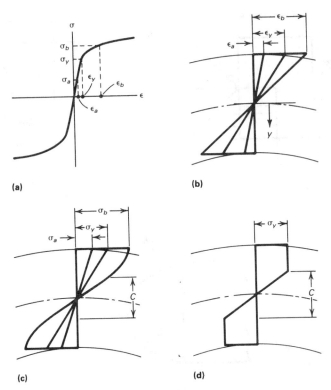

(a)

(b)

(c)

(d)

Stress-strain distribution in a beam. (a) Stress-strain curve. (b) Strain distribution. (c) Stress distribution. (d) Stress distribution for elastic-perfect plastic material.

For a beam with the stress-strain curve shown in (a) above, the development of longitudinal strain and stress at different stages of deformation is shown in (b) and (c), respectively. In (b), a linear strain distribution given by $\epsilon_x = (-y/R_n)$ was used. When the strain at the outermost fiber exceeds 0.1, it is suggested that the true strain description be used. For bending to a radius of curvature equal to R_n, the strain distribution and the subsequent stress distribution (from the σ-ϵ curve or from a known constitutive equation for the σ-ϵ dependence) can be found. The moment, M, required to produce R_n is:

$$M = -\int_A (\sigma_x dA)y$$

The thickness of the elastic core, C (in (c) above), is:

$$C = 2R_n \left(\frac{\sigma_y}{E}\right)$$

Therefore, in bending plates of the same material to the same radius of curvature R_n, the fractional thickness of the elastic core $C/2h$ becomes smaller as the thickness increases.

For elastic-plastic bending, a general equation for the relationship between R_n and M does not exist. For the simple case of an elastic-perfect plastic material (see (d) above), the following equation is obtained:

$$R_n = \frac{E}{\sigma_y} \left[3 \left(\frac{\sigma_y bh^2 - M}{b\sigma_y} \right) \right]^{1/2}$$

The predictions of this equation at large plastic deformations ($C/2h \le 0.02$) are not reliable.

Source: Metals Handbook, Ninth Edition, Volume 8, Mechanical Testing, American Society for Metals, Metals Park OH, 1985, p 121

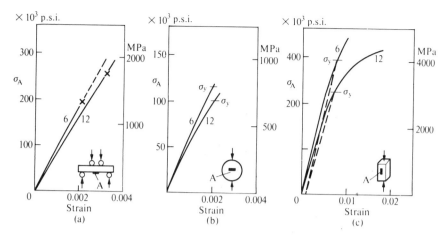

Stress-strain curves. (a) Bending (uniaxial tensile stress). (b) Disc test with strain gage mounted transverse to load at center of disc. (c) Compression test with strain gage mounted parallel to load. X marks the point of fracture and σ_y the yield stress. (After Takagi and Shaw, 1981.)

This figure shows stress-strain results for typical bending, disc, and compression tests. The strain readings are from very-small-gage-length strain gages made from foil having an unusually high yield strain since they were to be used to measure strains extending slightly beyond the yield strain of the tungsten carbides employed. The gages were mounted as shown by the insets at the lower right of each graph. The curves are slightly nonlinear all the way to the origin (zero applied load), and it is difficult to identify an exact yield point.

Source: Milton C. Shaw, Metal Cutting Principles, Clarendon Press, Oxford, 1984, p 127

1-49. Dependence of Shear Stress and Axial Stress on Effective Strain

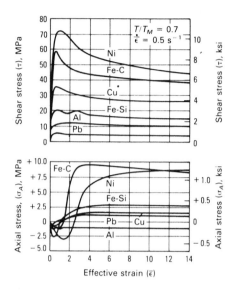

Dependence of shear stress and mean axial stress on effective strain in fixed-end torsion tests at high temperatures.

Triangular or hexagonal ends are most common when tests are to be run in stroke control to prevent changes in gage section geometry that may result from the axial extension that characterizes the torsion testing of many metals. Frequently, threaded ends can also be used in such situations; however, triangular or hexagonal ends have the added advantage of allowing rapid specimen removal for purposes of quenching, which is beneficial in subsequent metallurgical analysis. The possible development of axial stresses and their effect on flow and failure response in fixed-end tests should be carefully considered, because these stresses can be a significant fraction of the torsional shear stress (see figure above).

The final step in the preparation of torsion specimens for determination of workability is the application of a fine axial line on the gage section surface. This may be done using a felt-tip pen for room-temperature tests or a fine metal scribing instrument for elevated-temperature tests. After twisting to failure, the line may still be straight (i.e., it will form a helix at a fixed angle to the torsion axis), indicating fracture-controlled failure, or it will exhibit a "kink" at a larger angle to the torsion axis than the remainder of the line, indicating flow-localization-controlled failure.

Source: Metals Handbook, Ninth Edition, Volume 8, Mechanical Testing, American Society for Metals, Metals Park OH, 1985, p 157

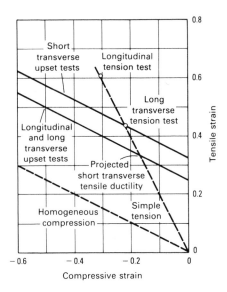

Comparison of the fracture strains from tension tests with those from upset tests. Data points are tension test results for zero-gage-length true strain at fracture.

When the value of ϵ_f is measured for the long transverse tension test, it plots on the figure above at the point where a line with slope 2 intersects the fracture-limit line for the short transverse upset test orientation. This leads to the prediction that the short transverse zero-gage-length fracture strain for the plate, which is difficult to measure because of thickness limitations, should be where the simple tension line intersects the fracture-limit line for longitudinal and long transverse upset tests.

Source: Metals Handbook, Ninth Edition, Volume 8, Mechanical Testing, American Society for Metals, Metals Park OH, 1985, p 581

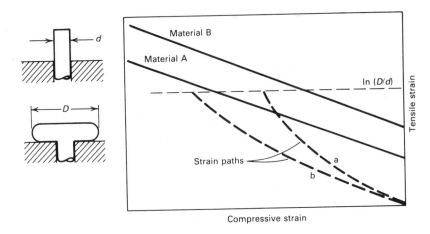

Comparison of strain paths and fracture locus lines.

A fracture-limit diagram can be used as a forming-limit diagram. For example, consider a simple bolt-heading process such as that illustrated above. If material A is used for the product, strain path a will cross the fracture line on its way to its position in the final deformed geometry, and cracking is likely to occur. Defects can be avoided in two ways: material B, which has a higher forming-limit line, can be used, or the strain path can be altered to follow path b, which in this case is accomplished through improved lubrication. In other deformation processes, the strain path can be altered by changing the preform design.

Source: Metals Handbook, Ninth Edition, Volume 8, Mechanical Testing, American Society for Metals, Metals Park OH, 1985, p 581

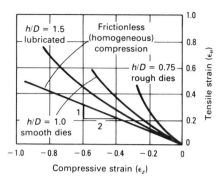

Strain paths in upset test specimens.

Free-Surface Strains. At the free surfaces of the compressed cylinders, the strains consist of circumferential tension and axial compression. For frictionless (homogeneous) compression of cylindrical specimens, the tensile strain is equal to one half of the compressive strain. With increasing frictional constraint, bulge severity increases, the tensile strain becomes larger, and the compressive strain decreases. Effects of the bulged profile, which develops naturally in cylindrical compression specimens, are imposed artificially through tapered and flanged compression tests.

The figure above summarizes the effects of friction, aspect ratio, and specimen profile on the measured free-surface strains at midheight. Measured strain paths are shown in terms of circumferential (tensile) strain versus axial (compressive) strain. Beginning with the strain ratio of one half for homogeneous deformation, the strain-path slope increases with increasing friction. For a given value of friction, a decreasing aspect ratio increases the strain-path slope slightly. Tapered compression specimens further increase the strain-path slope, and flanged compression specimens result in strain paths that lie nearly along the circumferential tensile strain axis.

Source: Metals Handbook, Ninth Edition, Volume 8, Mechanical Testing, American Society for Metals, Metals Park OH, 1985, p 580

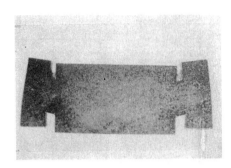

Statically deformed double-notch shear specimen.

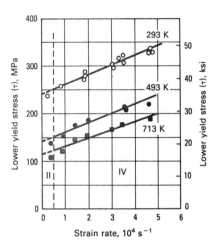

Variation of lower yield stress with strain rate.

In double-notch shear testing, the output bar is replaced by a tube, into which the input bar can slide. The lower end of the input bar and the upper end of the output tube are slotted to accommodate the thin plate specimen, into which two pairs of notches have been cut. With an effective gage length in this specimen of 0.84 mm (0.033 in.), a maximum shear strain rate of 40 000 s^{-1}, an order of magnitude greater than that reached in the standard Kolsky bar apparatus, has been achieved.

The principal disadvantage of this technique is that at shear strains greater than about 20% the specimen ceases to deform in pure shear. A double-notch shear specimen statically deformed to an apparent shear strain of about 60%, determined from the displacement across the loading surface, is shown in the figure above, left.

As shown, the end pieces have rotated with respect to the center and have deformed in compression. A calibration of the actual shear strain in the region under the notches yields an average value of about 20%. Reliable results can be obtained only at relatively low values of shear strain. Because about 20 μs is required for a constant strain rate to become established, results obtained at low strains can only be associated with an average value of strain rate.

Despite these problems, measurements can be taken at strain rates in excess of 10^4 s^{-1}. In this region, the lower yield stress, as shown above, is often found to be directly proportional to strain rate, rather than to the logarithm of strain rate. This implies a new region of mechanical response (region IV in the graph above), often considered to be controlled by a viscous damping mechanism, in contrast to region II, where thermally activated processes are usually assumed to control deformation.

Source: Metals Handbook, Ninth Edition, Volume 8, Mechanical Testing, American Society for Metals, Metals Park OH, 1985, p 229

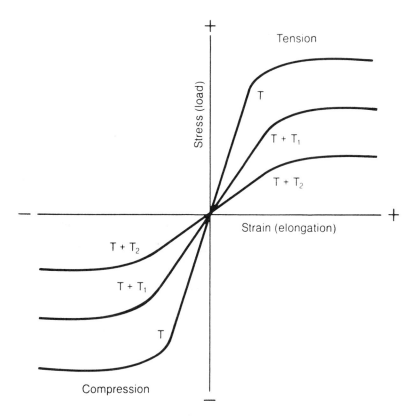

Effect of elevated temperatures T_1 and T_2 on tensile and compressive properties of a typical metal.

The modulus of elasticity is reduced when a metal at elevated temperature is in compression as it is when in tension at an elevated temperature. This is shown schematically above. Temperature T represents an arbitrarily selected base temperature, such as room temperature, while T_1 and T_2 represent elevated temperatures. Note not only the decrease in the modulus of elasticity (the slope of the straight-line portion) but also the decrease in yield and tensile (compressive) strengths with increasing temperature.

The preceding paragraph becomes exceedingly important in the understanding of residual (internal) stresses caused by thermal means, such as in welding. These stresses are the reason for weldment distortion and/or fracture resulting from tensile residual stresses caused by shrinkage during solidification and cooling of the weld.

Source: Donald J. Wulpi, Understanding How Components Fail, American Society for Metals, Metals Park OH, 1985, p 46

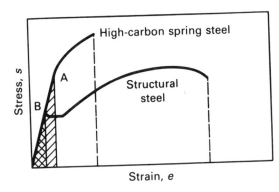

Comparison of stress-strain curves for high- and low-toughness steels. Cross-hatched regions in this curve represent the modulus of resilience, U_R, of the two materials. The U_R is determined by measuring the area under the stress-strain curve up to the elastic limit of the material. Point A represents the elastic limit of the spring steel; point B that of the structural steel.

The toughness of a material is its ability to absorb energy in the plastic range. The ability to withstand occasional stresses above the yield stress without fracturing is particularly desirable in parts such as freight-car couplings, gears, chains, and crane hooks. Toughness is a commonly used concept that is difficult to precisely define. Toughness may be considered to be the total area under the stress-strain curve. This area, which is referred to as the modulus of toughness, U_T, is an indication of the amount of work per unit volume that can be done on the material without causing it to rupture.

The figure above shows the stress-strain curves for high- and low-toughness materials. The high-carbon spring steel has a higher yield strength and tensile strength than the medium-carbon structural steel. However, the structural steel is more ductile and has a greater total elongation. The total area under the stress-strain curve is greater for the structural steel; therefore, it is a tougher material. This illustrates that toughness is a parameter that comprises both strength and ductility.

Several mathematical approximations for the area under the stress-strain curve have been suggested. For ductile metals that have a stress-strain curve like that of the structural steel, the area under the curve can be approximated by:

$$U_T \approx s_u e_f$$

or

$$U_T \approx \frac{s_0 + s_u}{2} e_f$$

For brittle materials, the stress-strain curve is sometimes assumed to be a parabola, and the area under the curve is given by:

$$U_T \approx \frac{2}{3} s_u e_f$$

Source: Metals Handbook, Ninth Edition, Volume 8, Mechanical Testing, American Society for Metals, Metals Park OH, 1985, p 23

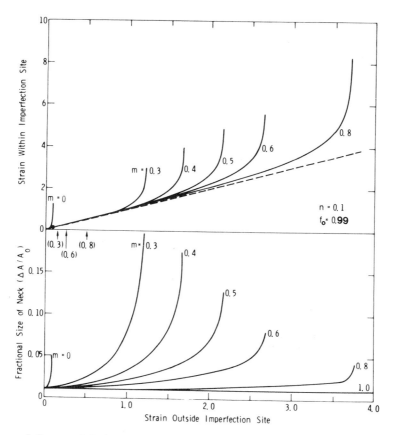

Influence of strain-rate sensitivity, m, on the growth of geometric (machining) imperfections. Strain with the imperfection as well as neck size increase from the outset. Considère strain (i.e., strain for $\gamma = 1$) is indicated by the solid circle, and the Hart strains for various values of m are indicated by the vertical arrows.

The following formula relating the strain away from and at the imperfection (where necking occurs) can be obtained:

$$\int_0^{\epsilon_h} \epsilon_h^{n/m} \exp(-\epsilon_h/m)d\epsilon_h = f_0^{1/m} \int_0^{\epsilon_i} \epsilon_i^{n/m} \exp(-\epsilon_i/m)d\epsilon_i$$

Because n and m are arbitrary numbers, the above equation cannot be integrated directly. However, it should be noted that it does not contain an explicit strain-rate dependence, except as $\dot{\epsilon}$ may affect m or n.

By evaluating this equation numerically, Ghosh has demonstrated the effect of variations in n, m, and f_0 on the rate of strain localization in a tensile bar. The effect is particularly strong with regard to m (see figure above). This dependence can also be seen from the following equation, with n set equal to zero. In this case, the equation is integrable with the result:

$$\exp(-\epsilon_h/m)\big|_0^{\epsilon_h} = f_0^{1/m} \exp(-\epsilon_i/m)\big|_0^{\epsilon_i}$$

Source: S. L. Semiatin and J. J. Jonas, Formability and Workability of Metals, American Society for Metals, Metals Park OH, 1984, p 162

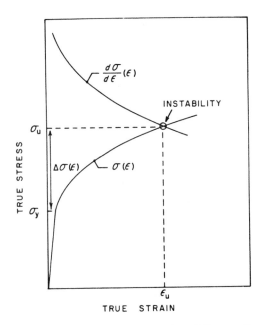

Schematic diagram showing the instability condition, $\sigma = d\sigma/d\epsilon$.

$$\frac{d\sigma}{d\epsilon} = \sigma \text{ when } \epsilon = \epsilon_u \qquad \text{(Eq 1)}$$

This condition is graphically illustrated for a typical parabolic stress-strain curve in the graph above, in which the strain-dependent functions for $d\sigma/d\epsilon$ and σ are plotted. The point of intersection defines, according to Eq 1, the maximum true uniform strain, ϵ_u. It is also apparent in these curves that the true stress at instability can be viewed as the sum of two terms: the yield stress, σ_y; and the change in stress due to strain hardening, $\Delta\sigma(\epsilon)$. Accordingly, Eq 1 is rewritten as

$$\frac{d\sigma}{d\epsilon}(\epsilon_u) = \sigma_y + \Delta\sigma(\epsilon_u) \qquad \text{(Eq 2)}$$

For the most general case, each term in Eq 2 will vary with microstructure, and $(d\sigma/d\epsilon)(\epsilon)$ and $\Delta\sigma(\epsilon)$ will be independent functions of strain.

Source: Deformation, Processing, and Structure, George Krauss, Ed., papers presented at the ASM Materials Science Seminar, 23 October 1982, St. Louis MO, sponsored by the Seminar Committee of the Materials Science Division of the American Society for Metals, Metals Park OH, 1984, p 64

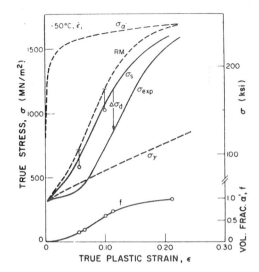

Experimental flow stress, σ_{exp}, and volume fraction martensite, f, vs plastic strain, ϵ, for metastable austenitic steel at -50 °C, $\dot{\epsilon}_1 = 2.2 \times 10^{-4}$ s^{-1}. Dashed curves represent the stable austenite flow stress, σ_γ, the martensite flow stress, $\sigma_{\alpha'}$, and the prediction of the rule of mixtures for two-phase hardening, RM. Solid curve, σ_s, is prediction of strain-corrected rule-of-mixtures model.

As in the case of stress-assisted transformation, dynamic softening is the dominant factor at low strains, causing the flow stress σ_{exp} of the transforming material in the figure above to fall below that of the stable austenite, σ_γ. The static hardening becomes dominant at high strains. The combined effect of these two factors delays the maximum hardening rate $d\sigma/d\epsilon$ to a higher strain than that where $df/d\epsilon$ is maximum. The maximum $d\sigma/d\epsilon$ arises from both the static hardening (proportional to $df/d\epsilon$) and the diminution of dynamic softening (proportional to $-d^2f/d\epsilon^2$) as $df/d\epsilon$ decreases. Again, the cessation of a softening phenomenon provides a major contribution to the net rate of hardening.

Source: Deformation, Processing, and Structure, George Krauss, Ed., papers presented at the ASM Materials Science Seminar, 23 October 1982, St. Louis MO, sponsored by the Seminar Committee of the Materials Science Division of the American Society for Metals, Metals Park OH, 1984, p 405

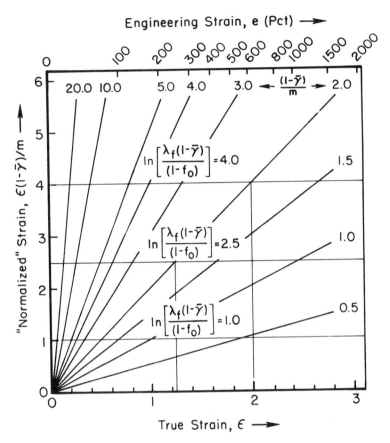

Diagram for determination of fracture strain given by the equation below. The fracture strain is given by the intersection of the horizontal line representing the value of $\ln[\lambda_f (1 - \bar{\gamma})/(1 - f_0)]$ with the locus $(1 - \bar{\gamma})/m = 0.5, 1.0, 2.0$, etc. For example, $(1 - \bar{\gamma})/m = 2.0$ can correspond to $\bar{\gamma} = 0$ and $m = 0.5$, $\bar{\gamma} = 0.25$ and $m = 0.38$, or $\bar{\gamma} = 0.5$ and $m = 0.25$. The intersections with the horizontals $\ln[\lambda_f (1 - \bar{\gamma})/(1 - f_0)] = 2.5$ and 4.0 correspond to fracture strain predictions of 1.25 and 2.0 (nominal strains of 250 and 640 percent, respectively). Note that for materials that fracture near the Considère strain, i.e., for low m metals tested at ambient temperatures, $\psi \gg 1$, so that the normalized strain must be replaced by $\epsilon\sqrt{mC}/m$ and the logarithmic term by $\ln[\lambda_f\sqrt{mC}/(1 - f_0)]$. Near ϵ_f, $\sqrt{mC}$ is generally less than 1 but greater than $(1 - \bar{\gamma})$ in such materials.

The effect of different experimental values of m, average γ or $\bar{\gamma}$, and $(1 - f_0)$ on the fracture strain predicted by the various formulas described above can be readily evaluated from the graph above. Here, the "normalized" true strain $[\epsilon(1 - \bar{\gamma})/m]$ is plotted as a function of strain ϵ for the usual ranges of m and $(1 - \bar{\gamma})$. Fracture can be considered to occur when:

$$\frac{\epsilon_f(1 - \bar{\gamma})}{m} = \ln\left[\frac{\lambda_f(1 - \bar{\gamma})}{(1 - f_0)}\right]$$

Source: S. L. Semiatin and J. J. Jonas, Formability and Workability of Metals, American Society for Metals, Metals Park OH, 1984, p 173

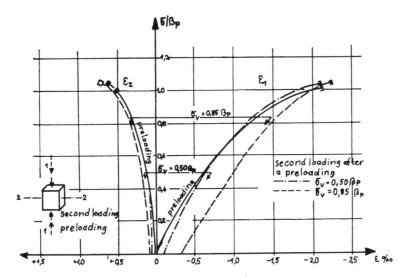

Stress-strain curves of uniaxial preloaded and reloaded specimens.

It is remarkable that the strains at failure were about the same as under loading without preloading even if high strains remained after unloading (see above).

Source: Proceedings of the Second International Conference on Mechanical Behavior of Metals, 16–20 August 1976, Boston MA, American Society for Metals, Metals Park OH, 1978, p 114

1-61. Stress-Strain Response With Residual Biaxial Compressive Stress

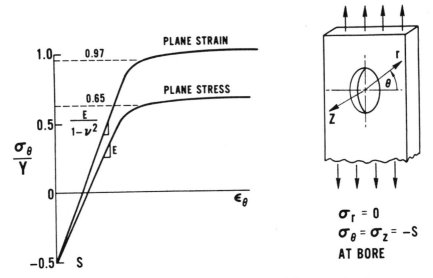

Stress-strain response of surface with residual biaxial compressive stress.

The development of constitutive models to predict the range and mean value of surface biaxial stress for complex cycles appears to be highly desirable for predicting crack initiation and early stages of growth, especially since many components are processed to impart a beneficial residual compressive stress. The relaxation of this stress during elevated-temperature cycling has been found to reduce fatigue resistance, and this effect should not be interpreted as a change in material properties. As a general precaution in analyzing surface-crack initiation, the surface residual stress and cold work from manufacturing should be closely reproduced in the laboratory or measured by X-ray diffraction and explicitly accounted for. Even subtle effects can be significant: the figure above illustrates the difference between stress-strain behavior in plane stress versus plane strain when a residual biaxial stress exists at the surface of a strain concentration.

Another practical requirement that is not now met by constitutive models is the prediction of the steady-state "shakedown" material response after many cycles of loading; currently this analysis involves integrating over successive half-cycles, which is extremely expensive.

Source: Fatigue and Microstructure, papers presented at the 1978 ASM Materials Science Seminar, 14–15 October 1978, St. Louis MO, sponsored by the Materials Science Division of the American Society for Metals, Metals Park OH, 1979, p 315

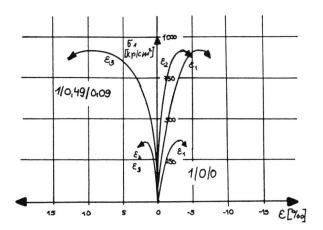

Triaxial stress-strain curve with brittle behavior.

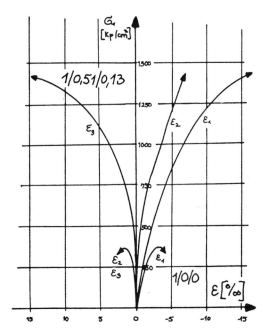

Triaxial stress-strain curve with ductile behavior.

Source: Proceedings of the Second International Conference on Mechanical Behavior of Metals, 16–20 August 1976, Boston MA, American Society for Metals, Metals Park OH, 1978, p 108

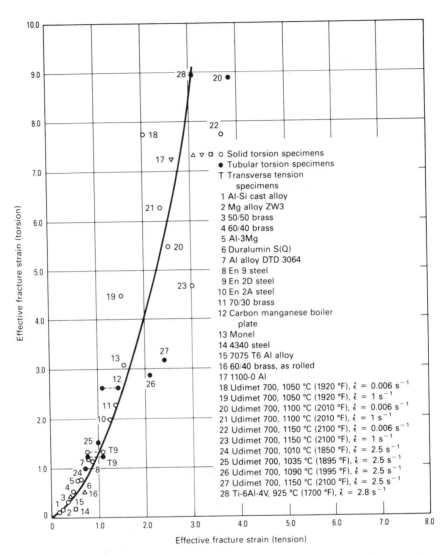

Relation between effective fracture strain from tension and from torsion tests for several alloys.

Correlation of Torsional Ductility Data. Because metalworking processes are not carried out under a state of pure torsional or shear loading, it is often necessary to convert the workability parameter measured in torsion to an index that is compatible with other deformation modes. Previously, attempts were made to correlate torsion, tension, and other types of data using the effective strain concept. Although a definite relationship exists between torsion and tension effective fracture strains (above figure), such a method is incapable of explaining how fracture can be totally avoided in homogeneous compression by preventing barreling. For fracture to occur, deformation must also involve tensile stresses to promote ductile fracture, wedge cracking, or some other failure mechanism.

Source: Metals Handbook, Ninth Edition, Volume 8, Mechanical Testing, American Society for Metals, Metals Park OH, 1985, p 168

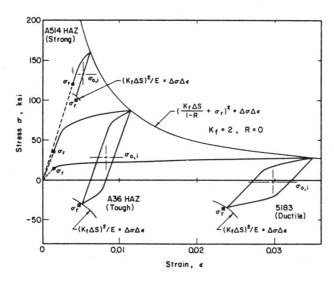

Set-up cycle stress-strain response for A514 HAZ (strong), A36 HAZ (tough), and 5183 WM (ductile) materials.

As shown above, the stabilized value of mean stress (σ_{os}) resulting from the residual stress (σ_r) may vary greatly depending upon the material. For many aluminums, the heat affected zone at the weld toe is in the zero-temper state; consequently, the notch-root plasticity in the first cycle results in $\sigma_{os} = 0$. Other materials, such as high strength steels, exhibit very little notch-root plasticity; consequently, σ_{os} may be larger than σ_r. The results obtained using the model agree with the experimentally observed behavior.

Source: Residual Stress for Designers and Metallurgists, Larry J. Vande Walle, Ed., proceedings of a conference sponsored by the American Society for Metals, Highway and Off-Highway Vehicles Activity of the Materials Systems and Design Division, 9–10 April 1980, Chicago IL, American Society for Metals, Metals Park OH, 1981, p 110

1-65. Flow Curves: Strain-Hardening Response

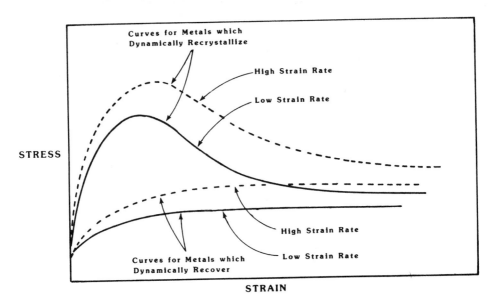

STRESS

Curves for Metals which
Dynamically Recrystallize

High Strain Rate

Low Strain Rate

High Strain Rate

Low Strain Rate

Curves for Metals which
Dynamically Recover

STRAIN

Flow curves showing strain-hardening response.

When the above processes are operative, the flow curves show negligible, or even negative, strain-hardening response (figure above). If dynamic recovery is the principal softening mechanism, the metal strain hardens to a certain level, after which it exhibits a steady-state flow stress. In metals in which dynamic recrystallization occurs at hot-working temperatures, the flow curve passes through a maximum at which recrystallization is initiated, "flow softens," and then achieves a steady-state flow stress. Both dynamic recovery and dynamic recrystallization are affected by the rate of straining. With an increase in strain rate, the deformations at which dynamic recovery leads to a steady-state flow stress, or at which dynamic recrystallization is initiated, both increase (see figure above).

Source: Forging Handbook, T. G. Byrer, S. L. Semiatin and Donald C. Vollmer, Eds., Forging Industry Association of America, Cleveland OH, 1985, p 89

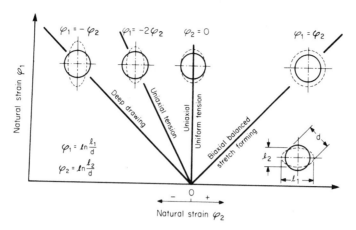

Strains in circular grid elements.

Circular grid patterns deform differently based on the type of loading. There exists a relationship between the distortion of the circle and the type of stressing (see above). By conducting a series of experiments it will be possible to find combinations of maximum strain (corresponding to the major axis of the ellipse) and minimum strain (perpendicular to the major strain and corresponding to the minor axis of the ellipse) for which neither necking nor fracture occurs.

Source: Handbook of Metal Forming, Kurt Lange, Ed., McGraw-Hill Book Co., New York, 1985, p 16.10

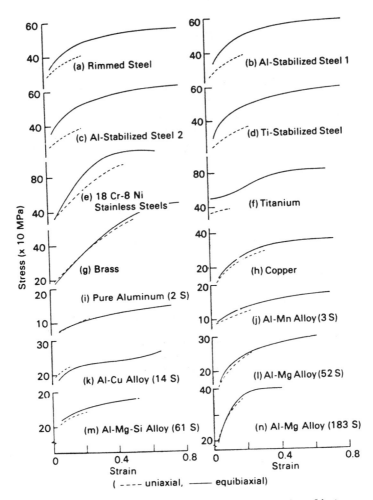

Stress-strain curves for various materials on the assumption of isotropy.

For the assessment of press formability, especially punch-stretchability, the stress-strain relation characterized by *n* value is very important and complicated as shown above. The stress-strain curves for different materials for deformation modes differ considerably from each other, and suggest the low possibility of single *n* value to express the work-hardening property for the entire measured strain range. Although we have usually used the averaged *n* value, the *n* value derived from a small range of strain may be required for more accurate prediction of the instability point and punch-stretching limit.

Source: Mechanics of Sheet Metal Forming: Material Behavior and Deformation Analysis, Donald P. Koistinen and Neng-Ming Wang, Eds., proceedings of a symposium sponsored by General Motors Research Laboratories, 17–18 October 1977, Warren MI, Plenum Press, New York, 1978, p 31

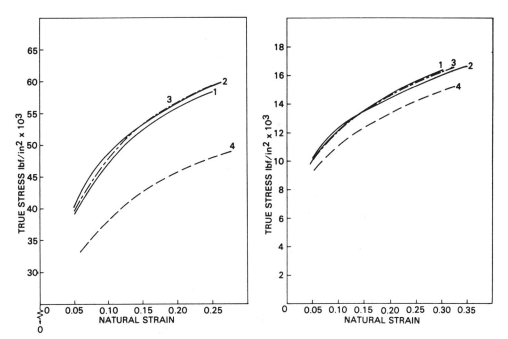

Estimation of *m*-values from work-hardening characteristics in simple and biaxial tension. (Left) Steel 'F', $r_{av} = 0.44$. (Right) Soft aluminum, $r_{av} = 0.72$. Curve 1: Experimental curve, simple tension, based on average of curves along 0°, 45°, and 90° to direction of rolling. Curve 2: Experimental curve, diaphragm test. Curve 3: Balanced biaxial tension curve predicted from Curve 1. For steel F, $m = 1.5$, for soft aluminum, $m = 1.8$. Curve 4: Balanced biaxial tension curve predicted from Curve 1, based on an average *r*-value.

Source: Mechanics of Sheet Metal Forming: Material Behavior and Deformation Analysis, Donald P. Koistinen and Neng-Ming Wang, Eds., proceedings of a symposium sponsored by General Motors Research Laboratories, 17–18 October 1977, Warren MI, Plenum Press, New York, 1978, p 70, 71

1-69. Idealized Shapes of Uniaxial Stress-Strain Curves

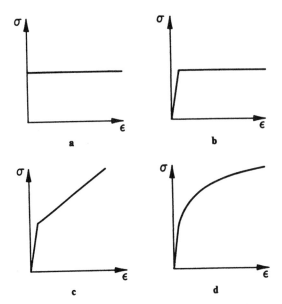

Idealized shapes of uniaxial stress-strain curve: (a) perfectly plastic; (b) ideal elastoplastic; (c) ideal elastoplastic with linear work hardening; (d) parabolic work hardening ($\sigma = \sigma_0 + k\epsilon^n$).

The plastic range of the uniaxial stress-strain curve is also assumed to have a simplified shape. One of the configurations displayed above can be assumed. Configuration (a) is called perfectly platic; the elastic strains are assumed zero. When the elastic deformation is assumed zero, the body is called rigid. When these assumptions cannot be made, configurations (b) and (c) are used. Configuration (b) is known as ideal elastoplastic. The volume of the material is assumed to be constant in plastic deformation. It is known that such is not the case in elastic deformation. The constancy in volume implies that

$$\epsilon_{11} + \epsilon_{22} + \epsilon_{33} = 0$$

or

$$\epsilon_1 + \epsilon_2 + \epsilon_3 = 0$$

and that Poisson's ratio is 0.5.

Source: Marc André Meyers and Krishan Kumar Chawla, Mechanical Metallurgy: Principles and Applications, Prentice-Hall, Inc., Englewood Cliffs NJ, 1984, p 73

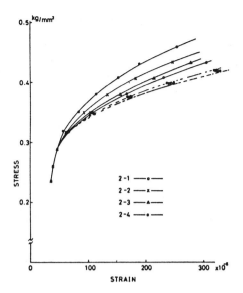

Stress-strain curves obtained by analyses.

Source: Proceedings of the Second International Conference on Mechanical Behavior of Metals, 16–20 August 1976, Boston MA, American Society for Metals, Metals Park OH, 1978, p 70

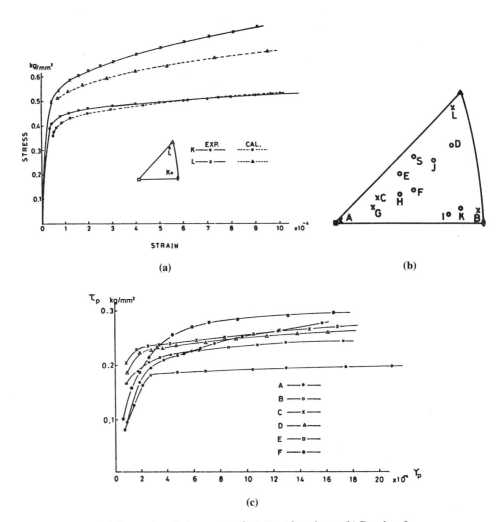

(a) Comparison between experiments and analyses. (b) Results of comparison, ○ denotes good agreement between experiments and analyses, x means bad ones. (c) τ_p-γ_p curves by experiments.

Source: Proceedings of the Second International Conference on Mechanical Behavior of Metals, 16–20 August 1976, Boston MA, American Society for Metals, Metals Park OH, 1978, p 65

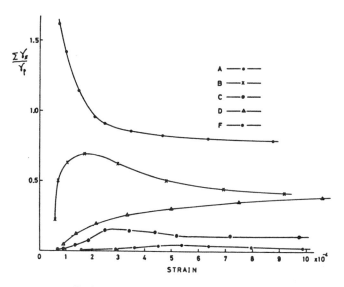

Ratios of secondary slip to primary slip.

Source: Proceedings of the Second International Conference on Mechanical Behavior of Metals, 16–20 August 1976, Boston MA, American Society for Metals, Metals Park OH, 1978, p 66

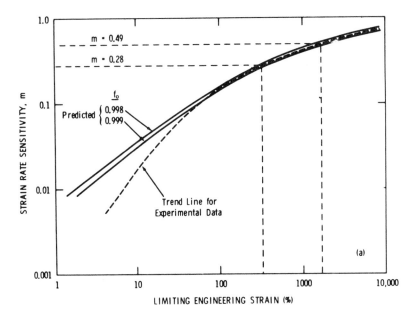

Limiting value of engineering strain as a function of m, predicted from the equation below, for two different inhomogeneity factors, $f_0 = 0.998$ and $f_0 = 0.999$. Comparison with trend line of experimental data is shown.

$$e_f(GD) = (1 - f_0^{1/m})^{-m} - 1$$

Source: S. L. Semiatin and J. J. Jonas, Formability and Workability of Metals, American Society for Metals, Metals Park OH, 1984, p 163

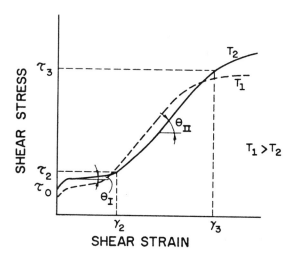

Generic shear stress-strain curves for FCC single crystals for two different temperatures.

Source: Marc André Meyers and Krishan Kumar Chawla, Mechanical Metallurgy: Principles and Applications, Prentice-Hall, Inc., Englewood Cliffs NJ, 1984, p 359

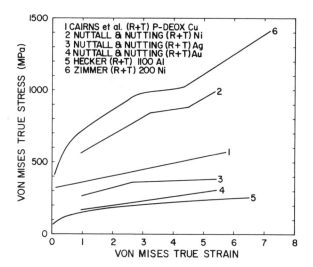

Von Mises effective stress-strain curves for various FCC metals of commercial purity. All curves are for rolling plus tension.

The flow curves for various commercial-purity FCC metals in the graph above show that continued hardening at large strains is very common. Few data are available on purity effects.

Source: Deformation, Processing, and Structure, George Krauss, Ed., papers presented at the ASM Materials Science Seminar, 23 October 1982, St. Louis MO, sponsored by the Seminar Committee of the Materials Science Division of the American Society for Metals, Metals Park OH, 1984, p 26

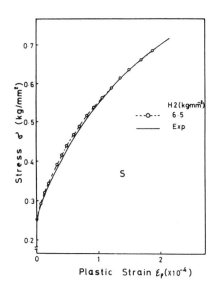
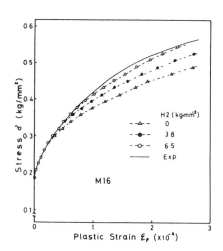

Calculated stress-strain curves of component single crystals.

	τ_{co} (kg·mm²)	γ_o	γ_i	H_1+H_{t} (kg·mm²)	H_2 (kg·mm²)
S	0·09046	0·83×10⁻⁶	0·83×10⁻⁶	16·432	6·5
M16	0·09046	0·13×10⁻⁶	0·13×10⁻⁶	7·617	6·5

Coefficients obtained from experimented and calculated analysis.

Source: Proceedings of the Second International Conference on Mechanical Behavior of Metals, 16–20 August 1976, Boston MA, American Society for Metals, Metals Park OH, 1978, p 58

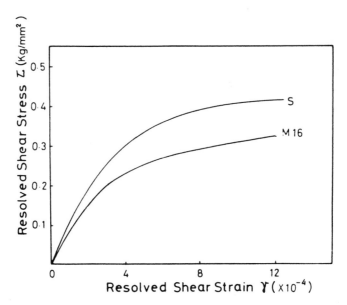

Resolved shear stress-strain curves of component single crystal.

Source: Proceedings of the Second International Conference on Mechanical Behavior of Metals, 16–20 August 1976, Boston MA, American Society for Metals, Metals Park OH, 1978, p 57

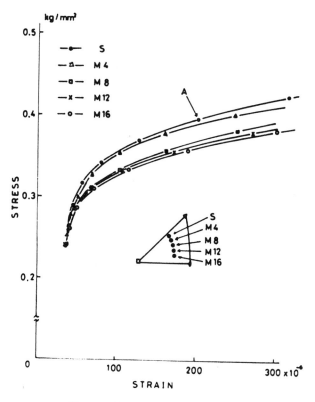

Stress-strain curves of single crystals.

Source: Proceedings of the Second International Conference on Mechanical Behavior of Metals, 16–20 August 1976, Boston MA, American Society for Metals, Metals Park OH, 1978, p 62

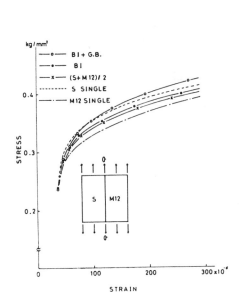

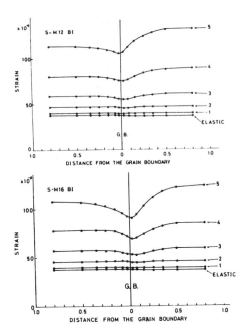

Analytical results of S-M16 bicrystal.

Strain distribution of S-M12, S-M16 bicrystal.

Source: Proceedings of the Second International Conference on Mechanical Behavior of Metals, 16–20 August 1976, Boston MA, American Society for Metals, Metals Park OH, 1978, p 68

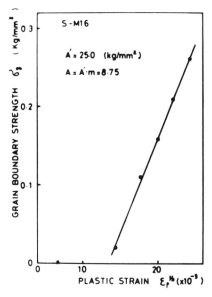

Determination of coefficient A.

Specimen size of polycrystal.

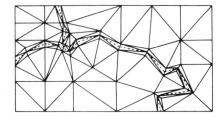

Meshing elements of polycrystal.

Source: Proceedings of the Second International Conference on Mechanical Behavior of Metals, 16–20 August 1976, Boston MA, American Society for Metals, Metals Park OH, 1978, p 59

1-81. Typical Stress-Strain Curve of a Pure FCC Metal Crystal

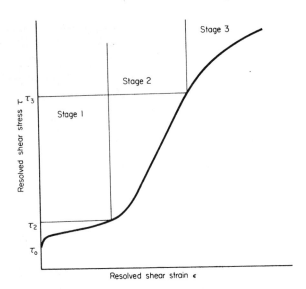

Typical stress-strain curve of a pure FCC metal crystal.

Source: R. W. K. Honeycombe, The Plastic Deformation of Metals, Second Edition, American Society for Metals, Metals Parks OH, 1984, p 31

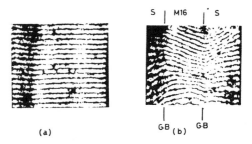

(a)　　　(b)

Moirè fringe pattern of tricrystals. (a) Before loading. (b) After loading.

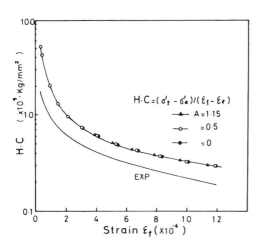

$$H \cdot C = (\sigma'_f - \sigma'_e)/(\varepsilon_f - \varepsilon_e)$$

A = 1·15
= 0·5
= 0

EXP

Relation between work-hardening rate and strain of polycrystal.

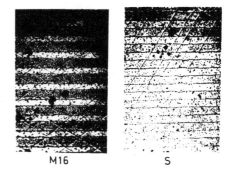

M16　　　S

Slip lines pattern on the surface of tricrystals.

Source: Proceedings of the Second International Conference on Mechanical Behavior of Metals, 16–20 August 1976, Boston MA, American Society for Metals, Metals Park OH, 1978, p 60

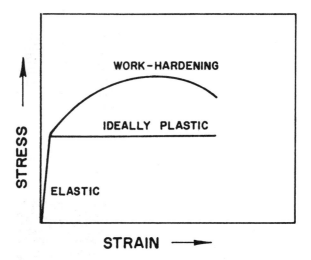

Stress-strain curves (schematic) for an elastic-ideally plastic solid and a work-hardening solid.

Hardening by plastic deformation (rolling, drawing, etc.) is one of the most important methods of strengthening metals, in general. Certain metals, in particular (e.g., copper), do not have many precipitation hardening systems but are ductile and can be appreciably hardened by cold working. If the relaxation times were short, the structure would return almost immediately to its state of equilibrium and a constant stress for plastic deformation would result independent of the extent of deformation. This is shown above in elastic–ideally plastic solid. However, when a real crystalline solid is deformed plastically, it turns more resistant to deformation and a greater stress is required for additional deformation (see figure above). This is called work hardening.

Source: Marc André Meyers and Krishan Kumar Chawla, Mechanical Metallurgy: Principles and Applications, Prentice-Hall, Inc., Englewood Cliffs NJ, 1984, p 354

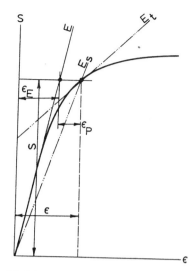

Modulus concepts for inelastic analysis.

To account for inelastic effects, the equation modifies to

$$S_{CR} = K_p \eta E \left(\frac{t}{b}\right)^2$$

Here η denotes the inelastic reduction factor, which is obtained by dividing the effective modulus by the elastic modulus. Many studies of this subject have been made and various formulas recommended. The approximate values for inelastic plate buckling in compression are given in the table below.

The relation of secant to tangent modulus is illustrated in the figure above. The secant modulus E_s is the relationship between stress and total strain at a particular point, consisting of elastic and plastic components ϵ_E and ϵ_p, respectively. The tangent modulus E_t may be regarded as a measure of instantaneous resistance of the material against increase in strain. This value diminishes quite rapidly as the total strain increases.

Approximate factors for inelastic buckling of plates

Edge conditions	Inelastic reduction factor, η
Both edges simply supported	$(E_t/E)^{1/2}$
Both edges fixed	$(E_t/E)^{1/2}$
One edge free, the other fixed	$(E_t/E)^{1/2}$
One edge free, the other supported	E_s/E

Source: Alexander Blake, Practical Stress Analysis in Engineering Design, Marcel Dekker, Inc., New York, 1982, p 90

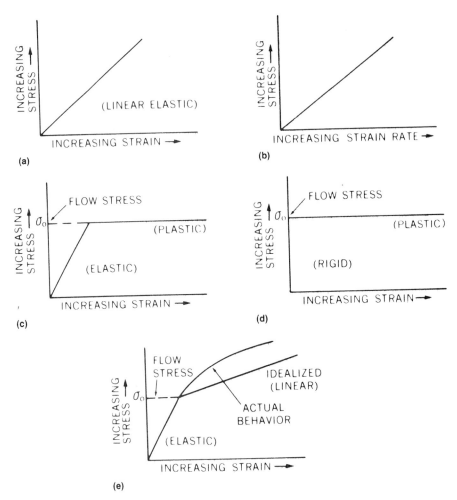

Mechanical behavior of classical bodies. (a) Hookean solid. (b) Newtonian fluid. (c) Perfectly plastic solid. (d) Rigid plastic solid. (e) Strain-hardening solid.

The mechanical behavior of real metals (represented above) is very complex, and no single model can adequately represent the phenomena of elasticity, elastic after-effect, hysteresis, plastic flow, and creep. The Hookean elastic solid and the Newtonian viscous fluid are the best-known models, which are mathematical abstractions. The most important model for large plastic deformations is the perfectly plastic solid (called the Tresca solid, or the Mises solid) in which elastic response of loading occurs up to a critical stress (flow stress) at which flow continues at constant stress.

Source: Forging Handbook, T. G. Byrer, S. L. Semiatin and Donald C. Vollmer, Eds., Forging Industry Association of America, Cleveland OH, 1985, p 87

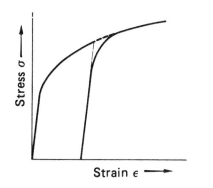

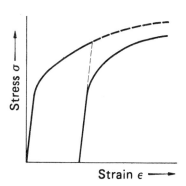

The plastic behavior of metals is determined by the arrangement and movement of dislocations. Therefore, any changes in properties by heating must be explained by changes of the structure of dislocations or of the dislocation density. The dislocation density rises during deformation at lower temperature. That is why strain hardening takes place. This strain hardening can be reduced by changing either the dislocation density or the arrangement of the dislocations.

Source: Handbook of Metal Forming, Kurt Lange, Ed., McGraw-Hill Book Co., New York, 1985, p 3.16

1-87. Schematic Presentation of the Change From Elastic to Plastic

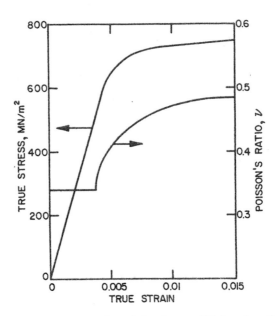

Schematic representation of the change of Poisson's ratio
as the deformation regime changes from elastic to plastic.

This figure shows that this assumption is reasonable and that
ν rises from 0.3 to 0.5 as deformation goes from elastic to
plastic.

However, prior to delving into the plasticity theories, we
have to know, for a complex state of stress, the stress level at
which the body starts to flow plastically. The methods developed to determine this are called flow criteria. The term "flow
criterion" will be used here rather than "yield criterion" or
"failure criterion." The term "failure criterion" has its histor-
ical origin in applications where the onset of plastic deformation indicated failure. However, in deformation-processing
operations this is obviously not the case, the plastic flow is
desired. The term "yield criterion" applies only to materials
that are in the annealed condition. It is known that when a
material is previously deformed by, for instance, rolling, its
yield stress increases due to work hardening. The term "flow
stress" is usually reserved for the onset of plastic flow of a
previously deformed material. For the reasons stated above, it
is felt that the term "flow criterion" is the most appropriate.

Source: Marc André Meyers and Krishan Kumar Chawla, Mechanical Metallurgy: Principles and Applications, Prentice-Hall, Inc. Englewood Cliffs NJ, 1984, p 74

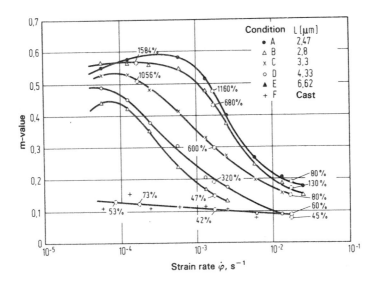

Dependence of *m*-value on strain rate φ and grain size *L* (metallographic mean free path), obtained for uniaxial tension tests at 20 °C (68 °F) of a eutectic tin-lead alloy, extruded and subjected to different heat treatments. Fracture strains based on constant starting strain rates φ are indicated on curves.

Metallurgical Processes in Superplasticity. Superplastic behavior of metallic materials requires $m \geq 0.3$. The relationship between m and fracture strain has been demonstrated with tests of titanium, zirconium, and tin-lead alloys. The curves above show the dependence of m on grain size or the metallographic mean free path in the microstructure. The curves $m = f(\dot{\varphi})$ of most superplastic materials display a distinct maximum for strain rates between 10^{-5} and 10^{-2} s^{-1}. This limited $\dot{\varphi}$-range, where m is sufficiently high for superplastic deformation, is known as stage 2. The curves above also underscore the fact that the high m-values required for superplasticity are obtained only with fine-grained microstructures. As the grain size decreases, m increases, and the maximum is shifted to higher strain rates. A fine-grained microstructure is especially easily obtained and maintained with eutectic and eutectoid alloys. For these materials different phases form simultaneously at subeutectic or subeutectoid temperatures and then precipitate in a very fine dispersion. It was demonstrated, however, that superplasticity is not restricted to two-phase structures, but can occur in fine-grained pure metals as well. A condition for high m-values is that $T_{def}/T_m \geq 0.5$. Only then is the microstructure nearly independent of the preceding deformation, and the flow stress is a function only of the initial microstructure, of $\dot{\varphi}$, and of T_{def} as a first approximation.

Source: Handbook of Metal Forming, Kurt Lange, Ed., McGraw-Hill Book Co., New York, 1985, p 30.6

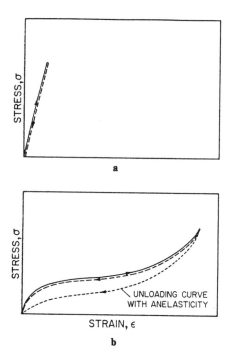

Stress-strain curves in elastic regime: (a) typical curve for metals and ceramics; (b) typical curve for rubber.

Metals exhibit some anelasticity, but it is most often neglected. Anelasticity is due to time-dependent microscopic processes accompanying deformation. An analogy that applies well is the attachment of a spring and dashpot in parallel. The spring represents the elastic portion; the dashpot represents the anelastic portion.

In 1678, Robert Hooke performed the experiments that demonstrated the proportionality between stress and strain. As was customary at that time, he proposed his law as an anagram: "ceiiinosssttuv," which in Latin is "ut tensio sic vis." The meaning is: "As the tension goes so does the stretch." In its most simplified form, we express it as

$$E = \frac{\sigma}{\epsilon}$$

where E is Young's modulus. For metals and ceramics it has a very high value. A typical value is 210 GPa for iron. E depends mainly on the composition, crystallographic structure, and nature of bonding of elements. Heat and mechanical treatments have little effect on E as long as they do not affect the former parameters. Hence, annealed and cold-rolled steel should have the same Young's modulus; there are, of course, small differences due to the formation of the cold-rolling texture.

Source: Marc André Meyers and Krishan Kumar Chawla, Mechanical Metallurgy: Principles and Applications, Prentice-Hall, Inc., Englewood Cliffs NJ, 1984, p 6

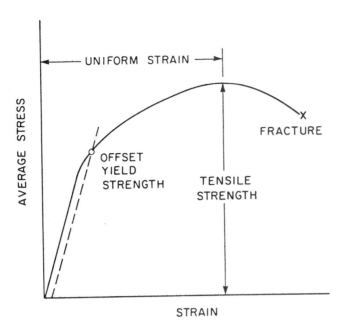

Design Against Excessive Yielding. A typical stress-strain curve for mild steel is shown above. At low stresses the material behaves in a linear elastic manner. Strain is directly proportional to stress and the deformation fully recovers as the load is released. Upon loading to higher stresses, the material plastically deforms and on release of the load, a permanent set is observed. An engineering boundary between elastic and plastic deformation is given by the 0.2% offset yield strength value, σ_{YS}. This is the stress required to produce a permanent extension of 0.2%. As the strain is increased beyond the yield point, the material strain-hardens and the stress must be increased to continue the deformation. As the tensile specimen elongates, its cross-sectional area decreases to maintain an essentially constant volume. At the maximum load point, the increase in strength due to strain hardening is balanced by the reduction in cross-sectional area. Now deformation can continue under a decreasing load until final fracture occurs. The stress at maximum load on the tensile specimen is the ultimate tensile strength of the material.

Source: Steel Castings Handbook, Fifth Edition, Peter F. Wieser, Ed., Steel Founders' Society of America, Rocky River OH, 1980, p 4-3

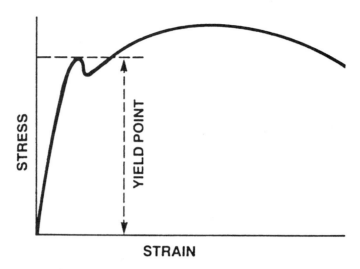

Stress-strain diagram showing material with a well-defined yield point.

In some materials, the yield point is well defined, such as shown above.

In many other materials, however, the first stages of plastic yielding are difficult to detect and the stresses corresponding to the apparent beginning of yielding depend on the sensitivity of the strain-measuring instrument used. In order to overcome this margin of potential error, the so-called offset method of determining yield strength is used. In this method, the stress at which a material exhibits a specified limiting permanent set is defined as its yield strength. The choice of the limiting amount of offset is to some extent arbitrary, but insofar as possible it should be based on that amount of plastic yielding that would be considered damaging in a statically loaded member of a structure. Generally, yield strength is based on 0.2% permanent set. The yield strength is shown in these cases with the offset value in parentheses, for example, Y.S. (0.2%) = 50,000 psi.

Source: Hamilton B. Bowman, Handbook of Precision Sheet, Strip and Foil, American Society for Metals, Metals Park OH, 1980, p 63

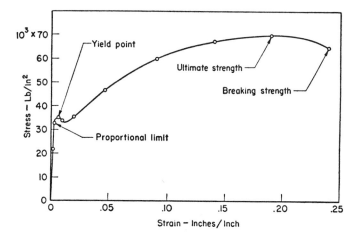

Stress-strain diagram for a low-carbon steel.

Some type of strain-measuring device may be attached to tensile specimens so as to permit accurate measurement of the strain that occurs over the desired gaging length as the applied load is increased. In this manner it is possible to obtain data from which a *stress-strain curve* may be plotted. Such a curve for a low-carbon steel is shown above. For this curve the stress is calculated from the loads and the original cross-sectional area of the specimen. It will be noted in this figure that up to a certain load the stress is directly proportional to strain. The stress at which this proportionality ceases to exist is known as the *proportional limit*. Up to the proportional limit, then, the material obeys *Hooke's law,* which states that within the elastic range of materials stress is proportional to strain.

Source: E. Paul Degarmo, Materials and Processes in Manufacturing, The Macmillan Co., New York, 1957, p 15

2-4. Low-Carbon Mild Steel: Results From Testing
a Ductile Material

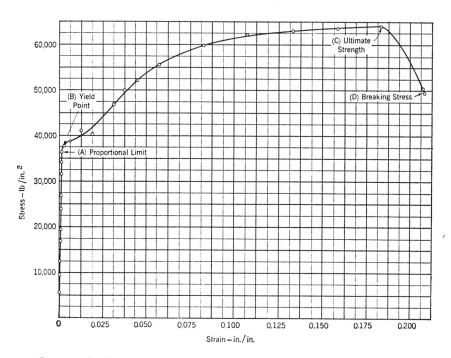

Stress-strain diagram for mild steel, which results from tension testing of a ductile material.

Source: Donald S. Clark, Engineering Materials and Processes, International Textbook Co., Scranton PA, 1962, p 33

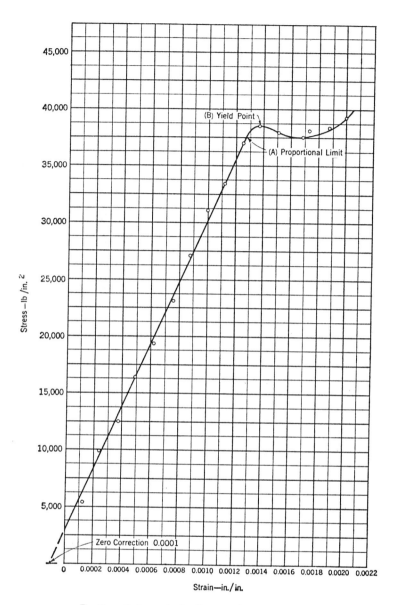

Portion of stress-strain diagram for mild steel.

Source: Donald S. Clark, Engineering Materials and Processes, International Textbook Co., Scranton PA, 1962, p 35

2-6. Low-Carbon Steel (0.18% C), Annealed

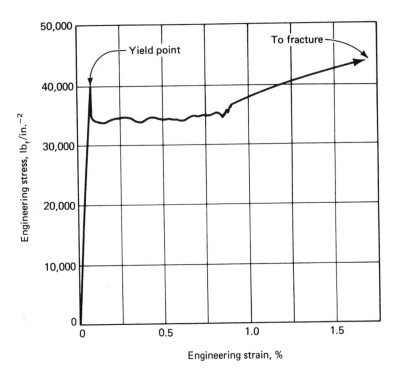

The stress-strain curve for an annealed 0.18% C steel, showing yield-point behavior.

Some materials, notably certain steels under certain heat treated conditions, have a stress-strain curve which shows a well-defined yield point (above curve). For such cases, the 0.2% yield strength is not used to define yielding.

Source: Charlie R. Brooks, Heat Treatment, Structure and Properties of Nonferrous Alloys, American Society for Metals, Metals Park OH, 1982, p 4

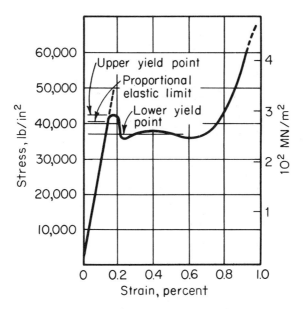

Yielding of annealed steel.

Source: Marks' Standard Handbook for Mechanical Engineers, Eighth Edition, Theodore Baumeister, Eugene A. Avallone and Theodore Baumeister III, Eds., McGraw-Hill Book Co., New York, 1978, p 5-4

2-8. Low-Carbon Steel: Yield Point Behavior, Showing Lüders Band

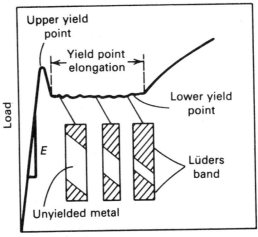

Typical yield point behavior of low-carbon steel. The slope of the initial linear portion of the stress-strain curve, designated by *E*, is the modulus of elasticity.

Many metals, particularly annealed low-carbon steel, show a localized, heterogeneous type of transition from elastic to plastic deformation that produces a yield point in the stress-strain curve. Rather than having a flow curve with a gradual transition from elastic to plastic behavior, metals with a yield point produce a flow curve or a load-elongation diagram similar to that shown above. The load increases steadily with elastic strain, drops suddenly, fluctuates about some approximately constant value of load, and then rises with further strain.

Source: Metals Handbook, Ninth Edition, Volume 8, Mechanical Testing, American Society for Metals, Metals Park OH, 1985, p 22

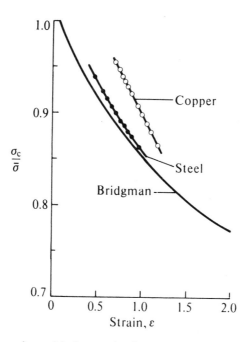

Bridgman's empirical correction factor versus true strain (ϵ).

Source: Milton C. Shaw, Metal Cutting Principles, Clarendon Press, Oxford, 1984, p 67

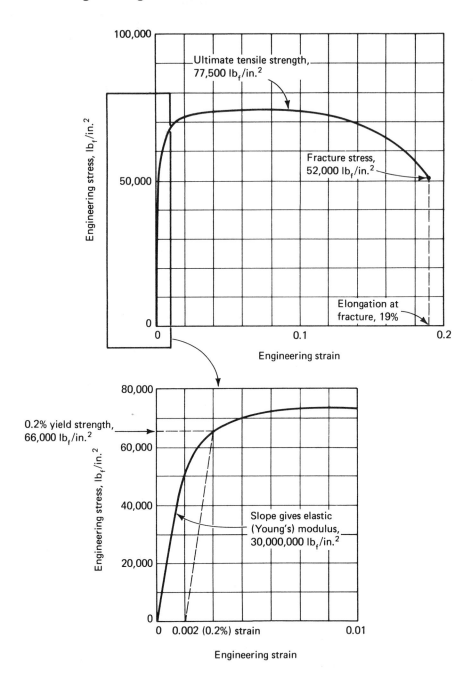

Engineering stress–engineering strain curve of 0.2% C plain carbon, cold worked steel, showing definition of mechanical property terms.

Source: Charlie R. Brooks, Heat Treatment, Structure and Properties of Nonferrous Alloys, American Society for Metals, Metals Park OH, 1982, p 2

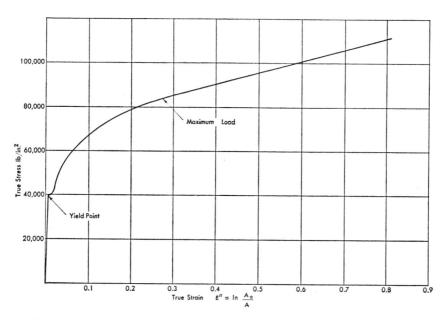

True-stress versus true-strain diagram or physical stress-strain diagram for mild steel.

Source: Donald S. Clark, Engineering Materials and Processes, International Textbook Co., Scranton PA, 1962, p 36

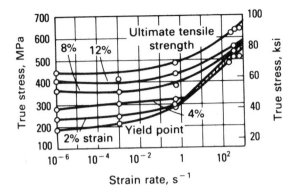

True yield stresses at various strains versus strain rate for a low-carbon steel at room temperature.

This figure illustrates true yield stress at various strains for a low-carbon steel at room temperature. Between strain rates of 10^{-6} s^{-1} and 10^{-3} s^{-1} (a 1000-fold increase), yield stress increases only by 10%. Above 1 s^{-1}, however, an equivalent rate increase doubles the yield stress. For the data in the figure, at every level of strain the flow stress increases with increasing strain rate. However, a decrease in strain-hardening rate is exhibited at the higher deformation rates.

Source: Metals Handbook, Ninth Edition, Volume 8, Mechanical Testing, American Society for Metals, Metals Park OH, 1985, p 39

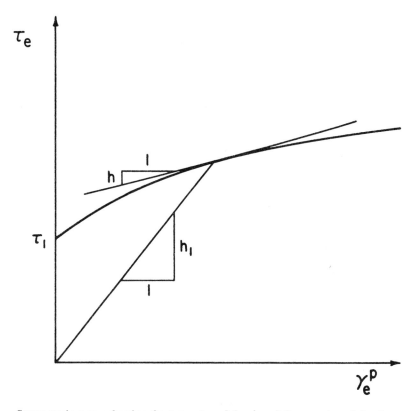

Stress-strain curve showing the tangent modulus h and the secant modulus h_1.

Here, $\mathbf{D}^{P\prime}$ is the deviatoric part of $\mathbf{D}^P$ and the plastic hardening modulus is given by

$$h = d\tau_e/d\gamma_e^{\,p} = \frac{1}{3}\,d\sigma_e/d\epsilon_e^{\,p}$$

as depicted above.

Source: Mechanics of Sheet Metal Forming, Material Behavior and Deformation Analysis, Donald P. Koistinen and Neng-Ming Wang, Eds., proceedings of a symposium sponsored by General Motors Research Laboratories, 17–18 October 1977, Warren MI, Plenum Press, New York, 1978, p 240

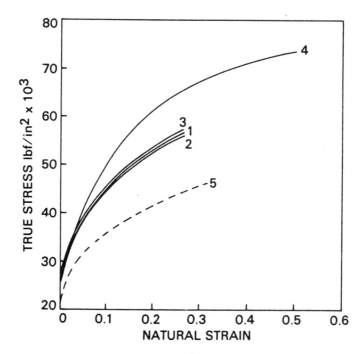

Work-hardening characteristics for annealed rim-steel: measured average r-value is 0.38. Curves 1, 2, and 3 are experimental curves in simple tension at 0°, 45°, and 90° to rolling direction respectively. Curve 4 is an experimental curve for balanced biaxial tension. Curve 5 is a balanced biaxial curve predicted from average r-value and corresponding work-hardening characteristic.

Pearce carried out similar experiments and made the same type of correlation for several materials including an annealed rim-steel having an average r-value of 0.38 (see curves above). It is seen that the experimental curve in biaxial tension is above the uniaxial curves, whereas the predicted curve falls well below the uniaxial curves. The difference between the two curves is far too great to be explained by experimental error, although the small diameter (112 mm) of the diaphragm testing machine might have given rise to a somewhat higher biaxial curve than if a larger machine had been used (see the work of Horta, Roberts and Wilson). It is now clear that the theory of orthotropic plasticity does not explain the behavior in biaxial tension where the r-value is less than unity. Dillamore, arguing from crystal plasticity, has concluded that the orthotropic theory is likely to give reasonable correlation only for r-values between 1 and 2.

Source: Mechanics of Sheet Metal Forming, Material Behavior and Deformation Analysis, Donald P. Koistinen and Neng-Ming Wang, Eds., proceedings of a symposium sponsored by General Motors Research Laboratories, 17–18 October 1977, Warren MI, Plenum Press, New York, 1978, p 66

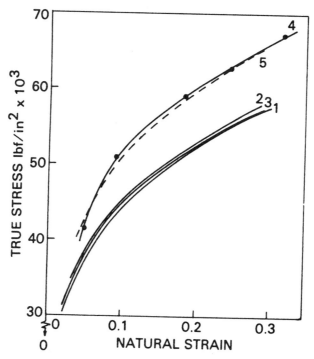

Work-hardening characteristics for killed steel: measured average r-value is 1.42. Curves 1, 2, and 3 are experimental curves in simple tension at 0°, 45°, and 90° to rolling direction respectively. Curve 4 is an experimental curve for balanced biaxial tension. Curve 5 is a balanced biaxial curve predicted from average r-value and corresponding work-hardening characteristic.

The r-value for a particular orientation was found to be constant up to the maximum measured strain of 0.2. Using the work-hardening characteristics obtained in simple tension, the theory of orthotropic anisotropy was used to predict the stress-strain curve in balanced biaxial tension. This was then compared with an experimental curve obtained from the diaphragm test. The correlation for one of the killed steels is shown above. The theory predicts correctly that the biaxial curve will be higher than the uniaxial curve and the correlation is considered to be good. Average r-values and an average uniaxial stress-strain curve were used in the correlation. If σ_b is the polar stress and ϵ the thickness strain then

$$\sigma_b = \left[\frac{1+r}{2}\right]^{1/2} \bar{\sigma} \; ; \; \epsilon = \left[\frac{2}{1+r}\right]^{1/2} \bar{\epsilon}$$

where r is the average r-value, $\bar{\sigma}$ is the average uniaxial stress, and $\bar{\epsilon}$ the average uniaxial strain.

Source: Mechanics of Sheet Metal Forming, Material Behavior and Deformation Analysis, Donald P. Koistinen and Neng-Ming Wang, Eds., proceedings of a symposium sponsored by General Motors Research Laboratories, 17–18 October 1977, Warren MI, Plenum Press, New York, 1978, p 65

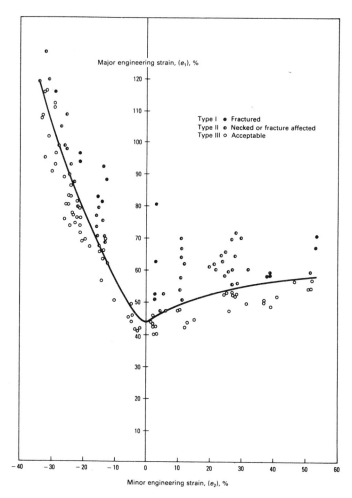

Strain measurements and forming-limit diagram for aluminum-killed steel.

The hemispherical punch method for determining forming limit diagrams uses circle-gridded strips of the test material ranging in width from 25.4 to 203 mm (1.0 to 8.0 in.) that are clamped in a die ring and stretched to incipient fracture by a 102-mm (4.0-in.) diam steel punch. The narrowest strip fractures at a minor-to-major strain ratio of about −0.5, which is comparable to that obtained in a tensile test. As the strip width is increased, the strain ratio increases to a slightly positive value for a full-width specimen. Further increases in the ratio to a maximum value of +1.0 (balanced biaxial stretching) are achieved by using progressively improved punch lubrication (oiled polyethylene, oiled neoprene) and by increasing thicknesses of polyurethane rubber.

The strains are measured in and around regions of visible necking and fracture. The forming-limit curve is drawn above the strains measured outside the necked regions and below those measured in the necked and fractured regions, as shown above.

Source: Metals Handbook, Ninth Edition, Volume 8, Mechanical Testing, American Society for Metals, Metals Park OH, 1985, p 566

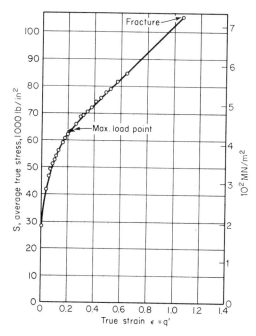

True stress-strain curve for 20 °C annealed mild steel.

The true stress-strain curve or flow curve obtained has the typical form shown in the figure above. In the part of the test subsequent to the maximum load point (UTS), when necking occurs, the true strain of interest is that which occurs in an infinitesimal length at the region of minimum cross section. True strain for this element can still be expressed as $\ln (A_0/A)$, where A refers to the minimum cross section.

Source: Marks' Standard Handbook for Mechanical Engineers, Eighth Edition, Theodore Baumeister, Eugene A. Avallone and Theodore Baumeister III, Eds., McGraw-Hill Book Co., New York, 1978, p 5-4

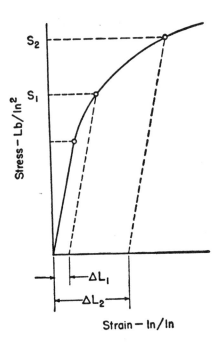

Stress-strain diagram for a material without a well-defined yield point, showing the "offset method" of determining yield strength.

Many materials do not have a well-defined yield point. Such materials have a stress-strain curve of the general type shown above. For such materials the *yield strength* is used. It is defined as the stress required to produce a given amount of permanent strain. The deformations used are usually 0.002 inch per inch (0.2 percent), or 0.001 inch per inch. Thus by using the "offset method" illustrated above the yield strength may be found when the stress-strain curve is known. However, it should be remembered that a statement of yield strength without reference to the corresponding permanent set is meaningless.

Source: E. Paul Degarmo, Materials and Processes in Manufacturing, The Macmillan Co., New York, 1957, p 15

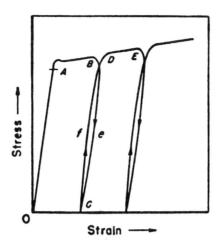

Stress-strain diagram obtained by unloading and reloading a specimen.

It is interesting to observe the manner in which a material reacts when subjected to more than one application of static loading. If a ductile steel is loaded and unloaded slowly, the stress-strain diagram as shown above is obtained. Unloading and reloading within the elastic range result in the stress-strain curve continuing to be a straight line.

Source: E. Paul Degarmo, Materials and Processes in Manufacturing, The Macmillan Co., New York, 1957, p 19

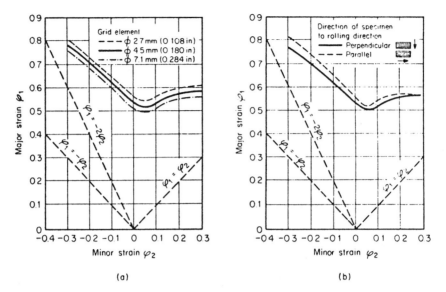

Influence of (a) grid size and (b) planar anisotropy on forming-limit curve. Material—RRSt 1403 (AISI 1006); sheet thickness—0.92 mm (0.036 in).

As shown in (a) above, the forming-limit curve is shifted to slightly lower strain values with increasing grid-circle diameter because the location of cracks or zones of necking can be determined more accurately with narrower grids. The influence of the direction of the test specimen to the strip-rolling direction—that is, the influence of planar anisotropy—is small but significant for steel sheet RRSt 1403 (AISI 1006) according to (b) above.

Source: Handbook of Metal Forming, Kurt Lange, Ed., McGraw-Hill Book Co., New York, 1985, p 18.13

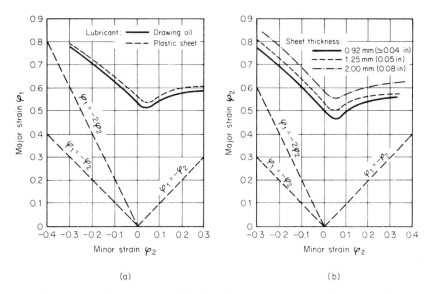

(a) (b)

Influence of (a) lubrication and (b) sheet thickness on forming-limit curve. Material—RRSt 1403 (AISI 1006); sheet thickness—0.92 mm (0.036 in). The influence of lubrication can be seen in these curves.

Source: Handbook of Metal Forming, Kurt Lange, Ed., McGraw-Hill Book Co., New York, 1985, p 18.13

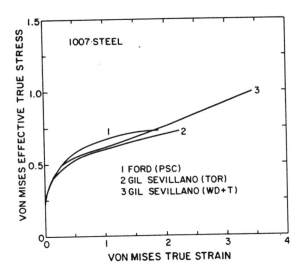

Comparison of stress-strain curves for low-carbon steels; 1008 steel deformed by plane-strain compression, and 1007 steel deformed by wire drawing plus tension and by torsion.

Source: Deformation, Processing, and Structure, George Krauss, Ed., papers presented at the ASM Materials Science Seminar, 23 October 1982, St. Louis MO, sponsored by the Seminar Committee of the Materials Science Division of the American Society for Metals, Metals Park OH, 1984, p 9

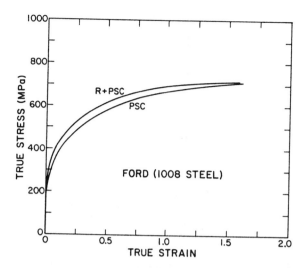

Comparison of stress-strain curves for monotonic plane-strain compression with rolling prestrain followed by plane-strain compression (1008 steel).

In some cases, prestraining and final deformation are carried out in similar stress states. For instance, in wire drawing plus tension, both stress states are axisymmetric. Ford compared the flow curves for low-carbon steel determined by plane-strain compression and by rolling plus plane-strain compression (see graph above). Here the stress states are very similar, and yet the rolling-plus-plane-strain compression curve is different. Ford explained this difference on the basis of redundant work, explaining that the curvature of the rolls causes some redundant shearing (not contributing to thickness reduction) and extra hardening.

Source: Deformation, Processing, and Structure, George Krauss, Ed., papers presented at the ASM Materials Science Seminar, 23 October 1982, St. Louis MO, sponsored by the Seminar Committee of the Materials Science Division of the American Society for Metals, Metals Park OH, 1984, p 10

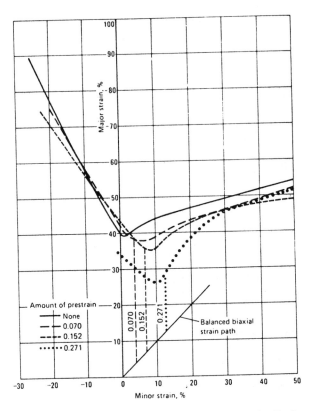

Effect of balanced biaxial prestrain on FLD: Forming limit diagrams for aluminum-killed steel as a function of balanced biaxial prestrain.

Several investigators have studied the effect of prior strain on the FLD. In one of these investigations, Ghosh and Laukonis subjected panels of steel sheet to uniform biaxial stretching and then determined the FLD for the stretched material. The prestrain shifted the FLD down and to the right, as shown in the chart above. This represents an apparent loss in formability that may be experienced in multistage forming operations.

Source: Metals Handbook, Ninth Edition, Volume 1, Properties and Selection: Irons and Steels, American Society for Metals, Metals Park OH, 1978, p 554

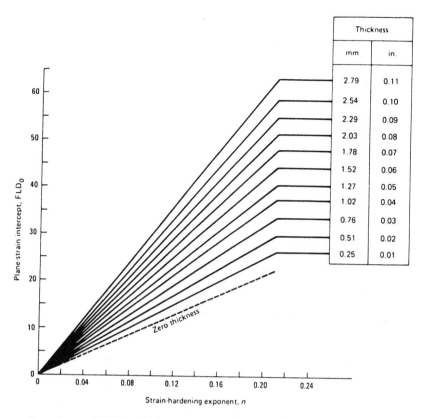

Thickness	
mm	in.
2.79	0.11
2.54	0.10
2.29	0.09
2.03	0.08
1.78	0.07
1.52	0.06
1.27	0.05
1.02	0.04
0.76	0.03
0.51	0.02
0.25	0.01

Dependence of FLD on thickness and n: Relationship between plane-strain intercept on FLD (FLD_0) and strain-hardening exponent as a function of thickness. FLD_0 depends only on thickness for values of n greater than 0.21.

Keeler and Brazier have shown that the FLD's of many steels have the same shape. Also, the FLD for any particular steel can be obtained merely by moving a curve of standard shape along the major strain axis to the appropriate point. They also showed that the intercept of the FLD on the major strain axis is a function of the sheet thickness and the strain-hardening exponent. Interrelations between these variables are shown in the chart above. Note that the ability of the steel sheet to withstand forming strains (as measured by the intercept of the FLD) increases as the thickness increases. The value of n may be obtained from the slope of the true stress/true strain diagram as plotted on logarithmic scales.

Source: Quality Control Source Book, A. K. Hingwe, Ed., American Society for Metals, Metals Park OH, 1982, p 220

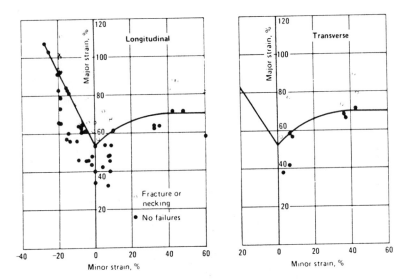

FLD for hot rolled steel sheet: Longitudinal and transverse forming-limit diagrams for hot rolled 1008 steel sheet, 2.03 mm (0.080 in.) thick. Yield strength group: 210 MPa (30 ksi).

Limitations to Use of FLD. Although the forming-limit diagram is a very useful tool for evaluating the formability of steel sheet, there are definite limitations to its use. In complicated parts, the FLD may be inaccurate because the paths of deformation may not be as well defined as those determined from relatively simple laboratory forming tests. A part may experience different modes of deformation at different stages of a single forming operation, or there may be several forming steps, which would not be reflected in the FLD.

The forming behavior of steel sheet depends only slightly on the rate of forming. A forming-limit diagram determined from low strain-rate laboratory tests almost always accurately predicts forming behavior at forming rates encountered in production stamping.

Source: Metals Handbook, Ninth Edition, Volume 1, Properties and Selection: Irons and Steels, American Society for Metals, Metals Park OH, 44073, 1978, p 551

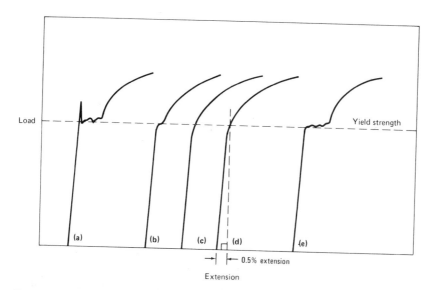

Load-extension curves for steel sheet having the same yield strength but different characteristic behavior. (a) Annealed dead soft rimmed or aluminum-killed steel. The yield strength is the average stress measured during yield point elongation. (b) Lightly temper rolled rimmed steel. The stress at the jog in the curve is reported as the yield strength. (c) and (d) Temper rolled low-carbon steel. May be rimmed, aluminum-killed, or interstitial-free steel with no detectable yield point. The yield strength is calculated from the load at 0.2% offset (c) or from the load at 0.5% extension (d). (e) Rimmed steel with a yield point elongation due to aging at room temperature for several months. The yield strength is the average stress measured during yield point elongation.

Source: Metals Handbook, Ninth Edition, Volume 1, Properties and Selection: Irons and Steels, American Society for Metals, Metals Park OH, 1978, p 548

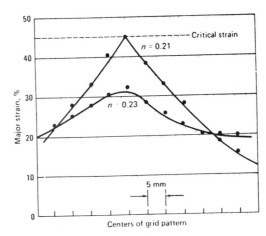

Effect of n on critical strain: The major strain ϵ_1 in the critical region of a formed part is more uniformly distributed for the steel having the higher value of n. One of these two parts (which are identical except for the n value of the steel selected) was strained to the point of excessive thinning; the other, made from steel with the higher n value, showed no inclination to fracture.

The strain-hardening exponent, n, is the slope of the true stress/true strain curve, when plotted on logarithmic coordinates. A significant portion of the curve is nearly a straight line for many low-carbon steels. The data are assumed to fit the equation

$$\sigma = K\epsilon^n$$

The n value will normally be around 0.22 for low-carbon steels used for forming. Higher values (up to 0.24) indicate improved ductility or biaxial stretchability. Freshly rolled rimmed steels generally have n values comparable to those of aluminum-killed steels. After aging, values of n for rimmed steels are less than for aluminum-killed steels. Some low-carbon steels that are not fully processed for formability, especially hot rolled grades, will have n values as low as 0.10, but most of the formable grades will have n values above 0.18. For other metals, these plots may not be linear, in this instance the slope of the curve developed for a given forming operation was used to obtain an estimate of the strain-hardening behavior.

The effects of different n values on strain distribution in critical regions of a specific formed part are illustrated in the figure above. Parts formed from steel sheet whose n value is low (for instance, 0.21) may undergo excessive thinning and fracture in critical regions. Identical parts formed from sheet with a higher strain-hardening exponent frequently will be strong enough in the critical areas to transfer strain to adjacent areas, thereby avoiding failure during forming.

Source: Metals Handbook, Ninth Edition, Volume 1, Properties and Selection: Irons and Steels, American Society for Metals, Metals Park OH, 1978, p 549

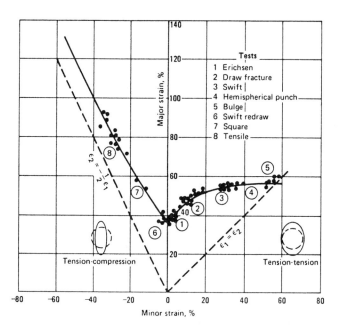

Schematic forming-limit diagram was constructed from several types of laboratory tests. Each type of test provides a different range of major strain/minor
strain values. A series of strips of various widths drawn over a hemispherical
punch can provide the full range of strain combinations. Ellipses in lower cor-
ners show appearance of circles on surface of sheet after deformation with
negative or positive minor strain.

Source: Metals Handbook, Ninth Edition, Volume 1, Properties and Selection: Irons and Steels, American Society
for Metals, Metals Park OH, 1978, p 550

2-30. Low-Carbon Steel Sheet

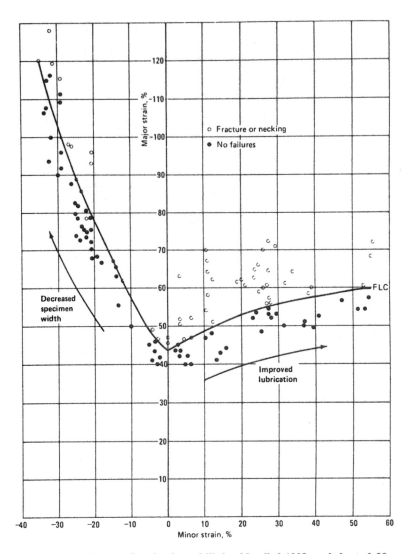

Forming limit diagram for aluminum-killed cold rolled 1008 steel sheet, 0.90 mm (0.035 in.) thick. Diagram was generated from analysis of specimens drawn over hemispherical punch 102 mm (4 in.) in diameter. Points representing negative minor strain obtained by forming strips ranging from 25 to 140 mm (1 to 5.5 in.) wide; points representing positive minor strain obtained by varying lubrication conditions.

Source: Metals Handbook, Ninth Edition, Volume 1, Properties and Selection: Irons and Steels, American Society for Metals, Metals Park OH, 1978, p 551

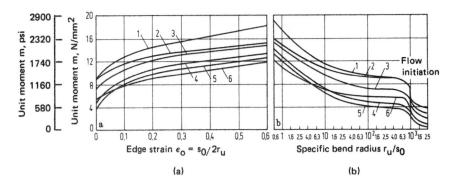

Unit-moment curves for various carbon steel sheet. 1—$S_u = 520$ N/mm^2 (74 ksi) normalized; 2—$S_u = 520$ N/mm^2 (74 ksi) spheroidized; 3—$S_u = 420$ N/mm^2 (61 ksi); 4—$S_u = 370$ N/mm^2 (54 ksi); 5, 6—$S_u = 340$ N/mm^2 (50 ksi). (a) Small bend radii. (b) Large bend radii.

Source: Handbook of Metal Forming, Kurt Lange, Ed., McGraw-Hill Book Co., New York, 1985, p 19.9

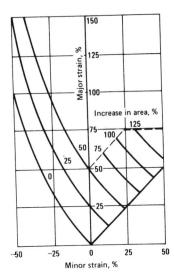

Theoretical severity curve for sheet metal forming.

Severity curves conveniently illustrate the nature and extent of deformation that occurs during forming (see graph above). By convention, the major strain ϵ_1, which is always positive, is plotted on the vertical axis of the graph. The minor strain ϵ_2, which is measured in a direction perpendicular to the major strain, may be either positive or negative and is plotted on the horizontal axis. A series of lines, each representing a particular increase in the surface area of a unit square on the sheet under various combinations of ϵ_1 and ϵ_2, can be constructed to represent the boundaries of ranges of conditions typically encountered in forming steel sheet. Increases in area resulting from drawing operations usually range from 0 to 50%. Decreases in area generally do not occur; the sheet wrinkles rather than thickening. Tearing will usually occur when increases in area exceed 50%. Increases in area above 50% can be achieved under stretching conditions; values of 200% have been observed. The occurrence of stretching conditions is limited because the major strain (by definition) must exceed the minor strain, and tearing or excessive thinning is frequently encountered when the major strain exceeds about 70%. These limits to deformation without tearing are maximums, assuming ideal forming conditions and steel sheet of the highest quality.

Source: Quality Control Source Book, A. K. Hingwe, Ed., American Society for Metals, Metals Park OH, 1982, p 213

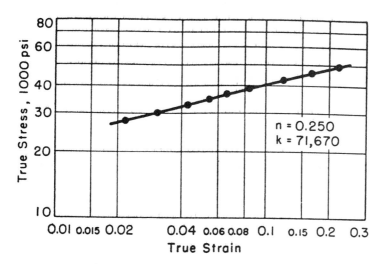

Typical logarithmic stress-strain curve for temper rolled, fully aluminum-killed, deep drawing steel sheet.

In the plot above, it can be seen that the conformity to a straight
line is very good. In a series of reproducibility tests conducted on
50 adjacent longitudinal specimens of 20-gage, fully aluminum-killed,
deep drawing steel sheet, it was found that the "n" values determined
by the above experimental technique only varied between 0.230 and
0.238.

Source: Source Book on Forming of Steel Sheet, American Society for Metals, Metals Park OH, 1975, p 217

2-34. 1015 Steel

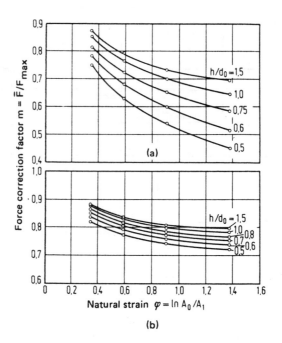

Force correction factor as a function of natural strain and relative deformation travel for solid forward extrusion. Material—Ck 15 (AISI 1015); annealed; lubricant—Bonderlube 235; surface ground. (a) Billet surface flat. (b) Billet chamfered.

Source: Handbook of Metal Forming, Kurt Lange, Ed., McGraw-Hill Book Co., New York, 1985, p 15.18

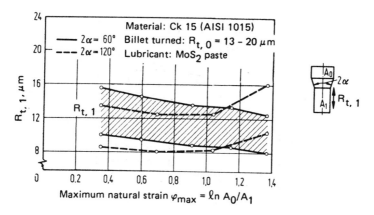

Measured surface roughness values in solid forward extrusion of steel.

Source: Handbook of Metal Forming, Kurt Lange, Ed., McGraw-Hill Book Co., New York, 1985, p 15.36

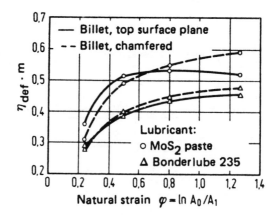

Product of deformation efficiency and force correction factor as a function of natural strain, lubricant, and shape. Material—Ck 15 (AISI 1015); annealed; surface peeled.

Source: Handbook of Metal Forming, Kurt Lange, Ed., McGraw-Hill Book Co., New York, 1985, p 15.18

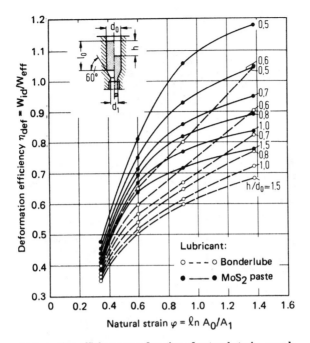

Deformation efficiency as a function of natural strain, punch travel, and lubricant. Material—Ck 15 (AISI 1015); annealed; peeled surface; plane billet surface.

Source: Handbook of Metal Forming, Kurt Lange, Ed., McGraw-Hill Book Co., New York, 1985, p 15.16

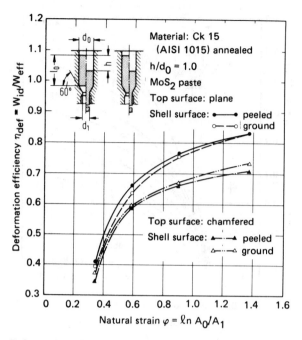

Deformation efficiency as a function of natural strain, billet shape, surface condition, and lubricant.

Source: Handbook of Metal Forming, Kurt Lange, Ed., McGraw-Hill Book Co., New York, 1985, p 15.16

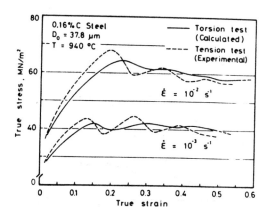

Comparison of *actual* **true stress–true strain curves determined in tension with** *calculated* **torsion curves. The latter were derived from tensile data determined at the strain rate appropriate to each cylindrical "shell" in the simulated torsion bar. The torque contributions of the individual shells were then summed and the effective stress–effective strain curves were deduced by the method of Fields and Backofen.**

The full lines represent the *estimated* torsion behavior as built up from a series of imaginary concentric shells, each of which follows the behavior expected for such a shell from a tension test of appropriate strain rate. The equivalent strain $\bar{\epsilon}$ in the simulated torsion test in the graph above is indeed that associated with the outer radius R of the specimen; however, the effective stress $\bar{\sigma}$ is *not* that of the outermost shell, but represents instead a kind of "average stress" that reflects only the over-all torque exerted by the individual layers, of which some are undergoing hardening and others are experiencing softening. The point of interest is that the peak strain ϵ_p for the composite torsion specimen is larger than ϵ_p for the tensile specimen, *even though both flow curves are based on the same set of tensile data.*

Source: Deformation, Processing, and Structure, George Krauss, Ed., papers presented at the ASM Materials Science Seminar, 23 October 1982, St. Louis MO, sponsored by the Seminar Committee of the Materials Science Division of the American Society for Metals, Metals Park OH, 1984, p 193

2-40. Cold Rolled 1018 Steel

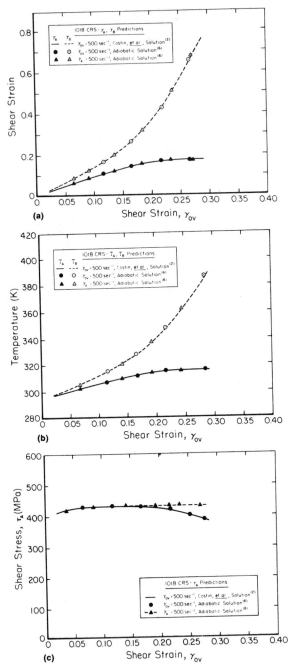

(a)

(b)

(c)

Comparison of torsion simulation results for 1018 CRS based on (1) heat-transfer model (Costin et al.) and (2) assumption of adiabatic heating. Predictions are in terms of (a) γ_A and γ_B versus γ_{ov}, (b) T_A and T_B versus γ_{ov}, and (c) τ_A versus γ_{ov}.

Source: S. L. Semiatin and J. J. Jonas, Formability and Workability of Metals, American Society for Metals, Metals Park OH, 1984, p 261

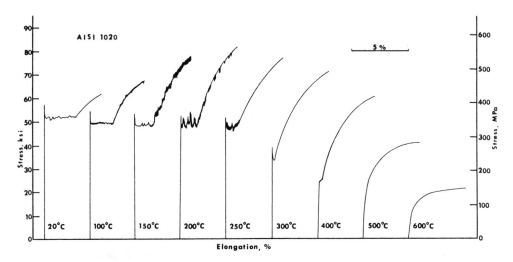

Stress-elongation curves of a carbon steel strained in tension, $\dot{\epsilon} = 1.75 \times 10^{-4}$ s^{-1}. Stressed at the various temperatures as shown.

Source: William C. Leslie, The Physical Metallurgy of Metals, McGraw-Hill Book Co., New York, and Hemisphere Publishing Corp., Washington, D.C., 1981, p 92

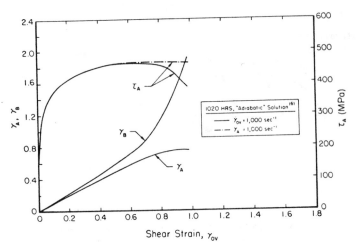

Predictions of γ_A, γ_B, and τ_A for high-speed torsion of 1020 HRS using boundary conditions of $\dot{\gamma}_A = 1000$ s^{-1} and $\dot{\gamma}_{ov} = 1000$ s^{-1}. Note that γ_A and γ_B are independent of the boundary condition.

At large strains of the order of $\gamma_{ov} = 0.9$ to 1.0, localization is indeed predicted to occur (see figure above). These strains are considerably in excess of those imposed in the experiments of Costin et al., suggesting that further torsion tests are warranted. However, comparison of the 1018 CRS and 1020 HRS results demonstrates that flow localization is delayed a considerable amount in the latter material relative to the former because of the stabilizing influence of larger n and m and smaller temperature sensitivity.

Source: S. L. Semiatin and J. J. Jonas, Formability and Workability of Metals, American Society for Metals, Metals Park OH, 1984, p 264

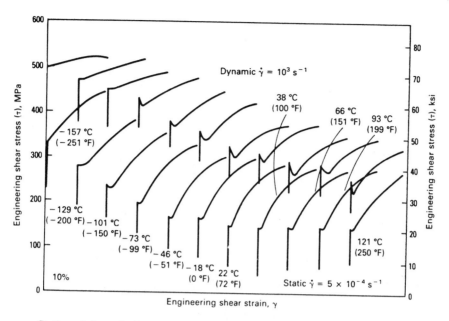

Static and dynamic shear stress/shear strain curves for hot rolled 1020 steel.

To obtain the shear strain in the specimen, the elastic rotation of the bar between the two differential transformers is subtracted from the total rotation. This elastic rotation is measured by cementing the loading bars together without a specimen and loading them quasi-statically. Typical test results obtained at a variety of temperatures using the Kolsky bar to test 1020 steel at a quasi-static strain rate of 5×10^{-4} s^{-1} are given in the figure above.

Source: Metals Handbook, Ninth Edition, Volume 8, Mechanical Testing, American Society for Metals, Metals Park OH, 1985, p 225

2-44. Carbon and HSLA Steels: 1020, 1522, 1035, A374, and Mn-Mo-Cb

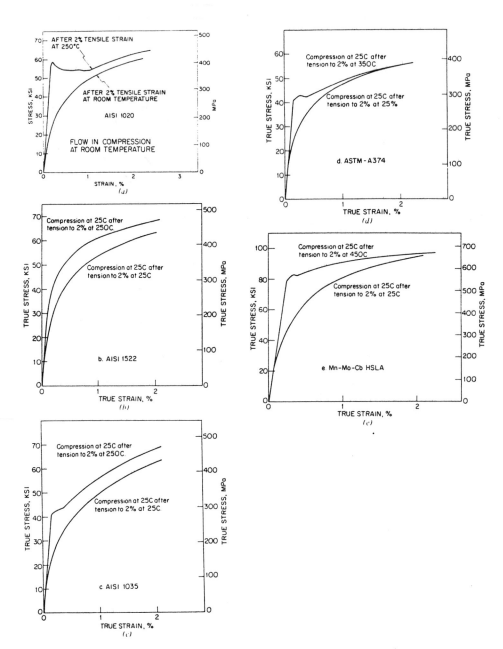

Ambient-temperature compression curves after prestraining in tension at 25 °C or at 250 °C. (a) AISI 1020, (b) AISI 1522, (c) AISI 1035, (d) A374, (e) Mn-Mo-Cb.

Source: C.-C. Li, J. D. Flasck, J. A. Yaker and W. C. Leslie, On Minimizing the Bauschinger Effect in Steels by Dynamic Strain Aging, Met. Trans. A, January 1978, American Society for Metals, Metals Park OH

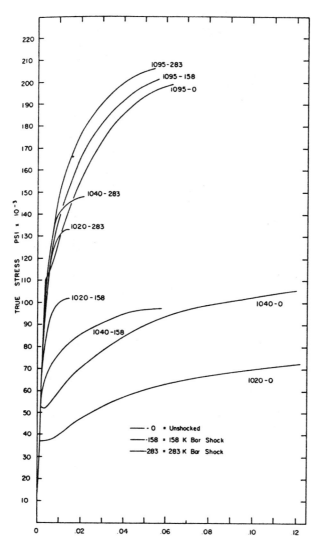

True stress–true strain curves illustrating the effect of shock-loading on 1020, 1040, and 1095 steels.

SAE 1020, 1040, and 1095 steels were shock-loaded to 158 and 283 kbar peak pressures by the mousetrap flyer plate technique. Post-shock mechanical properties, optical microscopy, and comparison with data from the literature indicated that shock pressure duration is critical to strengthening by the $\alpha \rightarrow \epsilon$ pressure induced transformation. It is suggested that the critical pressure duration is related to a time required for completion of the transformation process and that the required duration is inversely dependent on the volume percent carbide present in the steel.

Source: B. G. Koepke, R. P. Jewett, W. T. Chandler and T. E. Scott, Effects of Initial Microstructure and Shock Method on the Shock-Induced Transformation Strengthening of Carbon Steels, Met. Trans. A, August 1971, American Society for Metals, Metals Park OH, p 2045

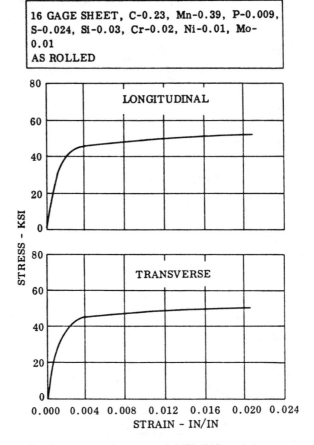

16 GAGE SHEET, C-0.23, Mn-0.39, P-0.009, S-0.024, Si-0.03, Cr-0.02, Ni-0.01, Mo-0.01
AS ROLLED

LONGITUDINAL

TRANSVERSE

STRESS - KSI

STRAIN - IN/IN

Tensile stress-strain curves of AISI 1023 steel sheet.

Source: Structural Alloys Handbook, Volume 1, Daniel J. Maykuth, Ed., Mechanical Properties Data Center, Battelle Columbus Laboratories, Columbus OH, 1980, p 28

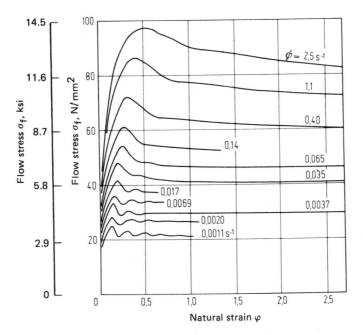

Flow-stress curves for carbon steel (0.25% carbon) at T = 1100 °C (2012 °F).

A carbon steel containing 0.25% carbon, for which the stress-strain curves at 1100 °C (2012 °F) are shown above, exhibits the described tendency even at low strains. The observation that at higher strains the flow stress is approximately constant is increasingly true at smaller strain rates $\dot{\varphi}$. The stress-strain curves shown above were obtained in hot torsion experiments.

Source: Handbook of Metal Forming, Kurt Lange, Ed., McGraw-Hill Book Co., New York, 1985, p 16.11

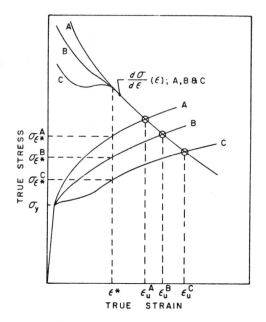

Schematic diagram showing the effect of strain hardening behavior at low strains on the maximum uniform strain for equivalent yield strength, and strain hardening at high strains.

An example illustrating the concepts shown in the graph above was presented by Pickering. He showed that in mild steels an increase in pearlite volume fraction, which does not alter strain hardening at high strains, causes a decrease in yield-point elongation, an increase in the rate of strength accumulation at low strains, and a corresponding decrease in the uniform strain at instability.

The preceeding discussion has shown that the true uniform strain at instability depends, in a complex way, on both the yielding and strain-dependent strain-hardening behavior. In attempts to simplify discussions of strain hardening, deformation behavior is commonly analyzed with the aid of idealized mathematical stress-strain equations and modified plotting techniques.

Source: Deformation, Processing, and Structure, George Krauss, Ed., papers presented at the ASM Materials Science Seminar, 23 October 1982, St. Louis MO, sponsored by the Seminar Committee of the Materials Science Division of the American Society for Metals, Metals Park OH, 1984, p 67

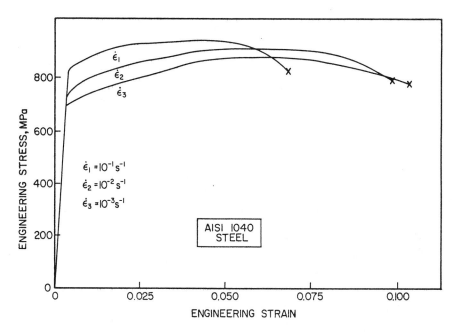

Effect of strain rate on the stress-strain curves for AISI 1040 steel.

This figure shows the effect of different strain rates on the tensile response of AISI 1040 steel. The yield stress and flow stresses at different values of strain increase with strain rate. The work-hardening rate, on the other hand, is not as sensitive to strain rate. This illustrates the importance of correctly specifying the strain rate when giving the yield stress of a metal. Not all metals exhibit a high strain rate sensitivity. Aluminum and some of its alloys have either zero or negative m. In general, m varies between 0.02 and 0.2, for homologous temperatures between 0 and 0.9 (90% of melting point in K). Hence one would have, at the most, an increase of 15% in the yield stress by doubling the strain rate.

Source: Marc André Meyers and Krishan Kumar Chawla, Mechanical Metallurgy: Principles and Applications, Prentice-Hall, Inc., Englewood Cliffs NJ, 1984, p 572

2-50. 1112 Steel

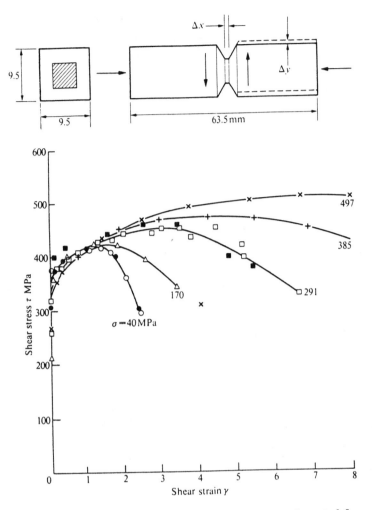

Shear stress-shear strain results for resulfurized low-carbon steel for specimen shown at top. σ = normal stress on shear plane. (After Walker and Shaw, 1969.)

From the figure above, it is seen that for a low value of normal stress on the shear plane of 40 MPa, strain hardening appears to go negative at a shear strain (γ) of about 1.5; that is, when the normal stress on the shear plane is about 10 percent of the maximum shear strain reached, negative strain hardening sets in at a shear strain of about 1.5.

Source: Milton C. Shaw, Metal Cutting Principles, Clarendon Press, Oxford, 1984, p 192

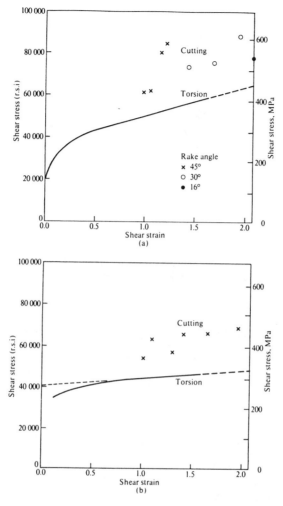

Comparison of shear-stress–shear-strain data for cutting and torsion tests. (a) Material: AISI B1112 steel cut at low speed using several rake angles and fluids. (b) Material: leaded free machining steel cut at low speed using several rake angles and fluids. (After Shaw and Finnie, 1955.)

When shear-stress–shear-strain orthogonal cutting data are compared with conventional torsional tests data for the same material, substantial differences are observed (see above). The cutting data do not lie on a single curve (appreciable scatter) and stresses for a given strain are considerably higher than for torsion. Likewise, when cutting data are compared with extrapolated uniaxial tensile data on the basis of equivalent stress and strain, the correlation is relatively poor.

Source: Milton C. Shaw, Metal Cutting Principles, Clarendon Press, Oxford, 1984, p 184

2-52. 1112 Steel

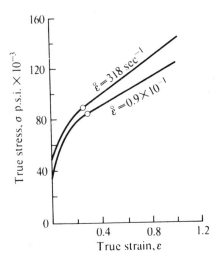

True stress-strain curves for AISI 1112 steel at different strain rates. Temperature = 70 °F (21 °C).

When metals are tested in tension at different strain rates, the flow stress corresponding to a given strain is found to increase with strain rate. The figure above illustrates this behavior. The following equation is frequently used to relate flow stress and strain rate at a given strain and temperature:

$$\sigma = \sigma_1 \dot{\epsilon}^m$$

where $\dot{\epsilon} = d\epsilon/dt$ and σ_1 and m are material constants. The exponent m (strain rate sensitivity) is found to increase with temperature, especially above the strain recrystallization temperature. In the hot working region, metals tend to approach the behavior of a Newtonian liquid for which $m = 1$.

Source: Milton C. Shaw, Metal Cutting Principles, Clarendon Press, Oxford, 1984, p 69

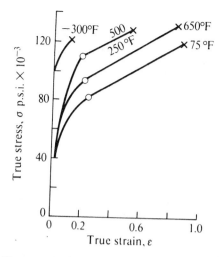

True-stress–true-strain tensile curves for AISI 1112 steel at different temperatures. Strain rate = 10^{-3} s^{-1}.

Temperature and Strain Rate Effects. When specimens are tested in tension at different temperatures, they are found to be more brittle (low strain at fracture) at low temperature than at elevated temperatures. However, in the vicinity of 550 °F (1022 °C) steels tend to have a higher yield stress and lower strain at fracture than at room temperature. This anomalous behavior, termed blue brittleness, is due to the migration of interstitial carbon and nitrogen to dislocations resulting in their immobilization. The figure above shows some typical tensile results for AISI 1112 steel tested at different temperatures but ordinary strain rate (10^{-3} s^{-1}). As a first approximation, the blue brittle anomaly is ignored and the yield stress is considered to decrease with increase in temperature.

Source: Milton C. Shaw, Metal Cutting Principles, Clarendon Press, Oxford, 1984, p 69

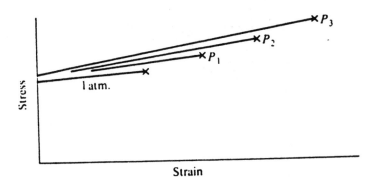

Influence of hydrostatic pressure on flow stress-strain curve and fracture stress (x)$P_1 < P_2 < P_3$.

The figure above illustrates these effects. Bridgman expresses the increased fracture strength (σ_f) with pressure in terms of a pressure coefficient of ductility (K):

$$\sigma_f = \sigma_{f0} + K\sigma_H$$

where σ_H is the hydrostatic pressure (+ for compression).

Source: Milton C. Shaw, Metal Cutting Principles, Clarendon Press, Oxford, 1984, p 68

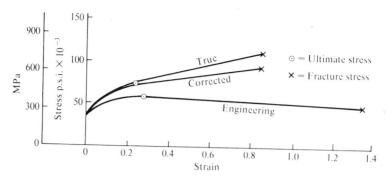

Relationship between engineering, true, and corrected tensile stress-strain curves for AISI 1112 steel.

The figure above shows the relationship between the so-called engineering stress-strain curve based on the original area, the true stress-strain curve, and the corrected true stress-strain curve where the stress plotted (σ_c) is the uniaxial tensile stress in the absence of the hydrostatic component introduced by curvature of the neck.

It is evident that interpretation of tensile test results is really quite involved despite the apparent simplicity of the test.

Source: Milton C. Shaw, Metal Cutting Principles, Clarendon Press, Oxford, 1984, p 67

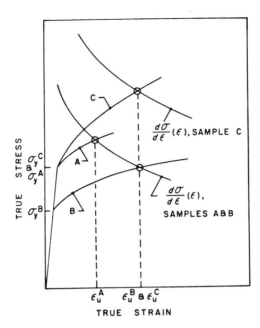

Schematic diagram showing the effect of strain dependency of strain-hardening rate on maximum uniform strain for equivalent yield strength (samples A and C), and the possibility of achieving the same maximum uniform strain in materials differing widely in deformation behavior (samples B and C).

The graph above illustrates the importance of the magnitude of the true strain-hardening rate at high strains. In this figure, samples A and C have the same yield strength, whereas samples A and B have the same dependence of strain-hardening rate on strain. Clearly, for constant σ_y, ϵ_u increases with $d\sigma/d\epsilon$ because the strain required to achieve the appropriate $\Delta\sigma$ also increases. However, as shown by a comparison of samples B and C, if both σ_y and $(d\sigma/d\epsilon)(\epsilon)$ are different, then it is possible for two samples that differ widely in deformation behavior, and thus in microstructure, to exhibit the same ductility.

Source: Deformation, Processing, and Structure, George Krauss, Ed., papers presented at the ASM Materials Science Seminar, 23 October 1982, St. Louis MO, sponsored by the Seminar Committee of the Materials Science Division of the American Society for Metals, Metals Park OH, 1984, p 66

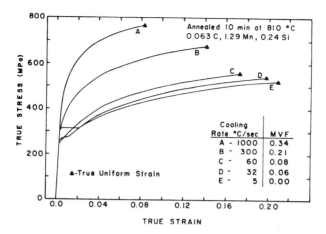

True stress–true strain curves for Fe-0.063C-1.29Mn-0.24Si steel intercritically annealed at 810 °C and cooled at different rates.

Source: Deformation, Processing, and Structure, George Krauss, Ed., papers presented at the ASM Materials Science Seminar, 23 October 1982, St. Louis MO, sponsored by the Seminar Committee of the Materials Science Division of the American Society for Metals, Metals Park OH, 1984, p 70

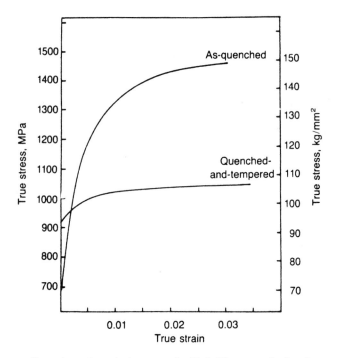

True stress–true strain curves for Fe-0.2C as-quenched and quenched-and-tempered lath martensite with packet size of 8.2 μm.

The figure above shows stress-strain curves that illustrate the changes in work hardening that develop with tempering of lath martensite in an Fe-0.2C alloy. In this case, the as-quenched martensite was obtained by quenching in a NaOH-NaCl solution and tempering was performed by heating in lead at 400 °C (750 °F) for 1 min. The work hardening rate in the as-quenched specimen was quite high, as shown by the rapid increase in stress with increasing strain, while the stress-strain curve for the tempered specimen was almost flat, indicating a very low rate of work hardening. This difference in work hardening behavior is attributed to interaction of dislocations with relatively coarse particles of cementite that form on tempering.

Source: George Krauss, Principles of Heat Treatment of Steel, American Society for Metals, Metals Park OH, 1980, p 194

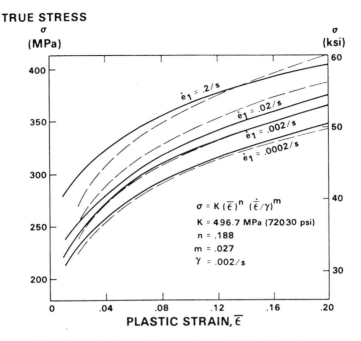

TRUE STRESS

Experimental values and power law model for AK steel.

It was attempted to match the experimental data with the equation below. However, it proved to be impossible to maintain accuracy at both high and low strains. Hence, we chose K and N to approximate the $\dot{e}_1 = 0.002/s$ curve, and then found m such that the stresses would be as accurate as possible at $\bar{\epsilon} = 0.1$. We found the values $K = 496.7$. MPa, $N = 0.188$, and $m = 0.027$. A comparison with the experimental data is shown in the figure above, where it is seen that the lower curves are matched quite well, but deviations are present for higher strain rates. However, it seems clear that the overstress model is more accurate. This is not surprising, of course, since more parameters were used.

Our computations are carried out using the model

$$\tau_e = K[(\dot{\epsilon} + \epsilon_0)^n + \eta \ln(1 + \dot{\epsilon}/\gamma)]$$

and sensitivity to strain rates are varied by means of the parameter η. However, neither of the numerical procedures are dependent on this representation, and, in principle, any well-behaved function could be used for $\tau_e(\bar{\epsilon}, \dot{\epsilon})$.

Source: Mechanics of Sheet Metal Forming: Material Behavior and Deformation Analysis, Donald P. Koistinen and Neng-Ming Wang, Eds., proceedings of a symposium sponsored by General Motors Research Laboratories, 17–18 October 1977, Warren MI, Plenum Press, New York, 1978, p 375

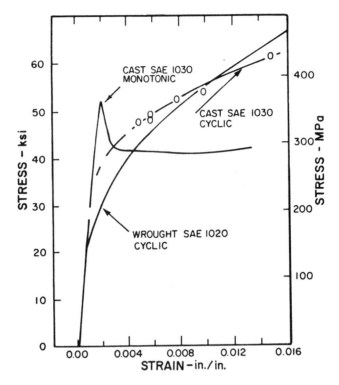

Monotonic tensile and cyclic stress-strain behavior of comparable wrought and cast carbon steel in the normalized and tempered condition.

The cyclic stress-strain characteristics in the graph above show a reduction of the strain-hardening exponent of the normalized and tempered cast carbon steel (SAE 1030) from $n = 0.3$ in monotonic tension to $n' = 0.13$ under cyclic-strain-controlled tests.

Source: Steel Castings Handbook, Fifth Edition, Peter F. Wieser, Ed., Steel Founders' Society of America, Rocky River OH, 1980, p 15-14

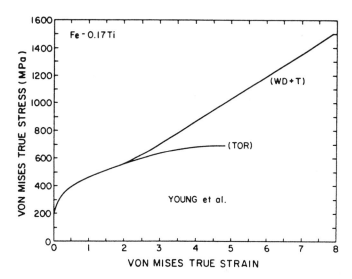

Comparison of stress-strain curves for Fe-0.17% Ti deformed by
torsion (solid rods) and by wire drawing plus tension.

Source: Deformation, Processing, and Structure, George Krauss, Ed., papers presented at the ASM Materials Science
Seminar, 23 October 1982, St. Louis MO, sponsored by the Seminar Committee of the Materials Science Division
of the American Society for Metals, Metals Park OH, 1984, p 7

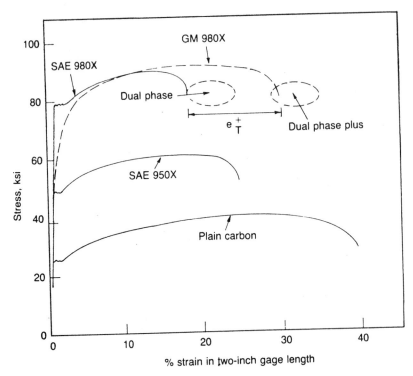

Stress-strain curves for plain carbon, SAE 950X, SAE 980X, and GM 980X. The GM 980X has been intercritically annealed and dual-phase microstructures produced. The two dashed ellipses indicate reported ranges of elongation for dual-phase steels.

The graph above shows the basis for three stages in the development of ferritic low-carbon steels. The lower stress-strain curve represents the deformation behavior of mild steel with ferrite-pearlite microstructures. The yielding is discontinuous and yield strengths are typically 30 ksi (207 MPa). SAE 950X and SAE 980X are HSLA steels with yield strengths of 50 ksi (345 MPa) and 80 ksi (562 MPa), respectively. The microstructures still consist of ferrite and pearlite but the ferrite grain size is highly refined because of controlled rolling and microalloying with vanadium. GM 980X is similar to SAE 980X but has been intercritically annealed to convert the pearlite to martensite. Rashid termed the resulting microstructure "dual-phase" to distinguish the ferrite-martensite microstructure from the ferrite-pearlite microstructure of conventionally treated mild steels of HSLA steels.

Source: George Krauss, Principles of Heat Treatment of Steel, American Society for Metals, Metals Park OH, 1980, p 242

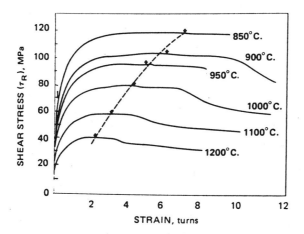

Effect of temperature on stress-strain curves for 0.16 C–0.04 Nb steel.

Source: Deformation, Processing, and Structure, George Krauss, Ed., papers presented at the ASM Materials Science Seminar, 23 October 1982, St. Louis MO, sponsored by the Seminar Committee of the Materials Science Division of the American Society for Metals, Metals Park OH, 1984, p 258

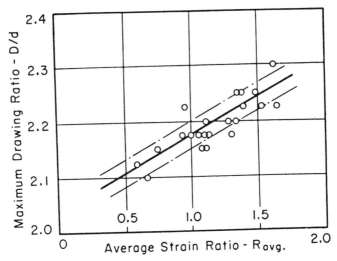

Effect of the average strain ratio, R_{avg}, on the maximum drawing ratio, D/d. The dashed lines represent the limits of experimental error.

Correlation of Drawing Ratio with Mechanical Properties

Mechanical Property	Coefficient of Correlation—r[1]	Significance of Correlation
Yield Strength	−0.12	Not significant
Tensile Strength	+0.07	Not significant
Yield-Tensile Ratio	+0.16	Not significant
Elongation in 2"	+0.42	Not significant
Olsen Ductility	+0.44	Not significant
Strain Hard. Exp.-n	+0.04	Not significant
Strain Ratio—R_{avg}.	+0.83	Highly significant[2]
Factor—R × n	+0.73	Highly significant[2]

[1] For perfect correlation r = ± 1, a negative coefficient indicating an inverse correlation.
[2] Probability that correlation is due to chance is less than 0.1%.

The normal directionality of each material was determined by averaging the strain ratios measured in all directions. The figure above shows the relationship between the drawing ratios and the average strain ratios of the materials. While the data show considerable scatter, a definite relationship exists between drawability and normal directionality. If we assume the experimental accuracy of the drawing ratio is ±0.025, a value suggested by Kemmis, a direct straight-line relationship is indicated between the drawing ratio, D/d, and the average strain ratio, R_{avg}.

The relative dependence of drawability on the strain ratio and other mechanical properties was determined by treating the data statistically to determine coefficients of correlation between the drawing ratio and each of the properties measured. The results of this statistical analysis are summarized in the table.

Source: Source Book on Forming of Steel Sheet, American Society for Metals, Metals Park OH, 1975, p 251

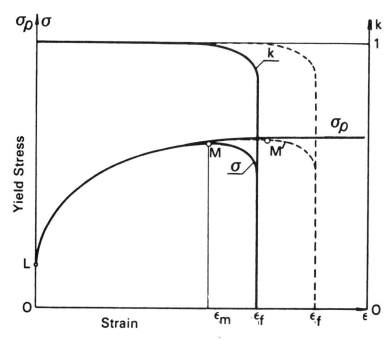

Yield stress σ as a product of strain-hardening function σ_p and strain-softening coefficient k.

Stress σ_p is treated here in a classical sense as a quantity expressing the hardening of the material depending on the strain, the strain path, and the temperature. On the other hand the coefficient k takes into consideration the softening of material caused by growth and coalescence of voids. Its value depends, in the first place, on the sign of stress prevailing in all the earlier stages of the process. For our purposes it is not necessary to specify the form of the functions σ_p and k. It is sufficient to assume that σ_p grows monotonically together with the strain and that k decreases from 1 to 0 all the faster, the higher are the tensile stresses compared to the compressive ones. As a result, the value of the product $k \cdot \sigma_p$ attains the maximum at point M (see figure above) and then falls to 0, which means a complete loss of material cohesion. The curve $\sigma(\epsilon)$ is thus not solely a characteristic of the material but depends also on the conditions of strain.

Source: Mechanics of Sheet Metal Forming: Material Behavior and Deformation Analysis, Donald P. Koistinen and Neng-Ming Wang, Eds., proceedings of a symposium sponsored by General Motors Research Laboratories, 17–18 October 1977, Warren MI, Plenum Press, New York, 1978, p 216

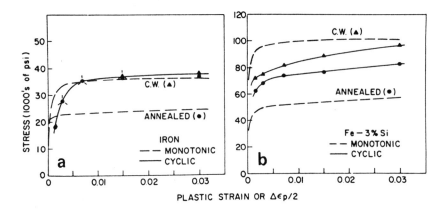

(a) History-independent cyclic stress-strain curve of iron compared to the monotonic curves of the uncycled material; (b) history-dependent response of Fe–3%
Si alloy compared to the monotonic response of the uncycled material.

Source: Fatigue and Microstructure, papers presented at the 1978 Materials Science Seminar, 14–15 October 1978,
St. Louis MO, sponsored by the Materials Science Division of the American Society for Metals, Metals Park OH,
1979, p 168

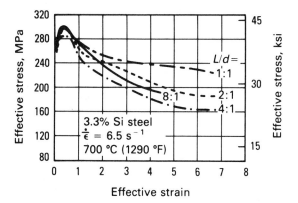

Stress-strain curves for solid torsion specimens of
3.3% silicon steel showing effect of gage length to
diameter ratio (L/d) on flow stress at high strain rates
when adiabatic heating occurs. The flow curves are
in terms of von Mises effective stress-strain ($\bar{\sigma}-\bar{\epsilon}$),
defined by $\bar{\sigma} = \sqrt{3}\tau$ and $\bar{\epsilon} = \Gamma/\sqrt{3}$, where τ-Γ is
the shear stress/shear strain curve obtained in tor-
sion testing.

In both solid bars and tubular specimens, the gage
length-to-diameter ratio may have a marked effect on the
actual specimen temperature during moderate-speed (Γ
= 10^{-2} to 10 s^{-1}) torsion tests because of the effects of
heat conduction. Because of this, flow curves derived
from data obtained at these rates tend to show a dependence on the length-to-diameter ratio (L/d). Flow curves
for large L/d specimens tend to fall below those for small
L/d ratios, in which most of the deformation heat is dis-
sipated into the shoulders (see above). Interpretation of
fracture strain data from such tests should take into account not only the nominal (initial) test temperature, but
also the temperature history during the test.

Source: Metals Handbook, Ninth Edition, Volume 8, Mechanical Testing, American Society for Metals, Metals Park
OH, 1985, p 157

2-68. 0.03% Carbon Rimmed Steel

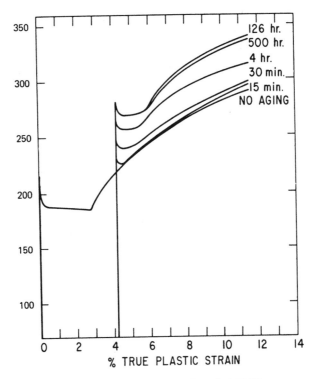

Strain aging of 0.03% C rimmed steel at 60 °C.

Strain aging in iron and steel at temperatures below about 100 °C is almost entirely due to nitrogen; at these temperatures the solubility of carbon is too low to produce any appreciable aging effects. Therefore, if nitrogen is removed from solution by precipitation as AlN, a steel can be made "nonstrain aging" even when the carbon content is normal, provided that the aging temperature is not much above ambient.

In static strain aging after prestraining in tension, the first change to occur is the reappearance of discontinuous yielding, as indicated in the figure above. With increased time the Lüders strain increases, up to about 0.02, and the yield stress increases to a maximum. After a considerable increase in the yield stress, the ultimate tensile strength and the work-hardening coefficient increase and the elongation to fracture decreases.

The rate at which these changes occur at a given aging temperature is a function of the interstitial solute content, as demonstrated by Wilson and Russell. The concentration of interstitial solute atoms in *solid solution* must be reduced to about 0.0001% (1 ppm) or less to eliminate the effects of strain aging. Strain-aging effects can approach their maxima at concentrations of only about 0.002%.

Source: William C. Leslie, The Physical Metallurgy of Metals, McGraw-Hill Book Co., New York, and Hemisphere Publishing Corp., Washington, D.C., 1981, p 89

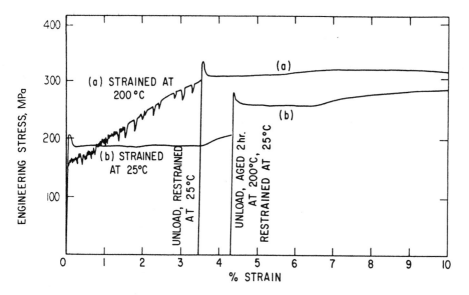

(a) Dynamic strain aging versus (b) static strain aging at 200 °C of 0.03% C rimmed steel.

Source: William C. Leslie, The Physical Metallurgy of Metals, McGraw-Hill Book Co., New York, and Hemisphere Publishing Corp., Washington, D.C., 1981, page 91

MATERIAL	F_{ty} MIN KSI
T-1, T-1 TYPE A, T-1 TYPE B	100
CON-PAC	80
EX-TEN 60	60
COR-TEN, TRI-TEN, EX-TEN 50	50
EX-TEN 42	42
ASTM A36	36

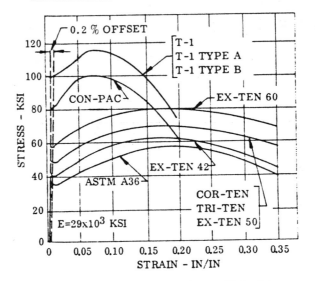

Comparison of stress-strain curves for specified minimum values.

Source: Structural Alloys Handbook, Volume 1, Daniel J. Maykuth, Ed., Mechanical Properties Data Center, Battelle Columbus Laboratories, Columbus OH, 1980, p 3

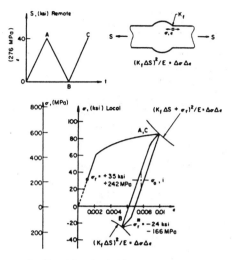

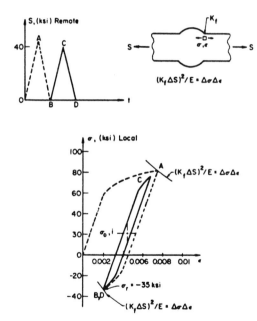

Computer-simulated local stress-strain response at the weld toe for a butt weld with tensile residual stresses (A36 HAZ material, $\sigma_r = +35$ ksi (242 MPa)).

Computer-simulated local stress-strain response at the weld toe for a butt weld with compressive residual stresses (A36 HAZ material, $\sigma_r = -35$ ksi (−242 MPa)).

Source: Residual Stress for Designers and Metallurgists, Larry J. Vande Walle, Ed., proceedings of a conference, 9–10 April 1980, Chicago IL, sponsored by the American Society for Metals, Metals Park OH, 1981, p 109

3-3. A-212B Structural Steel

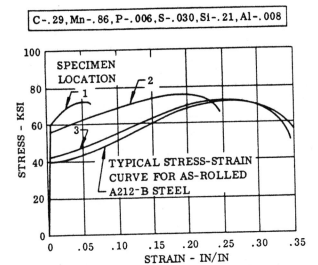

C-.29, Mn-.86, P-.006, S-.030, Si-.21, Al-.008

Stress-strain curves for hot-prestrained A-212B steel.

Source: Structural Alloys Handbook, Volume 1, Daniel J. Maykuth, Ed., Mechanical Properties Data Center, Battelle Columbus Laboratories, Columbus OH, 1980, p 8

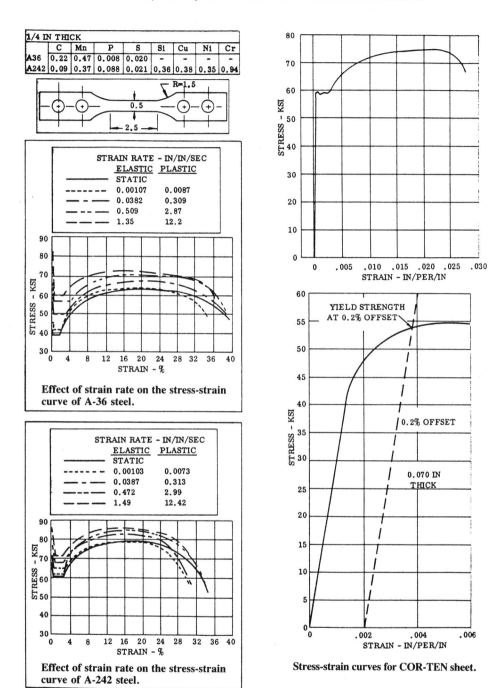

Effect of strain rate on the stress-strain curve of A-36 steel.

Effect of strain rate on the stress-strain curve of A-242 steel.

Stress-strain curves for COR-TEN sheet.

Source: Structural Alloys Handbook, Volume 1, Daniel J. Maykuth, Ed., Mechanical Properties Data Center, Battelle Columbus Laboratories, Columbus OH, 1980, p 9

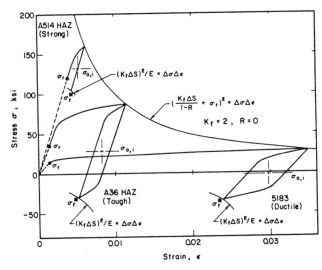

Set-up cycle stress-strain response for A514 HAZ (strong), A36 HAZ (tough), and 5183 WM (ductile) materials.

As shown in the figure above, the stabilized value of mean stress (σ_{0s}) resulting from the residual stress (σ_r) may vary greatly depending on the material. For many aluminums, the heat affected zone at the weld toe is in the zero-temper state; consequently, the notch-root plasticity in the first cycle results in $\sigma_{0s} = 0$. Other materials, such as high-strength steels, exhibit very little notch-root plasticity; consequently, σ_{0s} may be larger than σ_r. The results obtained using the model agree with the experimentally observed behavior.

Source: Residual Stress for Designers and Metallurgists, Larry J. Vande Walle, Ed., proceedings of a conference, 9–10 April 1980, Chicago IL, sponsored by the American Society for Metals, Metals Park OH, 1980, p 110

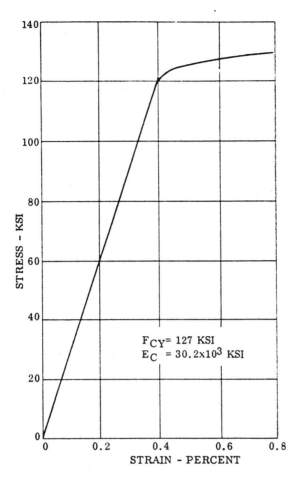

Typical compressive stress-strain diagram for T-1 steel at room temperature.

Source: Structural Alloys Handbook, Volume 1, Daniel J. Maykuth, Ed., Mechanical Properties Data Center, Battelle Columbus Laboratories, Columbus OH, 1980, p 11

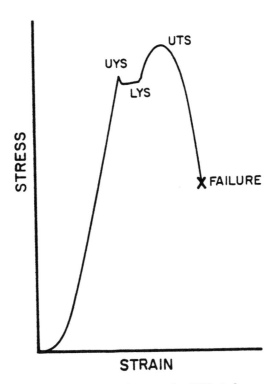

Typical stress-strain curve for A710 steel.

The stress-strain curves obtained for this alloy are typical of many steels, exhibiting upper and lower yield points followed by Lüders band propagation. Of interest is the magnitude of the upper yield strength.

Source: Engineering Properties of Steel, Philip D. Harvey, Ed., American Society for Metals, Metals Park OH, 1982

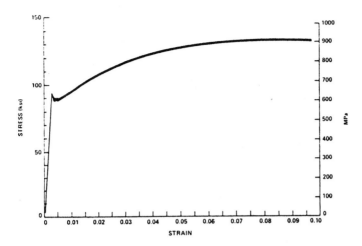

Monotonic stress-strain curve for SAE-1045 (260 HB) steel.

Above is shown a monotonic tension stress-strain curve for the SAE 1045 steel performed at a rate of 3×10^{-4} s^{-1}. Results of this test are given in the table below and are comparable to results for steels of similar hardness.

Monotonic Stress/Strain Results for SAE 1045 (260 HB)

```
Monotonic Tension Test of 1045-22
    Strain Rate              =  3.00000E-04

SUMMARY OF MONOTONIC TENSION PROPERTIES

    Elastic Modulus          =  28.96×10³ (199.7×10³)   ksi (MPa)

    0.002 Yield Stress       =  90.2 (621.9)            ksi (MPa)

    Ultimate strength        =  132.7 (914.9)           ksi (MPa)

    % Reduction in area      =  59.2

    True fracture ductility  =  0.90
    True fracture strength   =  258.7 (1783.6)          ksi (MPa)
        (corrected)

    Strain hardening exponent =  0.180
    Strength coefficient      =  233.04 (1606.4)        ksi (MPa)
```

Source: Proceedings of the SAE Fatigue Conference P-109, Fatigue Conference and Exposition, Dearborn MI, 14–16 April 1982, Society of Automotive Engineers, Inc., Warrendale PA, 1982, p 254

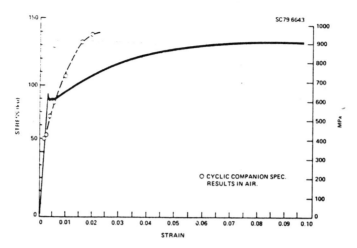

Comparison of monotonic and cyclic stress-strain curves for SAE 1045 (260 HB) steel.

Shown above is a comparison of the monotonic and cyclic stress-strain curve in lab air for this steel. The cyclic curve was obtained from the companion specimen fatigue results by plotting the steady-state stress response at 50% of the life to failure for the appropriate controlled strain amplitude. Note that cyclic strain softening occurs at strains less than approximately 0.007, while cyclic hardening occurs at greater strains. This response is also typical of steels of similar composition and hardness.

Source: Proceedings of the SAE Fatigue Conference P-109, Fatigue Conference and Exposition, Dearborn MI, 14–16 April 1982, Society of Automotive Engineers, Inc., Warrendale PA, 1982, p 256

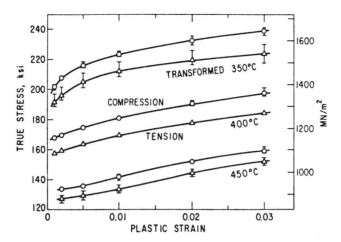

True stress-plastic strain curves in tension and in compression for lower, intermediate, and upper bainite in AISI 10B46 steel.

The SD effect has been found in all iron-carbon martensites for which data are available. To ascertain whether its occurrence in bainite is similarly common, a 10B46 steel was transformed in the lower, intermediate, and upper bainite ranges by the isothermal treatments. The resulting tensile and compressive stress-strain curves are shown in the curves above. It is apparent that the SD effect is present even in the upper bainite structure, in which the carbon has been in large part precipitated during isothermal transformation.

Source: G. C. Rauch and W. C. Leslie, The Extent and Nature of the Strength-Differential Effect in Steels, Met. Trans. A, February 1972, American Society for Metals, Metals Park OH, p 377

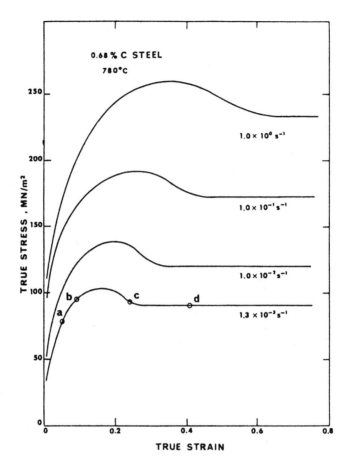

Flow curves for AISI C1060 carbon steel compressed at 780 °C and a series of constant true strain rates. The letters a, b, c, and d represent the interruption strains used in these experiments.

Source: R. A. P. Djaic and J. J. Jonas, Recrystallization of High Carbon Steel Between Intervals of High Temperature Deformation, Met. Trans. A, February 1973, American Society for Metals, Metals Park OH, p 622

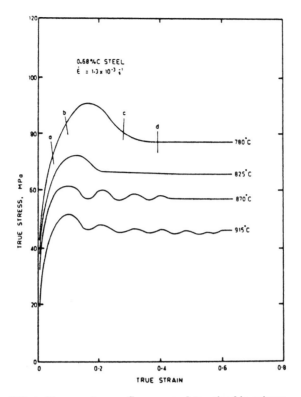

Effect of temperature on flow curves determined in axisymmetric compression on a 0.68% C steel at a strain rate of 1.3×10^{-3} s^{-1} After Petkovic et al.

Source: Deformation, Processing, and Structure, George Krauss, Ed., papers presented at the ASM Materials Science Seminar, 23 October 1982, St. Louis MO, sponsored by the Seminar Committee of the Materials Science Division of the American Society for Metals, Metals Park OH, 1984, p 188

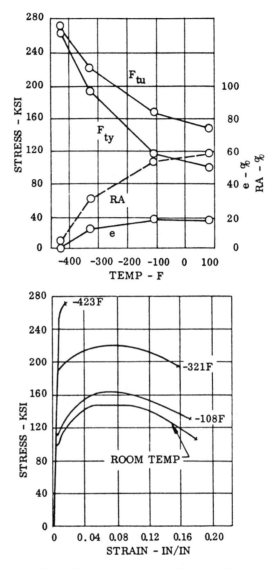

Effect of temperature on tensile properties.

Source: Structural Alloys Handbook, Volume 1, Daniel J. Maykuth, Ed., Mechanical Properties Data Center, Battelle
Columbus Laboratories, Columbus OH, 1980, p 7

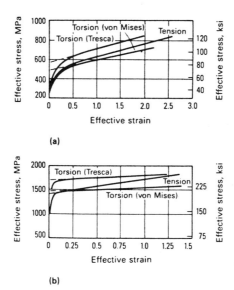

Flow curves determined via torsion testing and tension testing (following wire drawing) (a) 0.06% carbon steel. (b) 0.85% carbon steel (in pearlitic condition). Note that the torsion data are expressed in terms of both the von Mises and the Tresca effective stress-strain definitions.

The figure above illustrates results for a material with a crystal structure different from that of the face-centered cubic materials. These effective stress-strain curves are for body-centered cubic carbon steels that were tested in torsion and tension. To achieve the high deformation levels in tension, samples were prestrained by wire drawing. The von Mises flow curves for torsion lie below those for tension for both the low-carbon steel and the pearlitic, near-eutectoid high-carbon steel.

Also shown in this figure are effective stress-strain curves from torsion tests that were calculated on the basis of the Tresca criterion, in which the effective stress is equal to 2τ and the effective strain to $\Gamma/2$. A divergence between torsion and tension is still present.

Source: Metals Handbook, Ninth Edition, Volume 8, Mechanical Testing, American Society for Metals, Metals Park OH, 1985, p 165

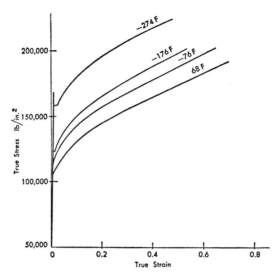

Effect of temperature on true-stress vs. true-strain relation for pearlitic steel.

In general, it may be said that the hardness and strength of a metallic material vary *inversely* with the temperature, while properties that depend on ductility vary *directly* with the temperature. The influence of temperature on the true-stress true-strain relation for a steel is shown in the figure above.

Source: Donald S. Clark, Engineering Materials and Processes, International Textbook Co., Scranton PA, 1962, p 67

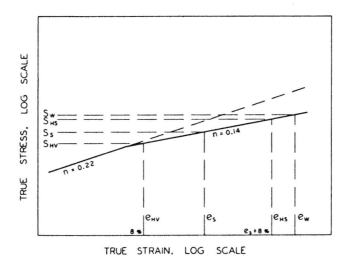

Sketch of the true stress-true strain relationship for a spheroidized 1.14% C steel. Adapted from Wilson and Konnan.

For the tempered steels, as for many other materials containing a dispersion of hard particles, the exponent of work hardening is quite small. Wilson and Konnan performed an accurate measurement on a spheroidized 1.14% C steel. Their results are sketched in the figure above. The value of n changes from 0.22 to 0.14 at a strain of 6 to 8%. Thus, a prediction of the abrasion resistance based on the bulk hardness and the higher value of n would be considerably in error in this case, yielding excessive values for R. The results of Wilson and Konnan substantiate this general relationship. They find that the true stress-true strain curves, for an almost pure iron and for a spheroidized 1.14% C steel, are parallel for strains above 6 to 8%. The stress levels are widely different, giving a smaller n value for the higher hardness connected with the 1.14% C steel. The same general correlation is substantiated by the work of other investigators.

Source: Source Book on Wear Control Technology, David A. Rigney and W. A. Glaeser, Eds., American Society for Metals, Metals Park OH, 1978, p 124

4-10. 1.3% C Steel

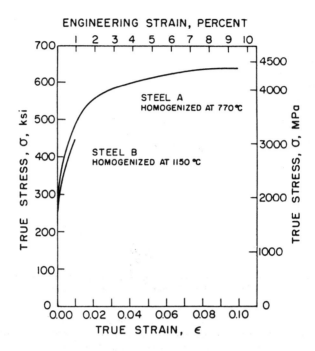

The influence of prior heat treatment on a UHC steel quenched from 770 °C is shown in the figure above. The hardness after quenching is about $R_c = 68$. On the graph the compression stress-strain curves of a 1.3% C steel after two different prior heat treatments are shown with accompanying optical micrographs. The martensite in steel A is so fine it is optically unresolvable and exhibits nearly 10% compression ductility with a fracture strength of 650 ksi.

Source: Superplastic Forming, Suphal P. Agrawal, Ed., proceedings of a symposium, 22 March 1984, Los Angeles CA, sponsored by the Los Angeles chapter of the American Society for Metals, Metals Park OH, 1985, p 38

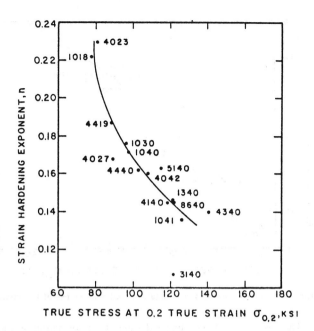

Variation of strain-hardening exponent with true stress at 0.2 true strain.

Source: Source Book on Cold Forming, American Society for Metals, Metals Park OH, 1975, p 142

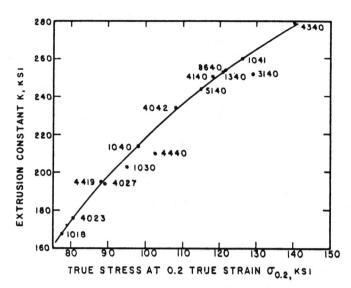

Relationship between extrusion constant k and true stress at 0.2 strain $\sigma_{0.2}$.

The resistance of the annealed steels to deformation as measured by compression and extrusion tests yield qualitatively similar results, i.e., increasing resistance to deformation by compression corresponded closely to increasing force required for extrusion. The relationship between the $\sigma_{0.2}$ values obtained by compression testing and the extrusion constant k is illustrated in the figure above.

Source: Source Book on Cold Forming, American Society for Metals, Metals Park OH, 1975, p 145

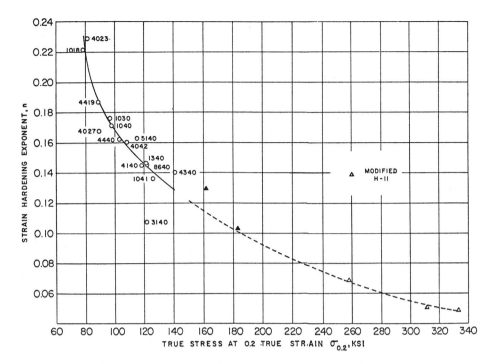

The influence of alloying elements on the cold deformation of steel.

Source: Source Book on Cold Forming, American Society for Metals, Metals Park OH, 1975, p 147

5-4. 4130 Steel

HOT ROLLED AND NORMALIZED SHEET
ALL THICKNESSES FROM 1 HEAT OF MATERIAL
C-.31, Mn-.50, P-.014, S-.15, Cr-.92, Mo-.19
AUSTENITIZE 1575F, OQ
TEMPER 1000F = 150 KSI NOMINAL STRENGTH LEVEL
TEMPER 830F = 180 KSI NOMINAL STRENGTH LEVEL
TEMPER 750F = 200 KSI NOMINAL STRENGTH LEVEL
0.064 IN SHEET
F_{tu} = 150 KSI
DATA SHOWN FOR VARIOUS EXPOSURE TIMES

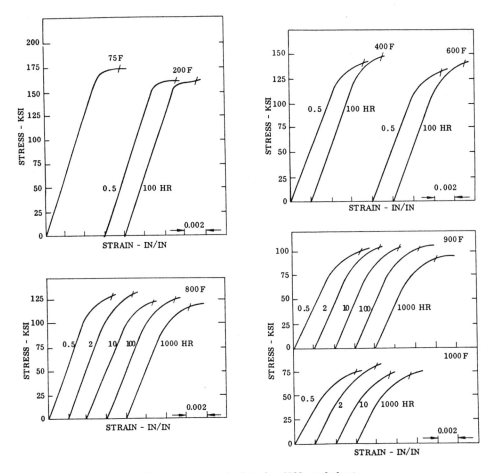

Compressive strain data for 4130 steel sheet.

Source: Structural Alloys Handbook, Volume 1, Daniel J. Maykuth, Ed., Mechanical Properties Data Center, Battelle
Columbus Laboratories, Columbus OH, 1980, p 31

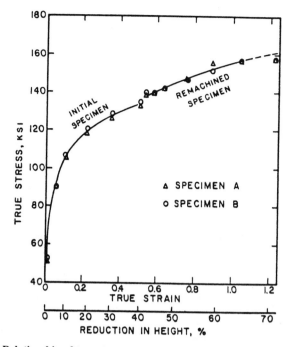

Relationship of true stress versus true strain obtained from compression tests of 4140 steel.

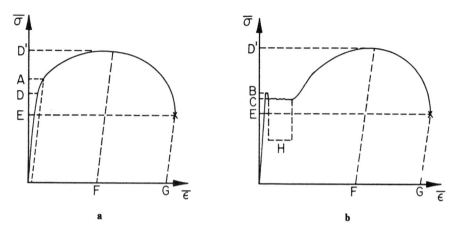

Engineering (or nominal) stress versus strain curves (a) without and (b) with yield point.

This figure shows two types of stress-strain curves. The first does not exhibit a yield point, while the second does. A number of parameters are used to describe the various features of these curves. First, the elastic limit; since it is difficult to determine the maximum stress for which there is no residual deformation, the 0.2% offset yield stress (point A in the figure above is commonly used instead; it corresponds to a residual strain of 0.2% after unloading. Actually, there is evidence of dislocation activity in a specimen at stress levels as low as 25% of the yield stress. This region (between 25 and 100% of the yield stress) is called the microyield region and has been the object of careful investigations. In case there is a yield drop, an *upper* (B) and a *lower* (C) *yield point* are defined. It should be emphasized that the lower yield point depends on the machine stiffness. A *proportionality limit* is also sometimes defined (D); it corresponds to the stress at which the curve deviates from linearity.

Source: Marc André Meyers and Krishan Kumar Chawla, Mechanical Metallurgy: Principles and Applications, Prentice-Hall, Inc., Englewood Cliffs NJ, 1984, p 565

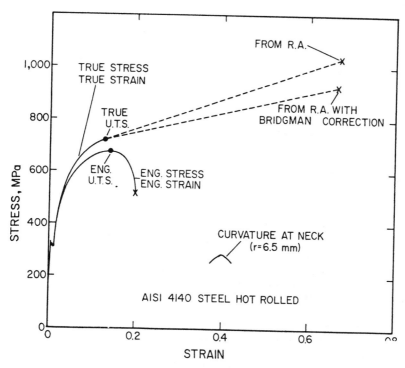

True and engineering stress versus strain curves for AISI 4140 hot rolled steel.

This figure shows engineering and true stress versus strain curves for the same hot rolled AISI 4140 steel. In the elastic regime the coincidence is exact because strains are very small (~0.5%). For $\epsilon = 0.20$ (a common value for metals) we have $\bar{\epsilon} = 0.221$. For this deformation the true stress is 22.1% higher than the nominal one. It can be seen that these differences become greater with increasing plastic deformation. Another *basic* difference between the two curves is the decrease in the engineering stress beyond a certain value of strain (~0.14 in the figure above).

Source: Marc André Meyers and Krishan Kumar Chawla, Mechanical Metallurgy: Principles and Applications, Prentice-Hall Inc., Englewood Cliffs NJ, 1984, p 565

5-8. 4320 Steel

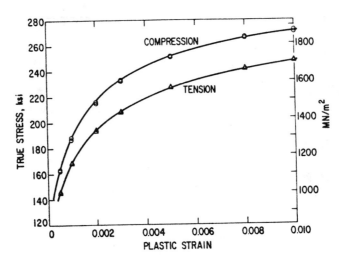

True stress-plastic strain curves in tension and in compression for as-quenched ultrafine-grained martensite (ASTM 14) in AISI 4320 steel. Tension-compression specimens.

The curves above show both tension and compression true stress-plastic strain curves for as-quenched martensite in 4320 steel, illustrating the SD effect. For these and similar stress-strain curves throughout this paper, the absolute value of compressive stress is plotted to facilitate comparison with the tensile data. The curves in the figure are for ultrafine-grained martensite (prior austenite grain size ASTM 14) produced by the thermal-cycling technique developed by Grange. The tension-compression specimen design was used for these tests. Coarser-grained specimens of 4320, with prior austenite grain size as large as ASTM 7, show a similar SD effect.

Source: G. C. Rauch and W. C. Leslie, The Extent and Nature of the Strength-Differential Effect in Steels, Met. Trans. A, February 1972, American Society for Metals, Metals Park OH, p 376

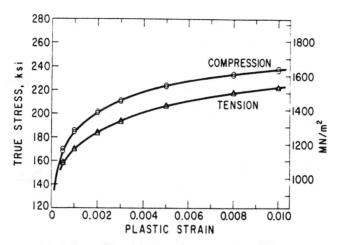

True stress-plastic strain curves in tension and in compression for lower bainite in AISI 4340 steel, transformed at 315 °C (588 °K) for 30 min. Tension-compression specimens.

The SD effect in martensite was known to exist before the present work began. The present study has shown that the effect exists in certain other austenite transformation structures as well. The figure above shows, for example, tensile and compressive stress-strain curves for lower bainite in a 4340 steel. As in the ultrafine-grained martensite, the absolute difference between tension and compression is virtually constant with strain greater than about 0.2%.

Source: G. C. Rauch and W. C. Leslie, The Extent and Nature of the Strength-Differential Effect in Steels, Met. Trans. A, February 1972, American Society for Metals, Metals Park OH, p 376

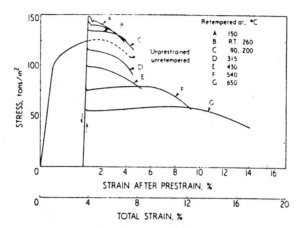

Strain tempering of AISI 4340 steel after hardening and tempering (1 h) at 205 °C. Engineering stress-strain curves show effect of prestraining 3% and retempering (1 h) as indicated.

The strengthening effect obtained by deforming tempered marten-site is similarly related to carbon content, but a strengthening effect is observed on tempering after straining at all carbon levels. If AISI 4340 steel in the martensitic condition and tempered at 205 °C, is plastically strained 3% in tension and then retempered at 205 °C, the yield strength is increased from 102 tons/in^2 to 145 tons/in^2 and the tensile strength from 127 tons/in^2 to 145 tons/in^2. The strengthening effect is attributed to the re-solution and reprecipitation of the car-bides in a much more finely dispersed form than can be achieved by ordinary tempering.

The stress-strain curves relevant to the above data are reproduced in the figure above, and it is clear that the applicability of this pro-cess is limited by the lack of tensile ductility shown by the strength-ened product. Other work has revealed that at very low strains the yield strength can be improved while still retaining appreciable duc-tility. It has also been shown elsewhere that the impact strength is impaired.

Source: Metallurgical Developments in High-Alloy Steels, proceedings of a joint conference on high-alloy steels held by the British Iron and Steel Research Association and the Iron and Steel Institute, 2–4 June 1964, London, England, Iron and Steel Institute, 1964, p 29

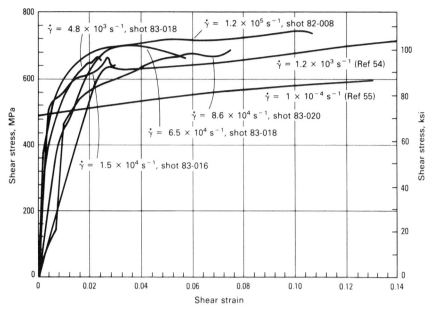

Dynamic stress-strain curves for vacuum arc remelted 4340 steel tempered at 600 °C (1110 °F).

Shot No.	Impact angle, θ, degrees	Specimen thickness, h		Projectile velocity, V_o	
		mm	in.	mm/μs	in./μs
82-008 . . . . 26.6		0.297	0.0117	0.205	0.0081
83-011 . . . . 18.4		0.194	0.0076	0.173	0.0068
83-016 . . . . 23.2		0.287	0.0113	0.145	0.0057
83-018 . . . . 23.2		0.255	0.0100	0.186	0.0073
83-020 . . . . 20.1		0.199	0.0078	0.201	0.0079

Stress-strain curves for several experiments on vacuum arc remelted 4340 steel specimens are compared in the figure above with stress-strain curves obtained from torsion experiments at lower strain rates. This comparison indicates that the flow stress in the pressure-shear experiments at strain rates of 10^5 s^{-1} is approximately 20% greater than in static torsion tests and less than 10% greater than in torsional Kolsky bar experiments at strain rates of 10^3 s^{-1}.

Systematic investigation of the effect of hydrostatic pressure indicates that there is no measurable effect for changes in pressure of as much as 1.0 GPa (145 ksi). The lack of a consistent trend in the pressure-shear experiments precludes a definitive conclusion on the strain rate sensitivity in this strain rate regime. Nevertheless, the trend is for the flow stress to increase with increasing strain rate over this regime, and the rate of increase is comparable to, and possibly slightly higher than, the logarithmic increase with strain rate that is observed at lower strain rates.

The weaker strain rate sensitivity of the flow stress in vacuum arc remelted 4340 steel, compared to that reported for 1100-O aluminum, is an indication that for the steel the mechanism controlling the rate of plastic flow is thermal activation of dislocations past obstacles, even at strain rates of 10^5 s^{-1}. This view is supported by the observation that the highest shear stresses obtained in the pressure-shear experiments (740 MPa, or 107 ksi) are below the values obtained in quasi-static torsion experiments at 80 K and well below the value of 900 MPa (130 ksi) obtained by extrapolation to 0 K of quasistatic flow stresses at 80 to 293 K.

Source: Metals Handbook, Ninth Edition, Volume 8, Mechanical Testing, American Society for Metals, Metals Park OH, 1985, p 237

5-12. 4620 Steel

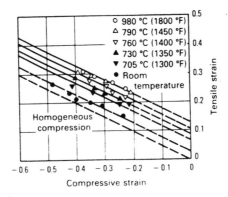

Fracture loci for elevated-temperature tests on sintered 4620 steel (80% of theoretical density) tested at $\dot{\epsilon} = 10 \ s^{-1}$.

For some materials, the fracture locus lines are straight and parallel with a slope of $-1/2$ (see above), while for other materials the fracture line is observed to have two slopes. In this case, a slope of unity is found at small strains using the tapered and flanged compression specimen.

Source: Metals Handbook, Ninth Edition, Volume 8, Mechanical Testing, American Society for Metals, Metals Park OH, 1985, p 583

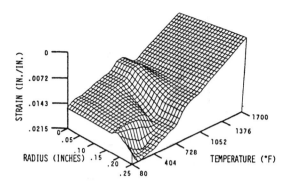

Dilatation curves were obtained on homogeneous carbon 86XXX steels with 0.18, 0.40, 0.60, 0.80, 1.0, and 1.2% carbon representing the carbon levels encountered in the layers of a carburized cylinder. It is well known that the various carbon levels within the steel have different temperatures on cooling at which the transformation of austenite to martensite occurs. Considering the carbon gradient depicted above, the combined phase transformation and thermal dilatation strains are shown in the figure above as a function of temperature and position/chemistry.

Source: Residual Stress for Designers and Metallurgists, Larry J. Vande Walle, Ed., Proceedings of a conference on Metals, Highway and Off-Highway Vehicles, activity of the Materials Systems and Design Division, 9–10 April 1980, Chicago IL, sponsored by the American Society for Metals, Metals Park OH, 1981, p 55

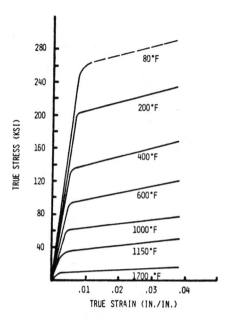

Typical stress-strain curve for 8660 steel.

An extensive number of tensile tests were run to determine the elastic-plastic and yield properties for the typical carbon levels of the 86XXX steel. Data for one of the homogeneous carbon levels (8660) are shown in the graph above. Similar data for 0.18, 0.40, 0.80, 1.00, and 1.20% carbon steels were also evaluated and used as data input to the finite element program.

Source: Residual Stress for Designers and Metallurgists, Larry J. Vande Walle, Ed., proceedings a conference on Metals, Highway and Off-Highway Vehicles, activity of the Materials Systems and Design Division, 9–10 April 1980, Chicago IL, sponsored by the American Society for Metals, Metals Park OH, 1981, p 56

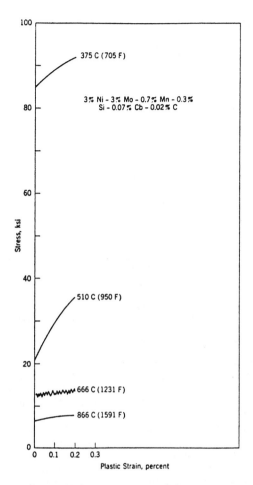

Appearance of stress-strain curves of 3Ni-3Mo alloy steel at various test temperatures.

The nature of the stress-strain curves obtained during the austenite deformation tests is also of interest. Those for the prototype steel are shown in the figure above. At the highest temperature, the form of the curve is descriptive of austenite deformation. At a test temperature, e.g., 666 °C (1231 °F), where the upper transformation would occur at long holding times, a serrated curve, indicative of stress-induced transformation in this alloy is obtained, which provides further support for the suggestion that the strength of the austenite is important. It would follow that if stress can induce transformation, some barrier to transformation, i.e., the strength of the austenite, must pre-exist.

In the temperature range of transformation to the shear dominated product, e.g., 510 °C (950 °F), rapid strain hardening is observed. This strain hardening may contribute to the phenomenon of transformation plasticity. Just prior to the onset of this region, a slight decrease in flow stress was sometimes detected, suggesting that the first shear-dominated transformation occurs readily but that the work-hardened austenite (and ferrite) provides a rapidly increasing resistance to transformation at a given temperature in this range.

Source: Transformation and Hardenability in Steels, a symposium sponsored by Climax Molybdenum Co. and the University of Michigan, 27–28 February 1967, Climax Molybdenum Co., Greenwich CT, p 187

5-16. Alloy Steels: Air vs Vacuum Melted

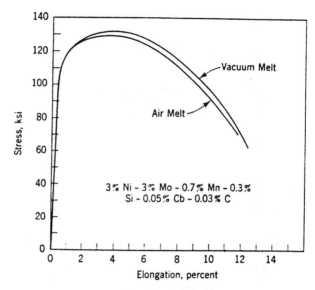

3% Ni – 3% Mo – 0.7% Mn – 0.3%
Si – 0.05% Cb – 0.03% C

Stress-strain curves for vacuum and air melted alloys with nominal composition of 3% Ni–3% Mo–0.7% Mn–0.3% Si–0.05% Cb–0.03% C.

The stress-strain curves of vacuum melted and air melted laboratory heats of this steel, in the as-rolled condition, are shown in the figure above. Here it is seen that melting procedure has little influence on these curves. The reduction in area for the air melt was 59%, that for the vacuum melt 65%. The yield strength of these alloys, 112,000 psi, is quite similar to those obtained for the 9% Ni–3% Mo alloy but a higher tensile strength of 130,000 psi and a greater extension to necking are observed, demonstrating a greater strain hardening capacity. If these alloys are furnace cooled at a very low rate of 1 °C/min (2 °F/min) the strength level drops to 69,000 psi, as a consequence of a loss of substructure hardening.

Source: Transformation and Hardenability in Steels, a symposium sponsored by Climax Molybdenum Co. and the University of Michigan, 27–28 February 1967, Climax Molybdenum Co., Greenwich CT, p 189

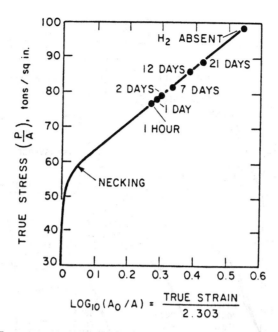

True-stress true-strain curves for a hydrogen-impregnated 3Cr-Mo steel after standing in air for varying periods. (After Hobson and Sykes.) Assume 1 tsi = 2.24 ksi = 15.5 MPa.

The predominant effect of hydrogen on the properties of steel is a decrease in ductility and true stress at fracture. This effect is manifested in tension tests with smooth specimens as a decrease in reduction of area and elongation, in tension tests with notched specimens as a decrease in notch tensile strength, and in bend tests as a decrease in bend ductility. The graph above illustrates the effects of hydrogen on ductility. Steels containing hydrogen characteristically fracture after the onset of necking but at lower than normal strain values, as illustrated in these data given by Hobson and Sykes for smooth tensile bars of an 0.26C-3Cr-Mo steel tempered to a strength level of 800 MPa ($\sim$116 ksi).

Source: Current Solutions to Hydrogen Problems in Steels, C. G. Interrante and G. M. Pressouyre, Eds., proceedings of the conference, 1–5 November 1982, Washington, D.C., sponsored by the American Society for Metals, Metals Park OH, 1982, p 10

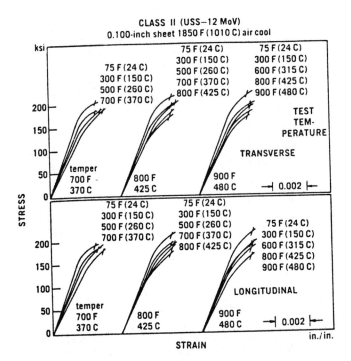

Compressive stress-strain curves.

Compressive Properties. Compressive characteristics are of prime interest in airframe design. While compressive strength is also a factor in some tool applications, the need for precise values is not as great as for airframes. Consequently, available data are confined to Class II grades that have been considered for airframes where much of the structure is loaded in compression. These data indicate the yield strength in compression is somewhat higher than that in tension, at least up to 1000 °F (540 °C). No data have yet been made available on Class III materials.

The ratio of compressive yield strength to density is often taken as a representative factor for depicting structural efficiency. This, of course, assumes a high degree of support of the compression sheet against structural instability. In the design range for a Mach 3 vehicle, the compressive yield strength/density ratio of Crucible 422 is comparable to that for PH 15-7 Mo and the 4% Al-3% Mo-1% V titanium alloy. This factor becomes important in the selection of honeycomb sections where one of the paramount reasons for a sandwich-type configuration is attainment of the desired structural stability at minimum weight. In tests on brazed honeycomb panels, the Class II Super 12% Cr steel detail had higher compressive strength than comparable panels of 17-7 PH and PH 15-7 Mo.

Source: Source Book on Materials for Elevated-Temperature Applications, Elihu F. Bradley, Ed., American Society for Metals, Metals Park OH, 1979, p 117

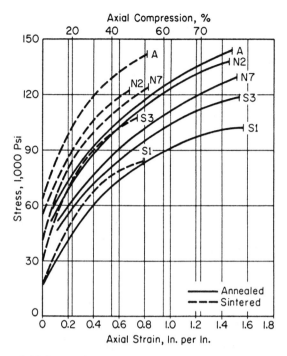

Cold forgeability of the following compositions is compared: (A) 0.27 C, 2.0 Ni, 0.5 Mo, bal Fe; (N2) 0.17 C, 2.7 Ni, 0.8 Cr, bal Fe; (N7) 0.24 C, 0.6 Ni, 0.5 Cr, 0.2 Mo, bal Fe; (S1) 0.01 C, bal Fe; (S3) 0.33 C, bal Fe.

Differences in the plastic deformability and resistance to deformation among preforms of various compositions and among sintered and annealed preforms are shown in the figure above.

Source: Source Book on Cold Forming, American Society for Metals, Metals Park OH, 1975, p 208

6-1. Quenched and Tempered 4130 Steel

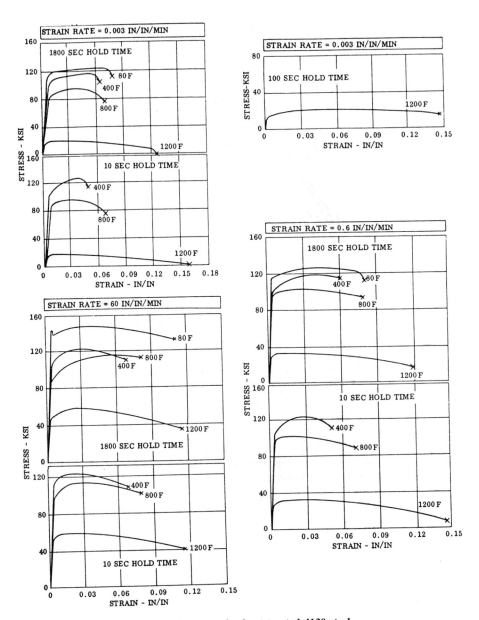

Stress-strain curves for heat treated 4130 steel.

Source: Structural Alloys Handbook, Volume 1, Daniel J. Maykuth, Ed., Mechanical Properties Data Center, Battelle Columbus Laboratories, Columbus OH, 1980, p 27

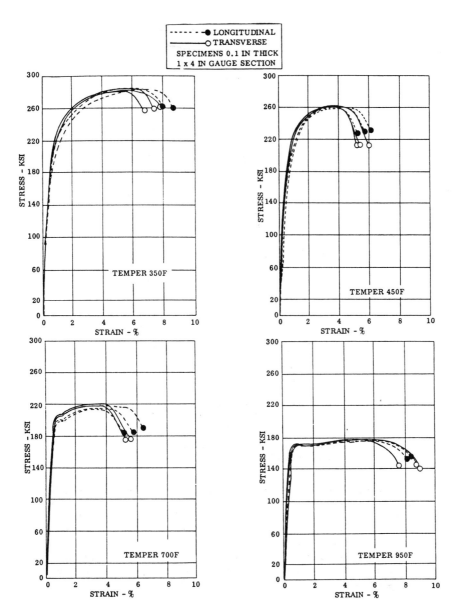

Comparison of longitudinal and transverse tensile stress-strain curves.

Source: Structural Alloys Handbook, Volume 1, Daniel J. Maykuth, Ed., Mechanical Properties Data Center, Battelle Columbus Laboratories, Columbus OH, 1980, p 42

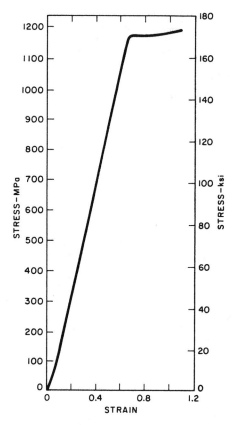

Stress-strain curve for AISI 4340 steel, quenched and tempered at 480 °C (900 °F).

A number of methods of determining the yield point are recognized by the American Society for Testing and Materials and are discussed in ASTM A 370. These include the drop-of-the-beam or halt-of-the-pointer (halt-in-gage) and autographic-diagram methods for material having a sharp-kneed stress-strain diagram; and the total-extension-under-load (total-strain) method for materials that do not exhibit a well-defined proportionate deformation that characterizes a yield point as measured by the drop-of-the-beam, halt-of-the-pointer, or autographic-diagram methods.

Source: The Making, Shaping and Treating of Steel, Tenth Edition, Wm. T. Lankford, Jr., Norman L. Samways, Robert F. Craven and Harold E. McGannon, Eds., Association of Iron and Steel Engineers, Pittsburgh PA, 1985, p 1401

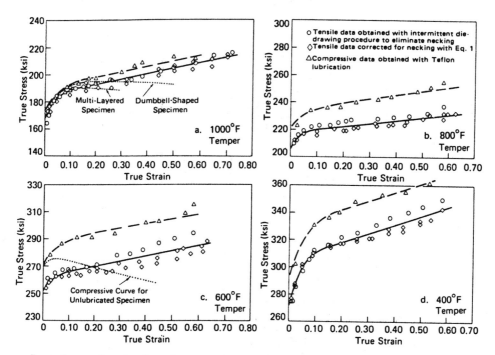

Room temperature tensile and compressive flow curves for 4340 steel quenched and then tempered at (a) 1000 °F (b) 800 °F, (c) 600 °F, and (d) 400 °F.

Source: R. Chait, Factors Influencing the Strength Differential of High Strength Steels, Met. Trans. A, February 1972, American Society for Metals, Metals Park OH, p 367

6-5. Quenched and Tempered 4340 Steel Compared With H-11 and Maraging Steels

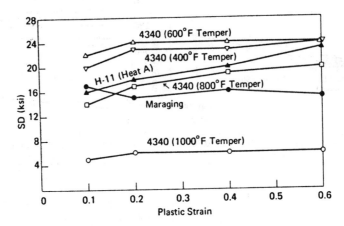

The effect of plastic strain on the strength differential (SD).

To obtain SD, the difference between the $(\sigma\text{-}e)_t$ and $(\sigma\text{-}e)_c$ curves was noted at various locations along the curve. As shown in the figure above, the SD of all materials remains at a relatively constant level despite large plastic deformation. Such behavior is particularly noteworthy as it bears on the applicability of some of the hypotheses that have been advanced to explain the SD.

Source: R. Chait, Factors Influencing the Strength Differential of High Strength Steels, Met. Trans. A, February 1972, American Society for Metals, Metal Park OH, p 368

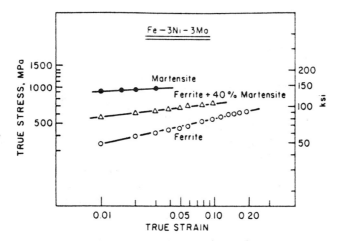

**A plot of log true-stress against log true-strain for three conditions
of Fe-3Ni-3Mo alloy showing that they obey a power law.**

Mileiko was the first to develop a theory relating the strength and
ductility of the two ductile components to the mechanical properties
of the composite; the theory is based on the application of plastic
instability criteria to the composite. There are two assumptions as to
materials properties in the theory. (1) The bond between fiber and
matrix is ideal, so that the strength of the interface is sufficient to
prevent the fiber necking without necking of the composite as a whole;
that is, the more stable matrix restrains the less stable fiber. The
interface in dual-phase steel is atomic in nature and therefore meets
this requirement. (2) The relationship between stress and strain for
both the composite and the components is a power law of the form:

$$\sigma = k\epsilon^n$$

Mechanical Properties. Plots of log true-stress against log true-
strain for the ferrite, martensite, and dual-phase structure (see figure)
are straight lines with slope n, confirming that all the structures obey
the equation above; thus one requirement of Mileiko's theory is ful-
filled. There were no Lüders strains or yield-point complications because all the specimens had smooth stress-strain curves; the small
vanadium addition effectively getters the residual carbon in the alloy.
Stress-strain data for all the structures were plotted as in the figure
above and n values were measured.

Source: R. G. Davies, The Mechanical Properties of Zero-Carbon Ferrite-Plus-Martensite Structures, Met. Trans. A,
March 1978, American Society for Metals, Metals Park OH, p 454

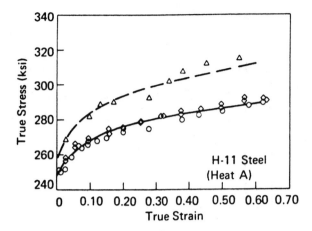

Room-temperature tensile and compressive flow curves for
H-11 steel.

Source: R. Chait, Factors Influencing the Strength Differential of High Strength Steels, Met. Trans. A, February
1972, American Society for Metals, Metals Park OH, p 367

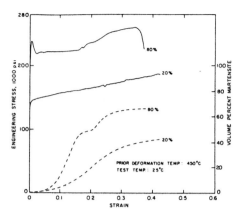

Engineering stress-strain (continuous curve) and volume percent martensite-strain curve (broken curve) for the alloys *a* and *d*, deformed at 450 °C.

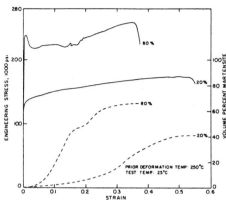

Same curves as shown at left, for alloys *b* and *c*, deformed at 250 °C.

Chemical Composition of the Alloys in the Graphs Above

Alloy Identification No.	Cr	Ni	Mn	Si	Mo	C	Fe
a	9	7.4	2.8	2.8	4.0	0.24	balance
a_1	9	7.4	2.8	2.8	4.0	0.25	balance
b	9	7.5	2.9	2.8	4.0	0.25	balance
b_1	9	7.5	2.9	2.8	4.0	0.25	balance
c	9	7.5	2.9	2.8	4.0	0.24	balance
c_1	9	7.5	2.9	2.8	4.0	0.25	balance
d	9	7.4	2.9	2.8	4.0	0.24	balance
d_1	9	7.4	2.9	2.8	4.0	0.24	balance
d_2	9	7.4	2.8	2.8	4.0	0.24	balance

Summary of Stress-Intensity and Tensile Data at 25°C for 0.075 In. Thick TRIP Steel Plate After Various Thermomechanical Processing

Alloy Identification No.	Processing	Yield Strength, ksi	Tensile Strength, ksi	Elongation, Pct	Critical Stress-Intensity Factor, K_c, ksi-in.$^{1/2}$	Magnetic* Characteristics Before Test	After Test
a	20 pct at 450°C	137	183	46		A	M
a_1	20 pct at 450°C	138	187	42	331†	A	M
b	20 pct at 250°C	126	174	55		A	M
b_1	20 pct at 250°C	119	160	49	279	A	M
c	80 pct at 250°C	237	237	36		A	M
c_1	80 pct at 250°C	237	258	38	296	A	M
d	80 pct at 450°C	233	248	34		A	M
d_1	80 pct at 450°C	231	243	32	334†	A	M
d_2	80 pct at 450°C	238	260	36		A	M

*A = Nonmagnetic. M = Magnetic.
†Tested at a cross-head speed of 0.2 ips. At 0.02 ips crack grew slowly giving no instability.

Tensile Test Results. Room-temperature tensile and fracture properties are listed below. The martensitic transformation was followed during tension testing by a "permeameter." The amount of martensite as a function of percent strain for four representative cases is shown in the two graphs above (broken lines). These figures also show the engineering stress-strain behavior for these cases (full lines). The tensile properties of TRIP steels have been discussed in detail by several investigators. As seen below, these steels have both high strength and elongation at room temperature.

Source: G. R. Chanani, Stephen D. Antolovich and W. W. Gerberich, Fatigue Crack Propagation in Trip Steels, Met. Trans. A, October 1972, American Society for Metals, Metals Park OH, p 2661, 2664

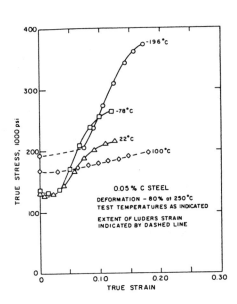

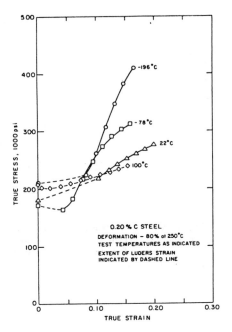

True-stress true-strain curves for the 0.05% C (TRIP) steel (deformed 80% at 250 °C) tested at several temperatures. Curves plotted to maximum load only—not to fracture.

True-stress true-strain curves for 0.20% C (TRIP) steel (deformed 80% at 250 °C) tested at several temperatures. Curves plotted to maximum load only—not to fracture.

Percentages of Alloying Elements in Steels						
C	Cr	Ni	Mo	Si	Mn	Fe
0.05	12.1	7.7	3.9	1.5	1.1	Bal
0.20	12.0	7.9	4.0	1.5	0.80	Bal

Source: G. R. Chanani, V. F. Zackay and Earl R. Parker, Tensile Properties of 0.05 to 0.20 Pct C TRIP Steels, Met. Trans. A, January 1971, American Society for Metals, Metals Park OH, p 134–135

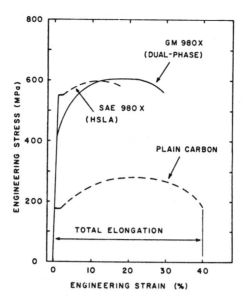

Comparison of the stress-strain curve for a dual-phase steel with those for a plain carbon steel and an HSLA steel.

The curves above compare the stress-strain curves for a dual-phase steel with those for a plain carbon steel and an HSLA steel. The compositions of the dual-phase and HSLA steels are identical (0.1 C, 1.5 Mn, 0.5 Si, 0.1 V). The dual-phase and HSLA steels have the same ultimate strength, but the dual-phase steel has significantly higher ductility. The conventionally processed plain carbon steel has much lower strength but higher ductility than the other two steels. Unique features of the stress-strain curve for the dual-phase steel are the absence of discontinuous yielding and a very high rate of strain hardening at low strains. Many combinations of strength and ductility may be developed in dual-phase steels, depending on processing conditions.

Source: Deformation, Processing, and Structure, George Krauss, Ed., papers presented at the ASM Materials Science Seminar, 23 October 1982, St. Louis MO, sponsored by the Seminar Committee of the Materials Science Division of the American Society for Metals, Metals Park OH, 1984

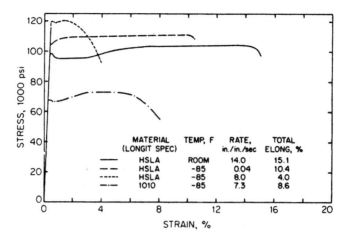

Change in strain of HSLA and 1010 steel with decreasing temperature.

The change in strain rate characteristics with low temperature is shown in the figure above. The data for HSLA steel are representative of a poorly performing high-strength steel; that is, other materials show greater ductility when tested under similar conditions.

Source: Source Book on Forming of Steel Sheet, American Society for Metals, Metals Park OH, 1975, p 125

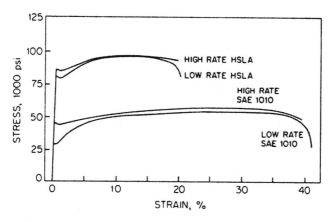

Typical high and low strain rate response of HSLA and low-carbon mild steel.

Many of the structures where HSLA steels are being considered for vehicles will be subject to impact-type damage. Thus, their performance under high strain rates and low temperatures is important. The figure above shows room-temperature data for low-carbon mild steel and a typical HSLA steel at high (10 in./in./s) and low (0.01 in./in./s) strain rates. The low rate would be considered quasi-static. The high rate is more realistic for impact conditions. On a percentage basis the mild steel is clearly more rate sensitive than the HSLA steel.

Source: Source Book on Forming of Steel Sheet, American Society for Metals, Metals Park OH, 1975, p 125

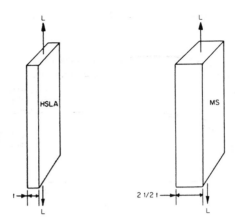

Comparative thickness of HSLA steel and mild steel under constant load and width.

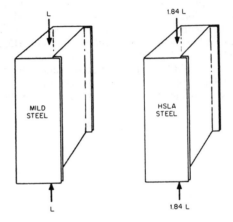

Incipient buckling load for HSLA and mild steel for same geometry and material thickness (from GM Engineering Staff.)

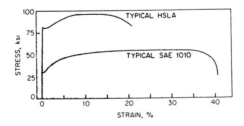

Stress-strain comparison of HSLA and SAE 1010 hot-rolled steel sheet.

Source: Source Book on Forming of Steel Sheet, American Society for Metals, Metals Park OH, 1975, p 124

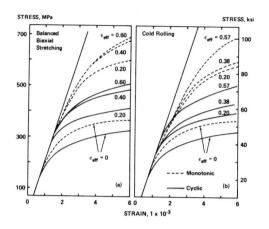

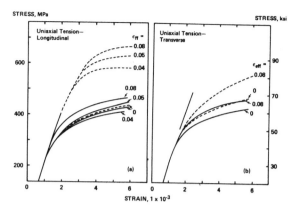

Fig. 1. AISI 1006 steel: monotonic and cyclic stress-strain curves before and after deformation.

Fig. 2. AISI 50XF steel: monotonic and cyclic stress-strain curves after deformation.

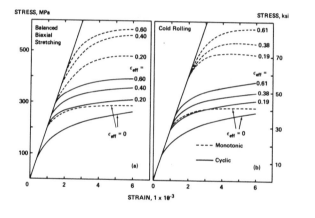

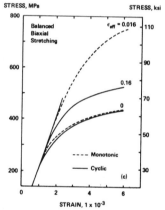

Fig. 3. AISI 80DF steel: monotonic and cyclic stress-strain curves before and after deformation.

Cyclic Stress-Strain Behavior. AISI 1006 STEEL: The cyclic stress-strain curves for 1006 for the various modes and amounts of deformation are shown in Fig. 1. Also shown in this figure for comparison are the corresponding monotonic stress-strain curves. As can be seen, both the cyclic and the monotonic 0.2% offset yield strengths (0.2CYS and 0.2MYS, respectively) increased as the amount of prior strain increased. Further, it can be seen that cyclic softening occurred for all conditions, including the undeformed material.

AISI 50XF STEEL: The cyclic and monotonic stress-strain curves for this steel are shown in Fig. 2. As is evident, the general behavior was the same as that of the 1006 steel.

AISI 80DF STEEL: The cyclic and monotonic stress-strain curves for this steel are shown in Fig. 3. The undeformed material is essentially cyclically stable in the longitudinal direction and exhibits minimum cyclic softening in the transverse direction; the 0.2CYS/0.2MYS ratio is 0.99 and 0.94 in these two directions, respectively.

Source: HSLA Steels: Technology & Applications, Proceedings of International Conference on Technology and Applications of HSLA Steels, 3–6 October 1983, Philadelphia PA, American Society for Metals, Metals Park OH, 1984, p 217–220

7-6. Effect of Deformation Mode on AISI 1006, 50XF, and 80DF Steels

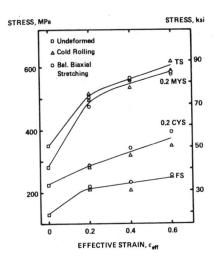

Fig. 1. AISI 1006 steel: mechanical properties after deformation as a function of prior effective strain.

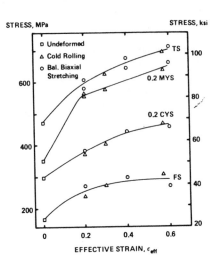

Fig. 2. AISI 50XF steel: mechanical properties after deformation as a function of prior effective strain.

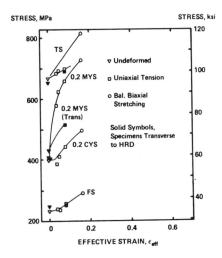

Fig. 3. AISI 80DF steel: mechanical properties after deformation as a function of prior effective strain.

AISI 1006 STEEL. Both the yield strength and the tensile strength are essentially the same for the two modes of deformation, and they both increase rapidly with the amount of deformation (Fig. 1). The strain-hardening capacity decreases rapidly initially, as evidenced by the rapid increase in the yield-to-tensile ratio, (Fig. 1). The uniform elongation also decreased rapidly between $\varepsilon_{eff} = 0$ and $\varepsilon_{eff} = 0.2$; the total elongation decreases more gradually with prior strain.

AISI 50XF STEEL. The effect of prior deformation on the postformed tensile properties was generally similar to that observed for 1006 steel (Fig. 2). Interestingly, the postformed uniform elongation was slightly larger in 50XF than in 1006 steel; the postformed total-elongation values for the two steels were nearly identical.

AISI 80DF STEEL. The postformed yield strength in the direction of prior tension, after uniaxial tension, appeared to be the same as after BBS (Fig. 3). In the direction normal to the prior tension, the yield strength was much lower, a Bauschinger-type effect. In contrast to the yield strength, the tensile strength exhibited no anisotropy after uniaxial tension; i.e., the yield strength was the same parallel and perpendicular to the prior tension. Interestingly, the tensile strength appeared to be much higher after BBS than after UT, even though the yield strength was the same.

Source: HSLA Steels: Technology & Applications, proceedings of International Conference on Technology and Applications of HSLA Steels, 3–6 October 1983, Philadelphia PA, American Society for Metals, Metals Park OH, 1984, p 215

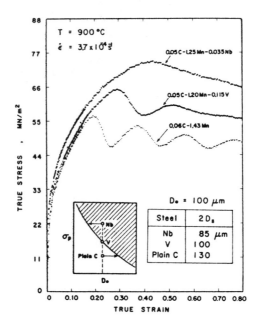

Effect of dynamic precipitation on the shape of the flow curve. The plain carbon steel exhibits cyclic recrystallization because $D_0/2D_s = 0.77$. Conversely, the niobium steel, with $D_0/2D_s = 1.18$, displays single-peak behavior. The vanadium steel, with $D_0/2D_s \simeq 1$, is at the transition between the two types of flow. The differences among the three steels are attributable to the occurrence of dynamic precipitation during compression of the two HSLA materials.

Source: Deformation, Processing, and Structure, George Krauss, Ed., papers presented at the ASM Materials Science Seminar, 23 October 1982, St. Louis MO, sponsored by the Seminar Committee of the Materials Science Division of the American Society for Metals, Metals Park OH, 1984, p 206

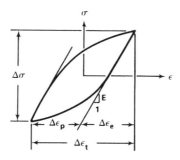

Stress-strain hysteresis diagram.
Testing Program Summary

Nominal Effective Strain Level	All Steels Undeformed, [1] As-Received Surfaces	AISI 1006 and AISI 50XF Steel		AISI 80DF Steel	
		Cold [1] Rolling (CR)	In-Plane Balanced [1] Biaxial Stretching (BBS)	Uniaxial Tension (UT)	In-Plane Balanced [1] Biaxial Stretching (BBS)
0.0	T,S	–	–	–	–
0.04 to 0.08	–	–	–	T,S	–
0.2	–	T,S	T,S	–	T,S
0.4	–	T,S	T,S	–	–
0.6	–	T,S	T,S	–	–

(1) All the test specimens were oriented parallel to the HRD except in AISI 80DF in the undeformed condition and after uniaxial tension, in which cases they were oriented parallel and perpendicular to the HRD, i.e., parallel and perpendicular to the direction of prior tension (see Figure 2).

T: Tension tests.
S: Strain-controlled fatigue tests.

Strain-controlled fatigue tests were conducted with the same specimen in accordance with ASTM practice E606-80 to obtain the strain-life curve corresponding to each combination of deformation mode, prior strain level, and specimen orientation (see table). The tests were performed under constant-amplitude, completely reversed, strain-controlled axial load using hydraulic, closed-loop testing machines and Wood's metal grips. Total strain, measured by an extensometer with a 7.6-mm (0.300-in.) gage length, was the control parameter. For each combination in the table above, tests were generally conducted on the duplicate specimens at the following nominal total-strain amplitudes: 0.75×10^{-2}, 0.5×10^{-2}, 0.25×10^{-2}, and 0.15×10^{-2}, or 0.125×10^{-2}. The extensometer was mounted with the knife edges against the machined sides of the specimen, not against the sheet surface. The strain-time waveform was sinusoidal with the first half-cycle positive. To minimize the strain-rate variation between the tests conducted at the different strain amplitudes, the sinusoidal frequency was increased from 0.2 to 3 Hz as the nominal strain amplitude was decreased, resulting in average strain rates ranging from 0.6 to 1.5 s^{-1}. For each specimen, the stress-strain response was recorded during the initial loading and during the first few strain cycles. Then hysteresis loops were recorded periodically throughout the test to monitor the cyclic response to obtain the half-life values of the various cyclic parameters.

For each specimen, from the hysteresis loop nearest half-life, the total-strain range, $\Delta\varepsilon_t$, was partitioned into its elastic and plastic components, $\Delta\varepsilon_e$ and $\Delta\varepsilon_p$, as shown in the figure above. The modulus of elasticity, E, and the stress range, $\Delta\sigma$, were also obtained from the same hysteresis loop.

The strain-life behavior was examined by plotting the total-strain amplitude, $\Delta\varepsilon_t/2$, versus log $(2N_f)$, the number of reversals to failure for each condition shown in the table, and these strain-life plots were compared visually to assess the effect of prior deformation.

Source: HSLA Steels: Technology & Applications, proceedings of International Conference on Technology and Applications of HSLA Steels, 3–6 October 1983, Philadelphia PA, American Society for Metals, Metals Park OH, 1984, p 213

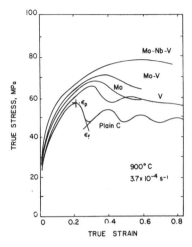

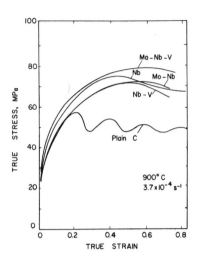

Fig. 1(a). Effect of molybdenum and vanadium addition on the flow curves of the first series of microalloyed steels.

Fig. 1(b). Effect of niobium addition on the flow curves of the first series of microalloyed steels.

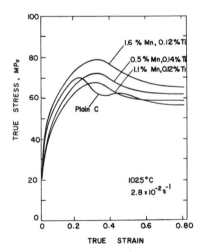

Fig. 1(c). Effect of manganese and titanium addition on the flow curves of the second series of steels.

Experimental Results. Typical sets of flow curves for the two series of steels are presented in Fig. 1. The curves for the first series of steels at 900 °C and a strain rate of 3.7×10^{-4} s^{-1} are shown in two parts for the sake of clarity (Fig. 1a and 1b). It is evident that the plain carbon steel recrystallizes the earliest, as seen by the smallest value of the first peak strain (strain at the maximum stress). In this material, several cycles of dynamic recrystallization occur, as indicated by the multiple peaks. (The end of the first cycle is depicted by ε_f in this figure). Similar behavior is displayed by the 0.115% vanadium steel, although fewer cycles of recrystallization occur, and the peak strains and stresses are greater. At this strain rate and temperature, the higher values are due in part to the effect of vanadium in solid solution and in part to the precipitation of VN. The addition of 0.30% Mo, on the other hand, produces a larger increase in the peak values,

Table Ia

Chemical Compositions (wt.%), Austenitization Temperatures and Austenite Grain Sizes for the Mo-Nb-V Steels Tested

Steel	C	Mn	Al	Si	Nb	V	Mo	T_{aus}'1' °C	Grain Size 1 μm	T_{aus}'2' °C	Grain Size 2 μm
Pl. C	.06	1.43	.025	.24	-	-	-	1030	110	1140	200
V	.05	1.20	.030	.25	-	.115	-	1045	100	1160	200
Nb	.05	1.25	.030	.27	.035	-	-	1100	130	1200	210
Mo	.05	1.34	.065	.20	-	-	.29	1060	110	1140	190
Mn-Nb	.06	1.90	.030	.23	.035	-	-	1100	120	- -	
Mo-Nb	.06	1.33	.025	.21	.040	-	-	1100	130	1200	200
Mo-V	.05	1.21	.050	.15	-	.115	.28	1100	120	1150	180
Nb-V	.05	1.18	.020	.24	.035	.115	-	1100	120	-	-
Mo-Nb-V	.04	1.16	.025	.20	.034	.115	.31	1100	115	1200	200

All: P= .006, N= .006, S= .012 Cr= .04

Table Ib

Chemical Compositions (wt.%), Austenitization Temperatures, Austenite Grain Sizes and Estimated Solution Temperatures for the Ti Steels

Steel	C	Mn	Al	Si	Ti	T_{aus} °C	Grain Size μm	Estimated Solution Temperature(*) °C	
								T_{min}	T_{max}
0.5 % Mn	.06	.48	.010	.12	.14	1260	120	1125	1160
1.1 % Mn	.10	1.07	.015	.29	.12	1260	110	1180	1195
1.6 % Mn	.10	1.60	.025	.24	.12	1260	110	1180	1210

All: N = .006, S = .008, P = .002

* Amount of Ti available for carbide formation:

Ti_{min} = Total % Ti - (48/32)% S - (48/14)% N

Ti_{max} = Total % Ti - (48/14)% N

but acts only as a solute. The combined presence of Mo and V produces an even larger increase in these values. The addition of niobium to this binary* steel (i.e. in the Mo-Nb-V steel) leads to a much larger increase than that attributable to either of the other two elements.

The effect of niobium when added alone is illustrated in Fig. 1(b), where the large increase in the peak stress and strain is evident. The addition of vanadium to this steel leads to a further increase in the peak strain. The somewhat lower peak stress is a result of the different grain sizes in these two materials (see tables above). The combined presence of Nb and Mo has an effect similar to that of Nb and V. The flow curve for the ternary steel is reproduced in Fig. 1(b) to illustrate the effect of the ternary addition relative to the other Nb steels. (The flow curves for the 0.42 and 1.90% Mn Nb steels have not been included for the sake of clarity.) There is a large difference between the peak values of the single V and single Mo steels on the one hand, and the binary (or ternary) steels containing Nb and V or Mn on the other. By contrast, the difference is relatively small between the single Nb and the binary (or ternary) Nb bearing steels.

Typical flow curves for the Ti series of steels are shown in Fig. 1(c) for 1025 °C and a strain rate of 2.8×10^{-2} s^{-1}. In comparison with the plain carbon steel, the addition of 0.12% Ti always leads to an increase in the peak strain, and

---

*In the present context, binary steel refers to two microalloy additions while ternary refers to three microalloy additions.

therefore, to a delay in the onset of dynamic recrystallization. This effect is enhanced in the presence of increased manganese. There is an unexpectedly small difference between the peak strain values of the 0.5 and 1.1% Mn steels. This can be attributed largely to the higher Ti concentration in the 0.5% Mn steel, and to a lesser extent to the grain size differences indicated in the tables.

Source: HSLA Steels: Technology & Applications, proceedings of International Conference on Technology and Applications of HSLA Steels, 3–6 October 1983, Philadelphia PA, American Society for Metals, Metals Park OH, 1984, p 150–152

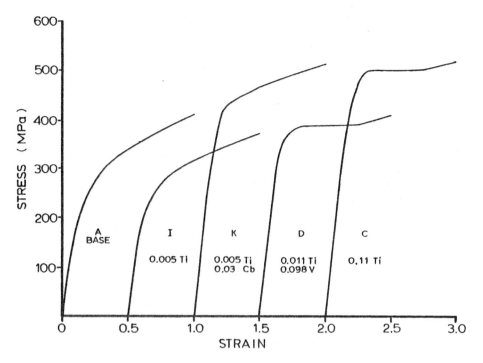

The effect of microalloy additions on strain behavior.

The base chemistry and base plus 0.005 and 0.010% titanium steels had continuous stress-strain curves with relatively low 0.5% underload yield and ultimate tensile strength (UTS). Columbium additions to both the base and base-plus-low-titanium chemistries significantly increased the yield and UTS level and altered the shape of stress-strain curves, as shown above. Vanadium additions to the columbium-titanium steels further increased strength.

Source: HSLA Steels: Technology & Applications, proceedings of International Conference on Technology and Applications of HSLA Steels, 3–6 October 1983, Philadelphia PA, American Society for Metals, Metals Park OH, 1984, p 885

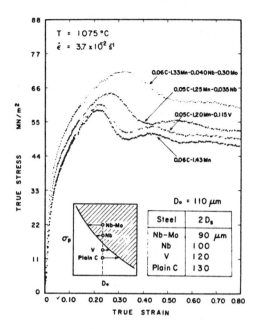

Effect of microalloy addition on the shape of the flow curve. The plain carbon and 0.115% V steels undergo cyclic recrystallization because $D_0/2D_s < 1$. The Nb and Nb-Mo grades display single-peak behavior because $D_0/2D_s > 1$. The transition in the character of the curves can be ascribed to the effect of the solute elements on the stable grain size produced at a given temperature and strain rate.

Source: Deformation, Processing, and Structure, George Krauss, Ed., papers presented at the ASM Materials Science Seminar, 23 October 1982, St. Louis MO, sponsored by the Seminar Committee of the Materials Science Division of the American Society for Metals, Metals Park OH, 1984, p 205

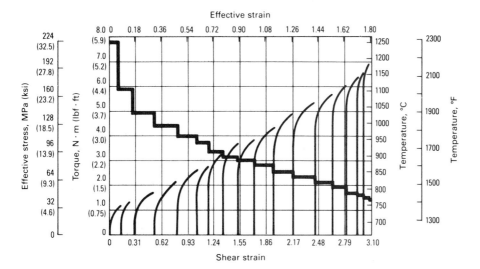

Flow behavior for a Nb-V microalloyed steel deformed in 17 passes in a torsion machine. The specimen temperatures are represented by the upper bold line.

Simulation of the rolling of high-strength low-alloy steels through the torsion test provides insight into how equipment requirements and final microstructures can be determined by this technique. In this instance, torsion tests were conducted in a servo-hydraulic test machine at a fixed strain rate. However, temperature was controlled to decrease continuously at a rate almost equivalent to that measured during actual production. Under these conditions, the torsion flow stresses increased rapidly, as shown in the figure above.

Source: Metals Handbook, Ninth Edition, Volume 8, Mechanical Testing, American Society for Metals, Metals Park OH, 1985, p 179

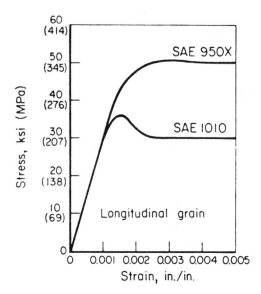

SAE 950X steel has a higher yield strength and a greater elastic range than SAE 1010 steel.

SAE 950X HSLA steel will perform much better than SAE 1010 steel in designs where it is important to delay the onset of inelastic buckling, as is the situation in automotive door beams and bumper reinforcements. Many highly stressed parts that do not require the ductility of low-carbon mild steels are prime candidates for conversion from mild steel to HSLA steel. Also, where safety or emission-control requirements in automobiles call for an increase in the thickness of mild-steel parts, HSLA steel of the same thickness as the existing part can be substituted. Only minor tool modification is necessary under those conditions.

Source: E. E. Fletcher, High-Strength, Low-Alloy Steels: Status, Selection and Physical Metallurgy, Battelle Press, Columbus OH, 1979, p 33

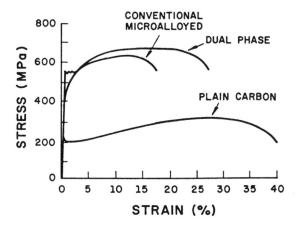

Stress-strain curves of dual-phase steel and a conventional microalloyed steel. The dual-phase steel shows a low yield stress, no yield drop, and has a high initial work-hardening rate.

The combinations of a low yield stress, high tensile stress, and good total elongation in dual-phase steels (see graph above) leads to good formability.

A high work-hardening rate is very important for good press formability. At any strain level, a high work-hardening rate results in a more uniform strain distribution throughout the formed piece. A high initial work-hardening rate also implies a low yield stress at low strains. This reduces springback and allows better shape control in large shallow panels (such as automobile skin panels).

Source: Marc André Meyers and Krishan Kumar Chawla, Mechanical Metallurgy: Principles and Applications, Prentice-Hall Inc., Englewood Cliffs NJ, 1984, p 527

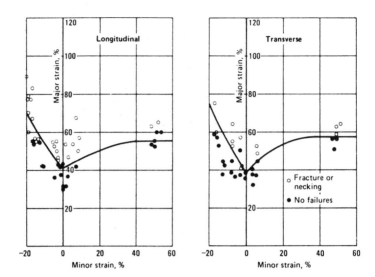

Longitudinal and transverse forming limit diagrams for hot rolled, semikilled, niobium-containing steel sheet, 2.11 mm (0.083 in.) thick. Yield strength group: 345 MPa (50 ksi).

Source: Metals Handbook, Ninth Edition, Volume 1, Properties and Selection: Irons and Steels, American Society for Metals, Metals Park OH, 1978, p 552

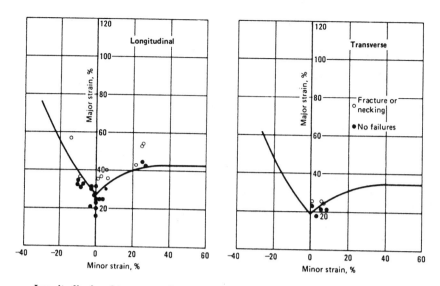

Longitudinal and transverse forming limit diagrams for hot rolled, fully killed, Si-Cr-Nb-Zr steel sheet, 2.13 mm (0.084 in.) thick. Yield strength group: 550 Mpa (80 ksi).

Source: Metals Handbook, Ninth Edition, Volume 1, Properties and Selection: Irons and Steels, American Society for Metals, Metals Park OH, 1978, p 552

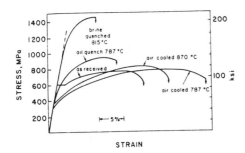

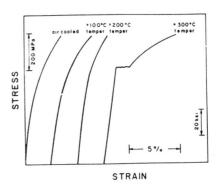

Stress-strain curves for V steel after various thermal treatments; the main features to note are the decreased yield stress and increased elongation of the a.c. 870 °C (ADP) and a.c. 787 °C (IDP) treated samples as compared to the as-received sample.

Stress-strain curves for tempered IDP samples showing absence of Lüders extension until 300 °C temper.

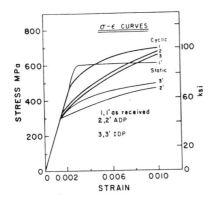

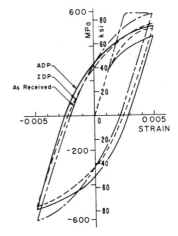

Cyclic and static stress-strain curves for V steel in the as-received, IDP, and ADP conditions.

Initial cyclic stress-strain curves for V steel in the as-received, IDP, and ADP conditions.

Compositions (Wt Pct) of the V Steels, Balance Fe

C	Mn	S	Si	V	Al	N	Thickness, mm (in.)	Use
0.15	1.50	0.015	0.36	0.12	0.075	0.019	1.65 (0.066)	Tensile
0.15	1.46	0.014	0.50	0.11	0.029	0.018	2.45 (0.102)	Fatigue
0.14	1.23	0.015	0.34	0.10	0.015	0.024	6.25 (0.250)	Impact

Source: R. G. Davies, The Deformation Behavior of a Vanadium-Strengthened Dual Phase Steel, Met. Trans. A, Volume 9A, January 1978, American Society for Metals, Metals Park OH, p 42, 46, 47

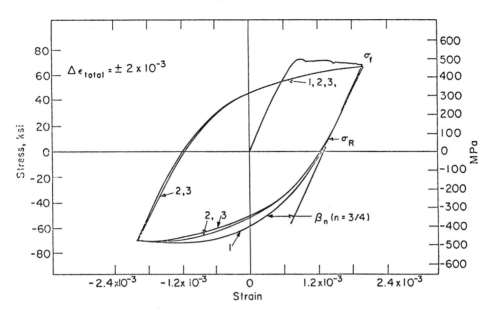

Stress-strain cycles 1, 2, and 3. HSLA steel (Van 5) with specimen axis parallel to the rolling direction, $\Delta\varepsilon_{total} = \pm 2 \times 10^{-3}$.

Source: Proceedings of the Second International Conference on Mechanical Behavior of Materials, 16–20 August 1976, Boston MA, American Society for Metals, Metals Park OH, 1978, p 37

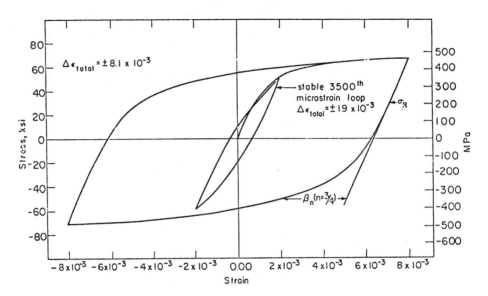

HSLA Steel (Van 6). First 3500 cycles were in preyield microstrain range, subsequent cycles were at $\Delta\varepsilon_{tot} = \pm 8.1 \times 10^{-3}$.

Source: Proceedings of the Second International Conference on Mechanical Behavior of Metals, 16–20 August 1976, Boston MA, American Society for Metals, Metals Park OH, 1978, p 38

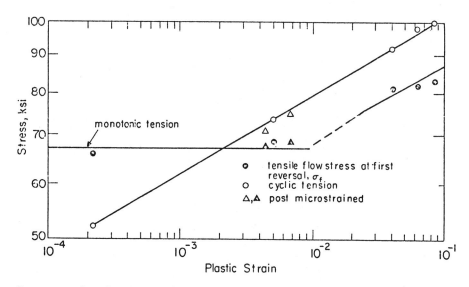

Log monotonic and cyclic stress/log plastic strain. HSLA steel (Van 6) with specimen axis parallel to the rolling direction.

Source: Proceedings of the Second International Conference on Mechanical Behavior of Metals, 16–20 August 1976, Boston MA, American Society for Metals, Metals Park OH, 1978, p 41

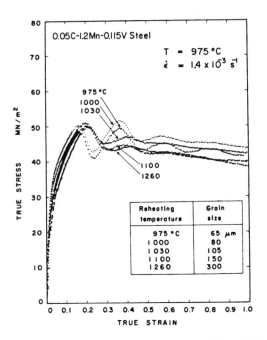

Critical experiment designed to test the model. A 0.05C-1.2Mn-0.115V steel was austenitized at a series of temperatures ranging from 975 to 1260 °C and selected to produce initial austenite grain sizes of 65 to 300 μm, respectively (see inset). Testing was carried out in axisymmetric compression at 975 °C and 1.4×10^{-3} s^{-1}, for which $2D_s \simeq 100$ μm. Cyclic σ/ε behavior was observed when $D_0 < 2D_s$; by contrast, single-peak behavior was obtained when $D_0 > 2D_s$.

Source: Deformation, Processing, and Structure, George Krauss, Ed., papers presented at the ASM Materials Science Seminar, 23 October 1982, St. Louis MO, sponsored by the Seminar Committee of the Materials Science Division of the American Society for Metals, Metals Park OH, 1984, p 203

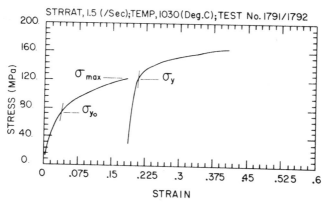

True stress data from a two-hit test on the cam plastometer. σ_{yo}, σ_y, σ_{max} are defined in the equation below.

The softening behavior of this Ti-N steel was investigated using two-hit tests. The fraction of softening at the start of the second compression can be expressed as the softening ratio S in the equation

$$S = \frac{\sigma_{max} - \sigma_y}{\sigma_{max} - \sigma_{yo}}$$

where σ_{max} is the maximum stress in the first compression, σ_{yo} is the yield stress (0.2% offset) in the first compression, and σ_y is the yield stress (0.2% offset) in the second compression. Test specimens were compressed once at the desired temperature, and after a time lapse of 15 s were compressed again at approximately the same temperature. This 15-s time interval simulates the interpass time in rolling practice. Strains impressed in the first and second compressions were 18% and 22%, respectively, at a nominal constant strain rate of 2 s^{-1}. Typical flow stress data from a two-hit test are shown in the figure above, where the stress parameters of the equation are schematically defined.

Source: HSLA Steels: Technology & Applications, proceedings of International Conference on Technology and Applications of HSLA Steels, 3–6 October 1983, Philadelphia PA, American Society for Metals, Metals Park OH, 1984, p 109

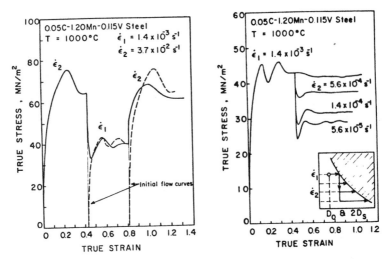

Stress-strain curves for the vanadium steel deformed at 1000 °C. (Left) The strain rate was cycled from 3.7×10^{-2} to 1.4×10^{-3} s^{-1} and then back again. The broken flow curves are for the annealed material. (Right) After a strain interval of 0.45 at an initial strain rate of 1.4×10^{-3} s^{-1}, the strain rate was decreased to a new rate in the range 5.6×10^{-5} to 5.6×10^{-4} s^{-1}.

The broken lines in this diagram are for purposes of comparison and indicate the shapes of the flow curves when *annealed* materials are tested at $\dot{\varepsilon}_1$ and $\dot{\varepsilon}_2$. In this case, following a decrease in strain rate by a factor of 26.4, multiple peaks are observed. Because $D_{s2} < D_{s1}$, it is apparent that cycles of grain coarsening accompany oscillatory flow, as in the deformation of dislocation-free specimens. It should be noted that, after the strain-rate decrease, the transient flow curve shows a rapid drop, with a stress minimum after an additional strain of 0.026. Beyond the minimum, the curve is similar to that for the annealed structure, but the oscillations tend to die out more quickly, and the peak strain is slightly larger than in the virgin material. The flow curve accompanying the strain-rate increase, on the other hand, differs considerably from that for the annealed condition, in that both the peak stress and the peak strain are smaller than those associated with the annealed structure. These differences are related to the presence of a highly inhomogeneous dislocation substructure in the "starting material," and therefore to the much larger distribution of nucleation strain in the strain-rate-change specimens than in the annealed material. The strain-rate increase and decrease transients are in sharp contrast to those observed when the deformation is controlled by dynamic recovery, a difference that can have an effect on the loads required to perform subsequent forming operations.

Source: Deformation, Processing, and Structure, George Krauss, Ed., papers presented at the ASM Materials Science Seminar, 23 October 1982, St. Louis MO, sponsored by the Seminar Committee of the Materials Science Division of the American Society for Metals, Metals Park OH, 1984, p 209

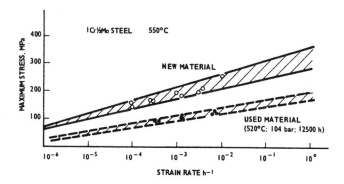

Results of constant strain-rate tests at 550 °C on 1% Cr-0.5% Mo steel.

A method suggested by Rajakovics involving constant strain-rate tests can give a rapid estimation of the degradation of the structure. However, tests carried out at laboratory strain rates (10^{-2}–10^{-4}/h) and service temperatures have to be extrapolated on a maximum stress-strain rate plot to typical service strain rates ($<10^{-8}$/h). An example of this is illustrated in the figure above, which gives results for a 1% Cr-0.5% Mo steel at 550 °C. We are thus faced with the problem of extrapolation in stress, so that while the technique provides a quick testing method, it has the same limitation as high stress-rupture testing.

Source: Flow and Fracture at Elevated Temperatures, Rishi Raj, Ed., American Society for Metals, Metals Park OH, 1985, p 313

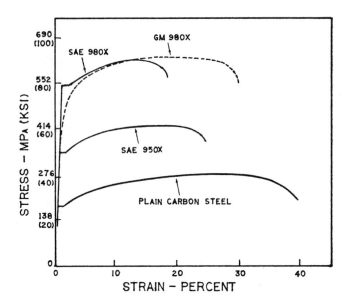

GM980X exhibits 10% higher total elongation than SAE980X. Also note that GM980X has no yield point elongation.

With its low yield strength, lack of yield point elongation, and improved total elongation compared to other HSLA materials, the GM980X looked as if it would provide the formability required to make the reinforcements (see curves above). The yield strength increase during forming and the strain aging treatment to 550 MPa from the initial yield strength of 400 MPa was important to obtain the yield strength required.

Source: HSLA Steels: Technology & Applications, proceedings of International Conference on Technology and Applications of HSLA Steels, 3–6 October 1983, Philadelphia PA, American Society for Metals, Metals Park OH, 1984, p 461

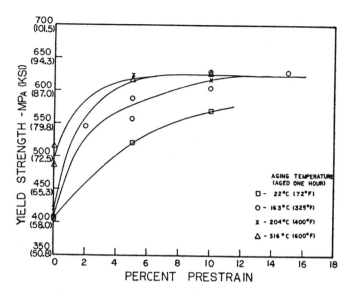

Yield versus prestrain at various aging temperatures.

The figure above shows how the yield strength varies with different amounts of prestrain—and at four different aging temperatures. You will note that with zero prestrain, aging has little effect, but small amounts of prestrain cause rapid increases in yield strength when it is followed by an aging process. Yield strength approaches maximum by prestraining at 10% and thermal aging at 163 °C. Maximum yield can be reached with less prestrain (5%) and using a higher aging temperature of 288 °C. No significant increase in yield strength can be gained by straining more than 10%.

Source: HSLA Steels: Technology & Applications, proceedings of International Conference on Technology and Applications of HSLA Steels, 3–6 October 1983, Philadelphia PA, American Society for Metals, Metals Park OH, 1984, p 463

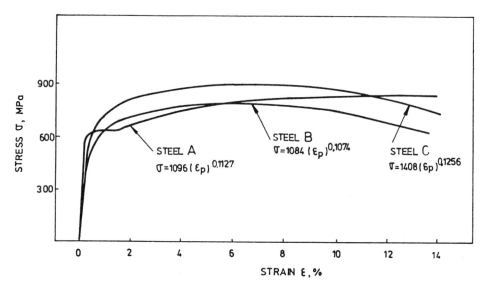

Stress-strain curves showing the strain-hardening exponent for the three steels.

Compositions of the Three Steels

Steel	Element C	Mn	Si	S	P	Al	V	Nb	Mo	Nppm
A	0.16	1.6	0.40	0.007	0.01	0.025	0.12	–	–	120
B	0.04	1.9	0.41	0.007	0.011	0.040	–	0.05	0.35	80
C	0.06	1.95	0.37	0.006	0.01	0.032	0.50	–	0.05	85

The stress-strain behavior of the three steels is presented in the curves above as an original plotting from the MTS recorder. While steel A has a yield plateau (with Lüders band formation), steels B and C are associated with continuous yielding. This behavior is significant in the U-O pipe forming process, providing different control of the Bauschinger effect (B;E.) and contributes to an increase in the yield strength from plate to pipe. Also indicated in the curves are the corresponding strain hardening exponents determined by the well-known relation:

$$\nu = K(\varepsilon_p)^n$$

Source: HSLA Steels: Technology & Applications, proceedings of International Conference on Technology and Applications of HSLA Steels, 3–6 October 1983, Philadelphia PA, American Society for Metals, Metals Park OH, 1984, p 793, 795

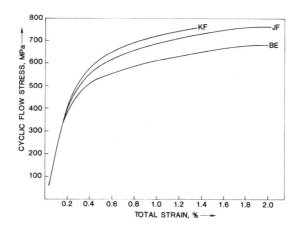

Comparison of the CSS curves for Cb(BE), Cb-V (JF), and Cb-V-Si (KF) steels as determined by the companion specimen method.

For clarity in presentation the actual data points in the curves above are omitted. It is clear that there are significant differences (in magnitude) between the CSS curves of the three steels although all three steels belong to the same commercial grade. This is an important conclusion in the discussion of the cyclic response of steels of the same grade.

Source: HSLA Steels: Technology & Applications, proceedings of International Conference on Technology and Applications of HSLA Steels, 3–6 October 1983, Philadelphia PA, American Society for Metals, Metals Park OH, 1984, p 586

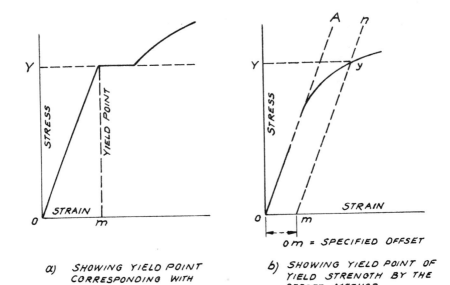

a) SHOWING YIELD POINT CORRESPONDING WITH TOP OF KNEE.

b) SHOWING YIELD POINT OF YIELD STRENGTH BY THE OFFSET METHOD.

Stress-strain diagrams showing discontinuous and continuous yield points.

Most commercially used steels exhibit definite yield point (see chart above at left). Materials with discontinuous yield usually develop visible stretcher strains, or Lüders bands, undesirable on high-quality finished products used by the automotive or appliance industry. Dual-phase steels on the other hand do not exhibit definite yield point (see chart above at right) and provide a better surface for finished products.

In addition to carbon and manganese, nitrogen is added to certain steels to increase strength and hardness. The addition of nitrogen allows reduction in carbon and manganese, thus increasing formability. The nitrogenized steel, however, is "aging" quickly after forming. This "strain aging" may increase yield strength as much as 140 MPa (20 ksi) accompanied by loss of ductility. This strain aging characteristic should be considered when secondary forming is planned after roll forming. Added phosphorus has similar effects as described above for nitrogen.

Source: HSLA Steels: Technology & Applications, proceedings of International Conference on Technology and Applications of HSLA Steels, 3–6 October 1983, Philadelphia PA, American Society for Metals, Metals Park OH, 1984, p 519

The deformation characteristics of the cold-rolled dual-phase steels were evaluated by comparing standard tensile properties and by analyzing the effects of strain on the strain-hardening behavior at low strains.

These data include a summary of the 0.2% offset yield strength, the flow stress at engineering strains of 3% and 5%, the ultimate tensile strength, and the total elongation. All the samples exhibited similar properties within the following ranges: yield strengths of 280 to 320 MPa, ultimate strengths of 550 to 650 MPa, and total elongation of 26 to 33%. Within these ranges, the differences between samples will be shown to result from variations in the strain-hardening behavior at low strains.

The shapes of the stress-strain curves observed on yielding and thus the strain-hardening behavior at low strains varied significantly with processing history and alloy content. The variation in yielding behavior is clearly shown in Fig. 1, which presents load versus engineering strain data recorded on the x-y plotter attached to the tensile machine. Figures 1(a) to (d) compare the initial yielding behavior of the four alloys of this study. These figures show the characteristic continuous yielding behavior associated with dual-phase steels. Note however, that the Cr steel (Fig. 1d) exhibits a well-defined inflection on yielding, a result that will be shown below to be a prerequisite for maximum ductility.

In Fig. 1(e) and (f), the deformation characteristics at the two ends of a coil of 1.5-mm-thick V steel are contrasted. Note that this is the same alloy shown in Fig. 1(a). Figures 1(e) and (f) show the change in yielding behavior from essentially continuous yielding in the lead end of the coil (Sample 9), where the stress-strain curve exhibits a well-defined inflection on yielding, to discontinuous yielding with a yield point in the trail of the coil (Sample 10). The different initial yielding characteristics presented in Fig. 1 have been shown to reflect variations in structure related to processing history (such as compositions, annealing temperature, or cooling rate), and furthermore, the inflection observed on yielding has been shown to be associated with the development of an incipient Lüders band which propagates with a positive

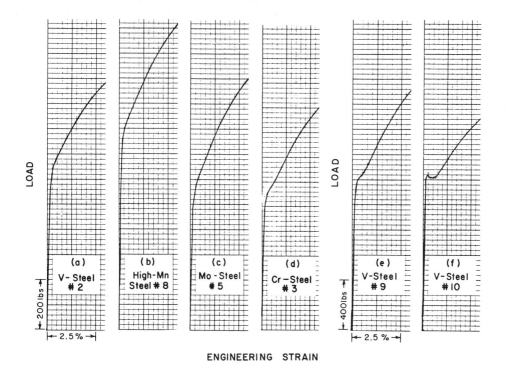

ENGINEERING STRAIN

Fig. 1. Direct traces of the load versus engineering strain data recorded on the MTS x-y plotter. (a) V steel #2, 0.7 mm thick, e_T = 27.4%. (b) High-Mn steel #8, 0.9 mm thick, e_T = 27.3%. (c) Mo steel #5, 0.7 mm thick, e_T = 26.4%. (d) Cr steel #3, 0.7 mm thick, e_T = 30.6%. (e) V steel #9, 1.5 mm thick, e_T = 33.5%. (f) V steel #10, 1.5 mm thick, e_T = 30.3%.

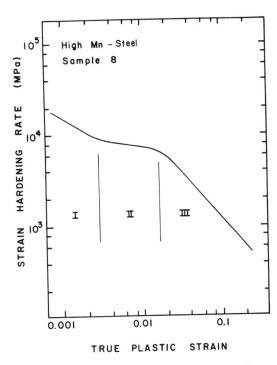

Fig. 2.   A Jaoul-Crussard Plot for Sample 2, from a 0.7-mm-thick V steel, showing the development of three distinct stages in strain hardening.

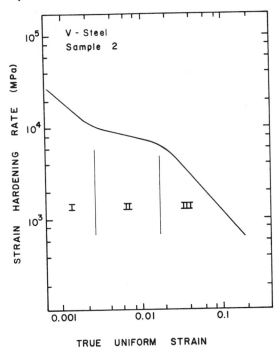

Fig. 3.   A Jaoul-Crussard plot for Sample 8, from a 0.9-mm-thick High-Mn steel, showing the development of three distinct stages in strain hardening.

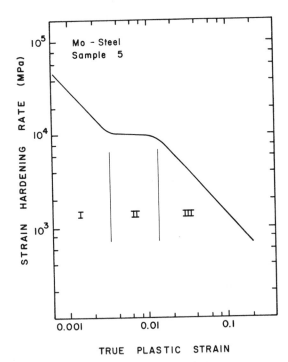

Fig. 4.   A Jaoul-Crussard plot for Sample 5, from a 0.7-mm-thick Mo steel, showing the development of three distinct stages in strain hardening.

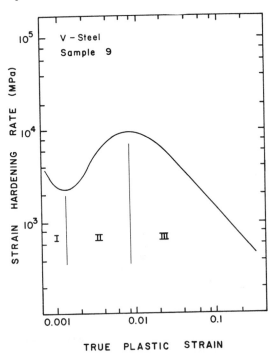

Fig. 5.   A Jaoul-Crussard plot for Sample 9, from a 1.5-mm-thick V steel, showing the characteristic dip associated with samples that exhibit a yielding inflection.

strain-hardening rate. Comparison of the curves for Sample 9 (Fig. 1e) with Sample 2 (Fig. 1a), both from the same steel, shows the transition to more continuous yielding with a decrease in sheet thickness, which in these continuously annealed steels is indicative of a difference in cooling rate.

Differences in the initial strain-hardening behavior are clarified with the Jaoul-Crussard plotting technique. This procedure plots the logarithm of the true strain-hardening rate versus the logarithm of the true plastic strain. The Jaoul-Crussard plots presented in Fig. 2, 3, and 4 show the presence of three well-developed "stages" in strain hardening. Variations in the slope of the second stage are also apparent. Figure 5 shows a characteristic minimum in the Jaoul-Crussard plot due to the yielding inflection. It should be noted that for a sample that exhibits true discontinuous yielding the strain-hardening behavior during Lüders band propagation is undefined on a Jaoul-Crussard plot. The stages in strain hardening that are amplified on a Jaoul-Crussard plot have been associated with strain-dependent changes in the basic requirements for strain hardening in dual-phase steels.

Source: HSLA Steels: Technology & Applications, proceedings of International Conference on Technology and Applications of HSLA Steels, 3–6 October 1983, Philadelphia PA, American Society for Metals, Metals Park OH, 1984, p 305, 306, 323–327

7-31. Effect of Constituents on the Stress-Strain Curve of Multiphase Steels

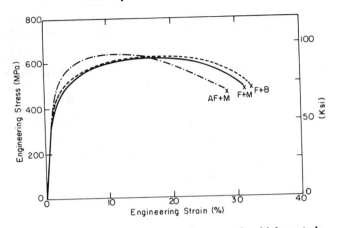

Effect of constituents on the stress-strain curves of multiphase steels. F: ferrite, AF: acicular ferrite, B: bainite, M: martensite.

Source: HSLA Steels: Technology & Applications, proceedings of International Conference on Technology and Applications of HSLA Steels, 3–6 October 1983, Philadelphia PA, American Society for Metals, Metals Park OH, 1984, p 283

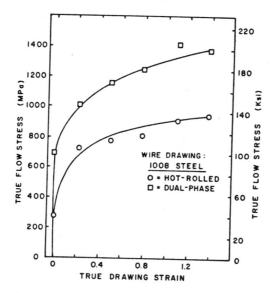

Comparison of flow curves obtained by wire drawing of 1008 steel with ferrite-pearlite (hot rolled) and ferrite-martensite (dual-phase) microstructures.

Source: Deformation, Processing, and Structure, George Krauss, Ed., papers presented at the ASM Materials Science Seminar, 23 October 1982, St. Louis MO, sponsored by the Seminar Committee of the Materials Science Division of the American Society for Metals, Metals Park OH, 1984, p 71

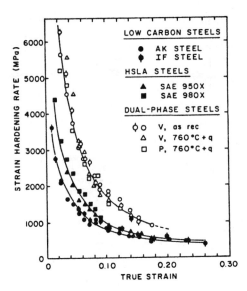

Strain-hardening rate versus strain for three classes of steel. The strain-hardening rates of the dual-phase steels are the highest at all levels of strain.

The curves above clearly indicate that strain-hardening rates are higher in dual-phase steels. However, it has also been well documented that dual-phase steels also exhibit distinct stages in strain hardening.

Source: Deformation, Processing, and Structure, George Krauss, Ed., papers presented at the ASM Materials Science Seminar, 23 October 1982, St. Louis MO, sponsored by the Seminar Committee of the Materials Science Division of the American Society for Metals, Metals Park OH, 1984, p 70

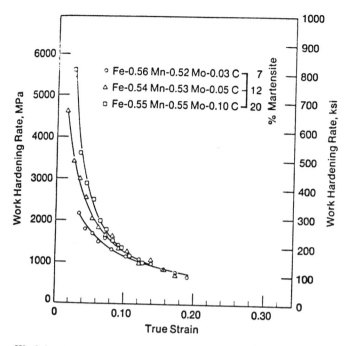

Work hardening as a function of strain for dual-phase steels containing various amounts of martensite.

The influence of varying the martensite content and the carbon content of the martensite on the work hardening of dual-phase steels was also investigated. Although increasing the martensite content from approximately 7 to 18% increased the work-hardening rate at low strains (see above), there is essentially no change in the work-hardening rate at strains above 0.12.

Source: Alloys for the Eighties, Robert Q. Barr, Ed., Climax Molybdenum Co., Greenwich CT, p 29

7-35. Modified C-Mn-Nb HSLA Steel

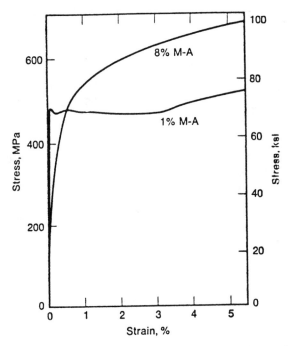

Change in stress-strain behavior with increasing amount of M-A constituent. Composition: 0.06C, 1.2Mn, 0.5Nb, 0.25Mo.

The formation of these lower transformation products has an important effect on the stress-strain behavior of these steels. When the amount of the M-A constituent exceeds 5%, the yield point is eliminated and a continuous stress-strain curve is obtained, as shown in the curves above. Steels with this type of stress-strain curve exhibit rapid work hardening when strained only a few percent. This behavior is desirable in plate product used to make UOE pipe because the pipe-forming strains can be utilized to increase the strength of the pipe over that of the plate. This increased strength can more than compensate for the Bauschinger effect that arises when tension test specimens cut from the pipe are flattened and tested.

Source: Alloys for the Eighties, Robert Q. Barr, Ed., Climax Molybdenum Co., Greenwich CT, p 189

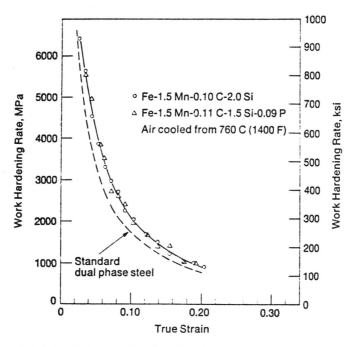

Work hardening as a function of strain for high-silicon and silicon-plus-phosphorus dual-phase steels showing enhanced work hardening over the standard dual-phase steel.

From these results, it appears that the properties of the ferrite are of primary importance in determining the work-hardening rate of dual-phase steels. This is demonstrated in the figure above, where the addition of 2% Si or 1.5% Si-0.1% P to a dual-phase steel leads to an increase in the work-hardening rate over that observed in the usual dual-phase steels containing only about 0.5% Si or 0.07% P.

Source: Alloys for the Eighties, Robert Q. Barr, Ed., Climax Molybdenum Co., Greenwich CT, p 29

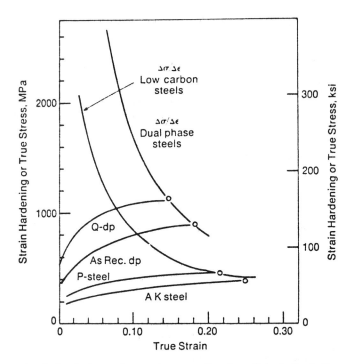

Strain hardening and true stress as a function of strain for both dual-phase and low-carbon steels. The circles indicate the extent of uniform elongation.

Since common use consists of at least 80% ferrite, it is to be expected that the properties of this ferrite will have a major influence on the work hardening of the steel; the other 20% or less of the structure is a high-carbon (~0.6% C) martensite that may contain a small amount of retained austenite. The ferrite in these dual-phase steels is usually fine-grained (<5μm), relatively free of interstitial elements and precipitates, and often strengthened by the addition of substitutional alloying elements such as Si and/or P.

Source: Alloys for the Eighties, Robert Q. Barr, Ed., Climax Molybdenum Co., Greenwich CT, p 28

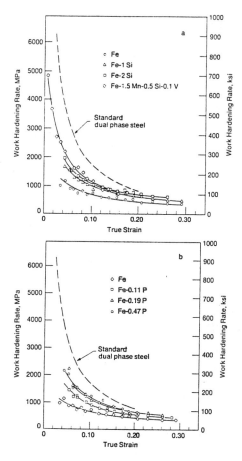

Work or strain hardening as a function of strain for iron and several alloys. (a) Iron-silicon alloys and a "dual-phase ferrite"; (b) Iron-phosphorus alloys.

The influence of Si and P on the work hardening of fine-grained ($\sim$10μm) pure iron alloys is shown in the figures above. It can be seen that the higher the alloy addition, the greater the work hardening at all strain levels. In (a) it can also be seen that the Fe-1.5 Mn-0.5 Si-0.1 V alloy (the approximate composition of the ferrite in the most common dual-phase steel) behaves similarly to the Fe-2% Si alloy, especially at high strains. Thus it is concluded that the 1.5 Mn and 0.1 V are having beneficial effects, similar to an extra 1.5 Si, on work hardening. Grain-size changes over the range of 10 to 30 μm in this synthetic dual-phase ferrite had no influence on the work-hardening rate.

Source: Alloys for the Eighties, Robert Q. Barr, Ed., Climax Molybdenum Co., Greenwich CT, p 28

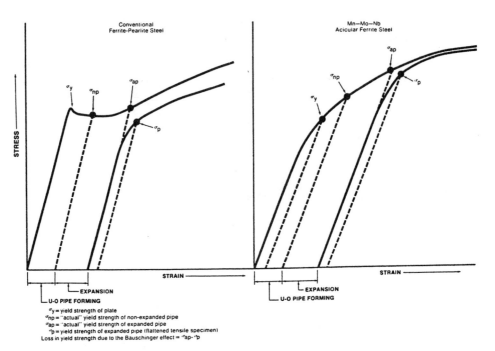

σ_y = yield strength of plate
σ_{np} = "actual" yield strength of non-expanded pipe
σ_{ap} = "actual" yield strength of expanded pipe
σ_p = yield strength of expanded pipe (flattened tensile specimen)
Loss in yield strength due to the Bauschinger effect = $\sigma_{ap} - \sigma_p$

Schematic representation of how the yielding behavior of ferrite-pearlite and acicular ferrite steels affects the yield strength of pipe (not to scale).

Pipe Fabrication: Strength Properties. One of the most important advantages of acicular ferrite Mn-Mo-Nb steels is their continuous yielding and rapid work hardening during the fabrication of plate into pipe. In contrast, conventional ferrite-pearlite pipe steels, in either the hot-rolled or the normalized condition, experience a sharp yield point followed by considerable strain at a constant stress. These steels often show no net increase in strength after pipe fabrication is complete.

In testing the strength of pipe, the usual method is to cut a circumferential strip specimen from the pipe and flatten it prior to test. The flattening introduces a Bauschinger effect that causes the measured yield strength to be lower than the "true" strength of the pipe as would be obtained from a ring expansion test. The difference in the measured yield strength between the starting plate (skelp) and the finished pipe is equal to the strength increase due to work hardening minus the Bauschinger effect.

Source: Molybdenum-Containing Steels for Gas and Oil Industry Applications, A State of the Art Review, Climax Molybdenum Co., Greenwich CT, 1977, p 25

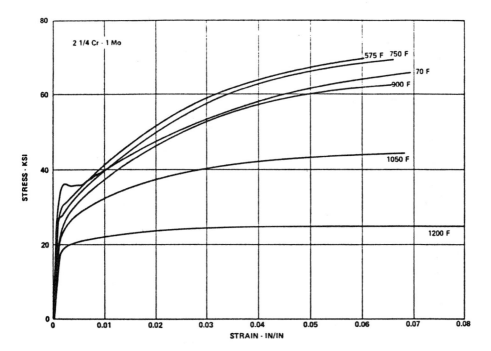

Stress-strain curves for annealed 2.25Cr-1Mo steel at various temperatures. (1 ksi = 6.895 MPa; F = 1.8 C + 32)

Source: Chrome Moly Steel in 1976, George V. Smith, Ed., proceedings of the Annual Winter Meeting of the American Society of Mechanical Engineers, 5–10 December 1976, New York, sponsored by the Metal Properties Council Inc., AIME, New York, 1976, p 5

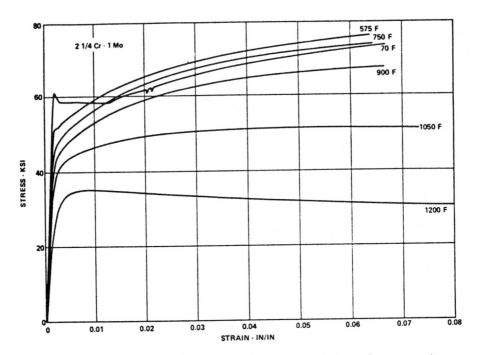

Stress-strain curves for normalized and tempered 2.25Cr-1Mo steel at various temperatures. (1 ksi = 6.895 MPa; F = 1.8 C + 32)

Source: Chrome Moly Steel in 1976, George V. Smith, Ed., proceedings of the Annual Winter Meeting of the American Society of Mechanical Engineers, 5–10 December 1976, New York, sponsored by the Metal Properties Council, Inc., AIME, New York, 1976, p 5

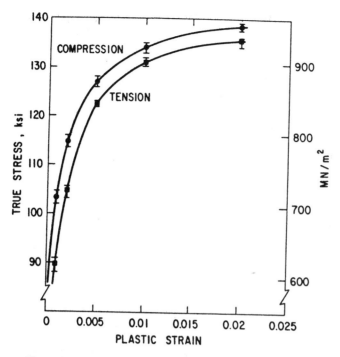

True stress-plastic strain curves in tension and in compression for as-quenched martensite in Fe-9.1 Ni-0.02 C steel.

The alloy in the figure above was tested in the as-quenched condition. In this material the SD effect was about 10% of the tensile stress when measured at 0.002 plastic strain, but diminished rapidly with further strain to about 2% at strains of 0.01 and greater, as shown in the figure above.

Source: G. C. Rauch and W. C. Leslie, The Extent and Nature of the Strength-Differential Effect in Steels, Met. Trans. A, February 1972, American Society for Metals, Metals Park OH, p 379

8-4. Nickel Steels (9.0 to 21% Ni)

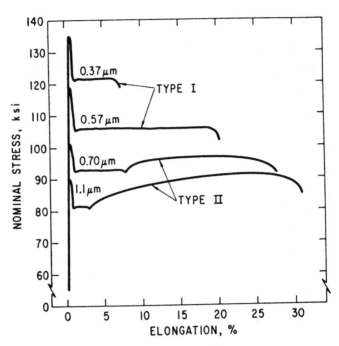

Two types of stress-elongation curves and the effect of grain size.

	Composition of Steels, Wt Pct								
Identifi-cation	Ni	Mn	C	Si	S	N	P	Cr	Mo
A	8.74	0.45	0.12	0.26	0.012	0.009	0.006	0.10	0.025
B	–	5.7	0.11						
C	16.3	–	0.053						
D	15.1	–	0.009						
E	21.0	–	0.053						

Effect of Grain Size on Tensile Elongation. Stress-strain curves exhibited a well-defined upper yield point followed by localized deformation at constant stress, similar to Lüders deformation. This type of yielding, which is characteristic of low-carbon steel, was observed in all the steels investigated regardless of the relative proportions of ferrite and austenite in the microstructure. However, plastic instability was observed in several of the steels. This instability is characterized by necking in the deformation band before the band traverses the entire gage section of the specimen. As a result, there is no uniform elongation and a very small total elongation.

Typical stress-elongation curves for these two types of specimens are shown in the figure above, designated as Type I and Type II. When the grain size is 0.7 μm or larger, deformation at constant stress (Lüders deformation) is followed, as in the normal case by a rise in stress (Type II) and total elongation values near 30% are obtained. However, when the grain size is 0.57 μm or smaller, the specimens neck and fracture during Lüders deformation (Type I) and total elongation values are as low as 7%.

Source: R. L. Miller, Ultrafine-Grained Microstructures and Mechanical Properties of Alloy Steels, Met. Trans. A, April 1972, American Society for Metals, Metals Park OH, p 910

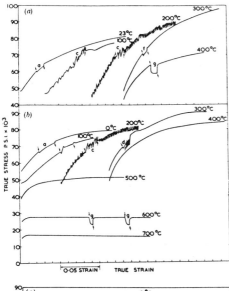

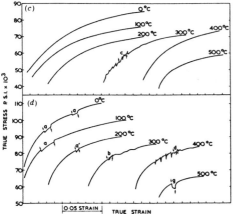

Typical true stress–true strain curves for as-tempered steels. All curves start from a strain of 0.002. (*a*) Plain-carbon steel tempered for 10 h at 700 °C. (*b*) 0.87% Cr tempered for 100 h at 700 °C. (*c*) 4.2% Cr tempered for 1000 h at 700 °C. (*d*) 11.7% Cr tempered for 10 h at 700 °C.

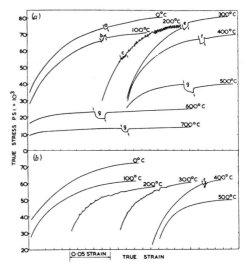

True stress-true strain curves for recrystallized steels. All curves start from a strain of 0.002. (*a*) 0.87% Cr tempered for 10 h, cold-worked 51% and tempered for a further 21 h. (*b*) 4.2% Cr tempered for 10 h, cold-worked 63% and tempered for a further 950 h.

A characteristic feature of the stress-strain curves at intermediate temperature was the occurrence of serrated yielding. This was associated with characteristic changes in stress-strain behavior when the strain rate was changed, as shown in the curves of this page. The different types of behavior have been labeled a to g for summary of the results. Type a results from work hardening only and shows a lower work-hardening rate at the lower strain rate. Type a′ is similar except that a small yield occurs on increasing the strain rate. Types b, c, and d are expected at the lower, middle, and upper temperatures in the dynamic strain-aging, but types e and f also show some effect of strain aging as faster work hardening occurs at the lower strain rate, and in the latter case yield effects also occur. Type g results when dynamic recovery occurs at a significant rate. It can be seen that, in general, equivalent behavior occurs at higher temperatures with increasing chromium content.

Source: T. Mukherjee and C. M. Sellars, Tensile Properties of Tempered Chromium Steels in the Temperature Range 0 Degrees to 700 Degrees, Met. Trans. A, April 1972, American Society for Metals, Metals Park OH, p 456–457

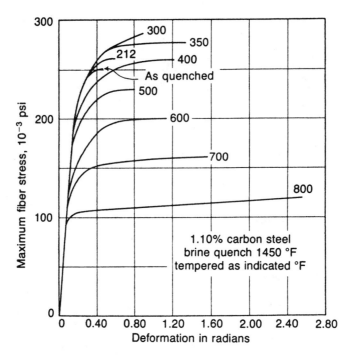

Torsion stress-strain curves for 1.1% C tool steel brine quenched from 1450 °F and tempered at a series of temperatures. (Data of Green and Stout.)

Source: George A. Roberts and Robert A. Cary, Tool Steels, Fourth Edition, American Society for Metals, Metals Park OH, 1980, p 320

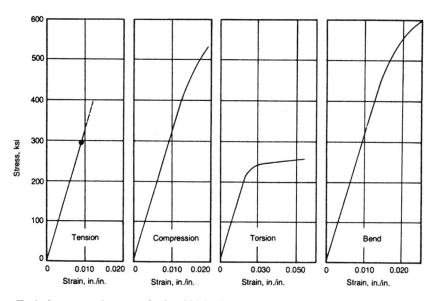

Typical stress-strain curves for hard high-alloy steel in tension, compression, torsion, and bend.

Source: George A. Roberts and Robert A. Cary, Tool Steels, Fourth Edition, American Society for Metals, Metals Park OH, 1980, p 61

9-3. T1 High Speed Tool Steel

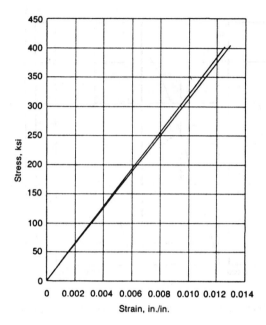

Stress-strain curve for hardened and tempered type 610 (T1) high speed steel. Hardness, Rockwell C 65.2 to 65.5. Three samples. (Grobe and Roberts.)

Source: George A. Roberts and Robert A. Cary, Tool Steels, Fourth Edition, American Society for Metals, Metals Park OH, 1980, p 696

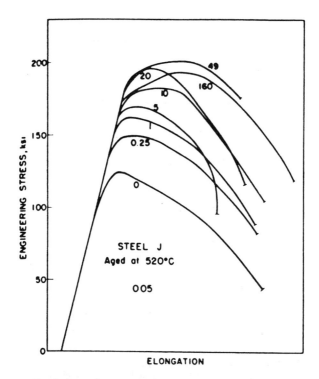

Load elongation curves after different aging times at 520 °C for the 12% Ni grade J.

At 520 °C where precipitation starts much before austenite reversion, uniform elongation did not increase until austenite started to form. This is seen in the figure above, which shows how the uniform strain varies with aging time at 520 °C. A slight decrease in the uniform strain at small aging times is found, followed by a rapid increase when austenite starts to form. A modest drop in the uniform strain is found at peak yield stress, followed by a rise presumably associated with the overaging of precipitates.

Source: C. A. Pampillo and H. W. Paxton, The Effect of Reverted Austenite on the Mechanical Properties and Toughness of 12 Ni and 18 Ni (200) Maraging Steels, Met. Trans. A, November 1972, American Society for Metals, Metals Park OH, p 2897

10-2. 18% Ni Maraging Steel

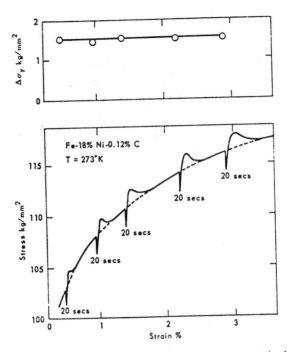

The high-carbon martensite strained at $\dot{\epsilon} = 8.3 \times 10^{-4}\ s^{-1}$ at 273 K and relaxed for 20 s at five successively larger strains. The yield increment $\Delta\sigma_\gamma$ is shown as a function of the strain at which the relaxation occurred.

Source: Proceedings of the Second International Conference on Mechanical Behavior of Materials, 16–20 August 1976, Boston MA, American Society for Metals, Metals Park OH, 1978, p 36

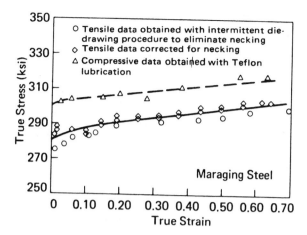

Room-temperature tensile and compressive flow curves.

Source: R. Chait, Factors Influencing the Strength Differential of High Strength Steels, Met. Trans. A, February 1972, American Society for Metals, Metals Park OH, p 367

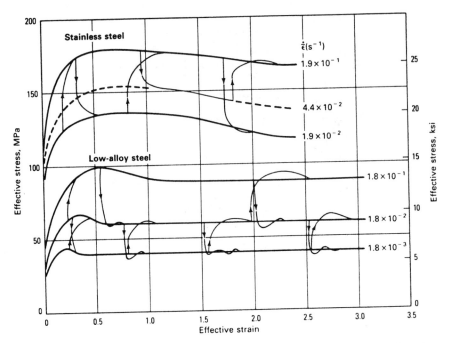

Typical stress-strain curves for tests involving instantaneous changes in strain rate for an austenitic stainless steel and a ferritic low-alloy steel. Note that the rate-change strain-rate sensitivity is either lower (stainless steel) or higher (low-alloy steel) than that based on continuous (constant strain-rate) torsion tests.

Source: Metals Handbook, Ninth Edition, Volume 8, Mechanical Testing, American Society for Metals, Metals Park OH, 1985, p 177

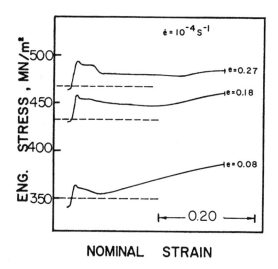

NOMINAL STRAIN

The figure above shows the result of experiments done with an annealed austenitic alloy of composition Fe-31% Ni-0.1% C. The tensile test was stopped three times, for 3 h each time, after three different strains: $\epsilon =$ 0.08, 0.18, and 0.27. The test was stopped by simply turning off the machine. Initially, the sample did not show a well-defined yield point. However, on reloading after the 3-h rest, the stress-strain curve showed clearly the appearance of a yield point followed by a plateau, a horizontal load drop region, and finally a return to the original trajectory. This can be seen clearly in the curves above. The dashed lines indicate the stress at which the test was stopped. Note that on reloading, the yield stress of the alloy increased for the three strains. This phenomenon is known as strain aging. The term "aging" is normally used when there is precipitate formation. However, this is not necessarily the case in the example given above. As the test and its interruption were carried out at ambient temperature, there occurs a migration of interstitial atoms to dislocations during the test stoppage, with its consequent dislocation locking. Thus, on reloading, these dislocations have to be unlocked with the appearance of the well-defined yield point. Meyers and Guimarães repeated these experiments under identical conditions but maintaining the test sample unloaded during 3 h. The well-defined yield point reappeared, but it was less marked. This indicates that the applied stress has an accelerating effect on the aging process.

Source: Marc André Meyers and Krishan Kumar Chawla, Mechanical Metallurgy: Principles and Applications, Prentice-Hall Inc., Englewood Cliffs NJ, 1984, p 398

11-3. Type 201 Stainless Steel

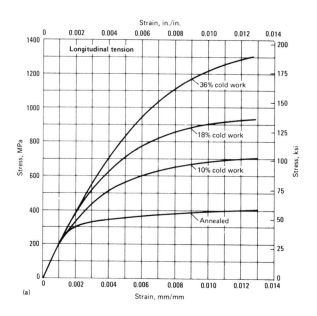

(a)

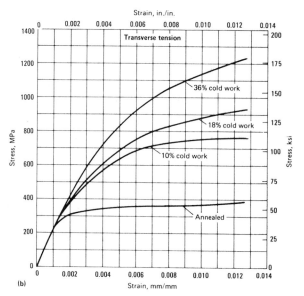

(b)

Effects of cold work on the stress-strain properties in longitudinal tension (top) and transverse tension (bottom).

Source: Engineering Properties of Steel, Philip D. Harvey, Ed., American Society for Metals, Metals Park OH, 1982, p 247

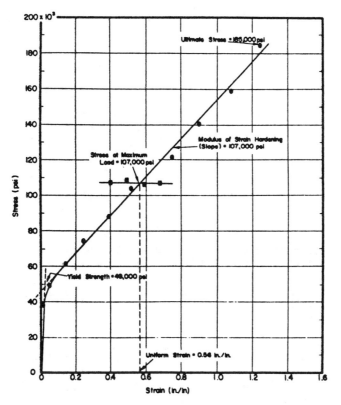

A true stress-true strain graph provides useful data for evaluating stretch forming operations.

A test procedure to evaluate press formability of stainless steel must be as simple and reproducible as possible. One test of value is the true stress-true strain tensile test. The figure above shows a plot of the stress applied to a specimen versus the specimen strain for each stress level. (True stress is the load on the specimen divided by the cross sectional area of the specimen at that load, and true strain is the deformation that occurs at each increment of load.)

This graph illustrates a true stress-true strain tensile curve for type 301 strip. The true stress-true strain tensile properties that are significant in an analysis of press formability (shown in the graph) include:

1. Yield Strength—The stress at which a specimen shows deviation from straight line proportionality of stress to strain.

2. Stress at Maximum Load—The stress at the highest load (in pounds) sustained by the specimen.

3. Maximum Uniform Strain—Maximum value of straining before uniform deformation ceases and localized deformation and necking take place. This is the strain at point of maximum load.

4. Modulus of Strain Hardening—Slope of the plastic region of the true stress-true strain curve. Modulus indicates rate of cold-work hardening.

Source: Source Book on Cold Forming, American Society for Metals, Metals Park OH, 1975, p 124

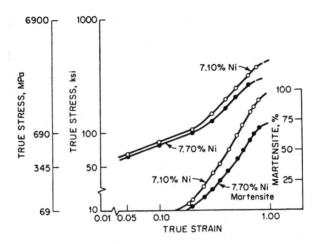

Relationship between alloy stability, strain-hardening, and martensite formation in type 301 stainless steel.

A simple representation of the true-stress–true-strain curve is given by

$$s = k\epsilon^n$$

where s = true stress, ϵ = true strain, and n = work-hardening exponent. This expression defines the straight-line segment at the beginning of the true-stress–true-strain curve. In order to define the full stress-strain curve, Griffiths and Wright developed a more complicated quadratic equation:

$$\log s = C_1 + C_2 \log \epsilon + C_3 (\log \epsilon)^2$$

They found all the constants were influenced by austenite stability as expressed by the modified Post and Eberly stability factor. This is shown above.

Source: Donald Peckner and I. M. Bernstein, Handbook of Stainless Steels, McGraw-Hill Book Co., New York, 1977, p 4–34

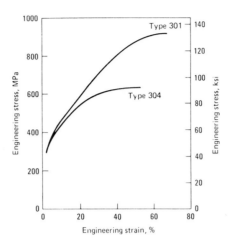

Typical stress-strain curves for types 301 and 304 stainless steel.

Certain austenitic stainless steels—the so-called metastable types—can develop higher strengths and hardnesses than other "stable" types for a given amount of cold work. In metastable austenitic stainless steels, deformation triggers transformation of austenite to martensite. The effect of this transformation on strength is illustrated in the figure above, which compares the stress-strain curve for stable type 304 with that for metastable type 301. The parabolic shape of the curve for type 304 indicates that strain hardening occurs throughout the duration of the application of stress, but that the amount of strain hardening for a given increment of stress decreases as stress increases. On the other hand, the stress-strain relationship for type 301 indicates an accelerated rate of strain hardening after an initial increment of 10 to 15% plastic strain. This accelerated strain hardening is a direct result of deformation-induced transformation to martensite.

Source: Metals Handbook, Ninth Edition, Volume 3, Properties and Selection: Stainless Steels, Tool Materials and Special-Purpose Metals, American Society for Metals, Metals Park OH, 1978, p 17

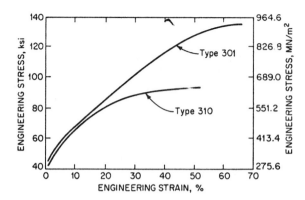

Stress-strain curves for types 301 and 310.

The austenitic stainless steels can be classified into two groups based on the stability of the austenite: stable austenitic stainless steels and metastable austenitic stainless steels. The stable austenitic stainless steels are those with microstructures that remain austenitic even after much straining. The metastable austenitic stainless steels are those with microstructures that transform readily to an acicular martensitic structure during straining. The difference between these two groups of steel is best illustrated by the stress-strain diagrams of two typical steels—one from each group (see above). Type 310, representative of the stable austenitic stainless steels, exhibits a normal stress-strain curve for the austenitic structure. Type 301, representative of the metastable austenitic stainless steels, exhibits a stress-strain curve in which the rate of strain hardening increases markedly at the onset of the martensite transformation that usually occurs after about 10 to 15% strain.

Source: Donald Peckner and I. M. Bernstein, Handbook of Stainless Steels, McGraw-Hill Book Co., New York, 1977, p 20–26

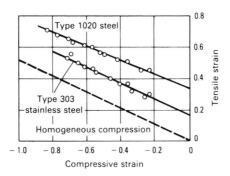

Fracture loci in cylindrical upset test specimens of two materials deformed at room temperature.

Upset Test Fracture Limits. A wide range of stress and strain states can be produced at the free surfaces of cylindrical, tapered, or flanged test specimens, which permits evaluation of the effects of variations in stress and strain states with fracture. The most convenient representation of the limits of fracture is a plot of the circumferential and axial strains that existed on the specimen surface at fracture. This plot, as shown in the figure above, is a fracture-limit line. At strain combinations below the line, the material has not fractured. For strains above the line, the material has fractured. The fracture-limit line is parallel to the line for homogeneous compression of a cylinder that has a slope of $-1/2$.

This relationship is appropriate for a wide range of materials. The intersection of ϵ_θ (tensile strain axis) corresponds to the fracture strain in plane-strain tensile testing.

Source: Metals Handbook, Ninth Edition, Volume 8, Mechanical Testing, American Society for Metals, Metals Park OH, 1985, p 580

11-9. Type 304 Stainless Steel

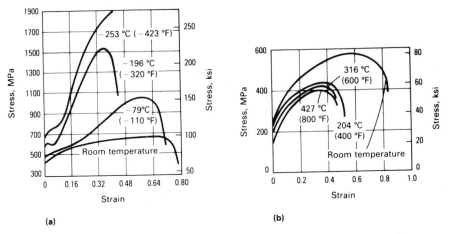

(a)

(b)

Stress-strain diagrams for type 304 stainless steel. (a) At low temperatures. (b) At elevated temperatures.

Elevated/low temperature stress-strain diagrams are similar in appearance to those determined at room temperature (see figures above). Relative to ambient temperature, materials become stronger, but less ductile, as temperature decreases.

Source: Metals Handbook, Ninth Edition, Volume 8, Mechanical Testing, American Society for Metals, Metals Park OH, 1985, p 35

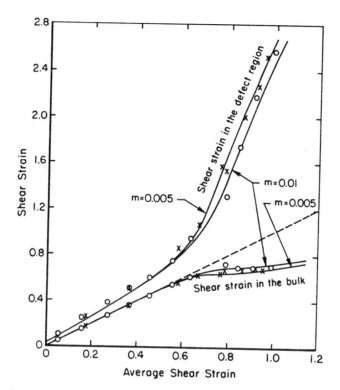

Comparison of experimentally observed localization (data points) in 304 stainless steel torsion tests at room temperature with results of the simulation corrected for radius change (solid lines). The simulations were run with two rate sensitivities, m = 0.01 and m = 0.005, whose values bounded those measured in torsion tests on specimens without geometric defects. Average surface shear strain rate was approximately 0.05 s^{-1} in both experiments and simulations.

Source: S. L. Semiatin and J. J. Jonas, Formability and Workability of Metals, American Society for Metals, Metals Park OH, 1984, p 145

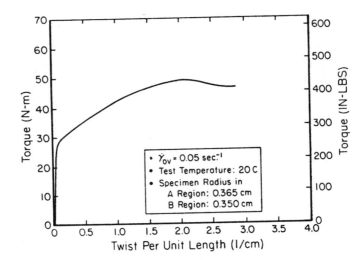

Experimental torque-twist curve for AISI 304 samples displaying an initial geometric defect at their centers.

Source: S. L. Semiatin and J. J. Jonas, Formability and Workability of Metals, American Society for Metals, Metals Park OH, 1984, p 263

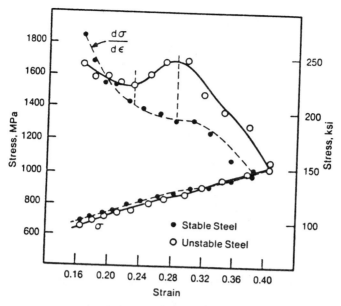

Stress-strain behavior of austenitic stainless type 304.

Effect of Percent Reduction on Strain Ratio and
Plane Anistropy Coefficient of Type 304 Stainless Steel

Percent Reduction	r_0	r_{45}	r_{90}	$\bar{r}$	Δr
69	0.87	1.06	0.88	0.97	−0.18
53	0.89	1.19	0.88	1.04	−0.31

The average strain ratio, $\bar{r}$, of austenitic stainless steel sheet is approximately 1, independent of composition, amount of cold-rolling reduction and annealing treatment. Only the plane anisotropy coefficient, Δr, can be affected by changing the amount of cold reduction. Low degrees of reduction between annealing treatments produce low values of Δr and earing, but high amounts of reduction result in higher Δr and more earing.

Source: Stainless Steel '77, Robert Q. Barr, Ed., symposium sponsored by the Climax Molybdenum Co., Greenwich CT, p 216

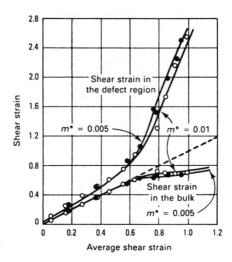

Comparison of experimentally observed localization kinetics (data points) with simulation results (solid lines) in type 304 stainless steel specimens. Specimens had premachined radius defects at the center of the gage section and were tested in torsion at room temperature. The simulations were run with two rate sensitivities: $m^* = 0.01$ and $m^* = 0.005$, whose values bounded those measured in torsion tests on specimens without geometric defects. Average surface shear strain rate was approximately 0.05 s^{-1} in both experiments and simulations.

For the type 304 specimen, the material coefficients required for the analysis were determined from low-speed tests at which localization does not occur. Using a specimen with a premachined 8% defect in radius at the center of the gage section, localization occurred during tests at $\dot{\Gamma} \sim 0.05$ s^{-1}. The localization rate (measured using scribe lines) showed good agreement with the localization simulation based on material parameters, when the additional effect of geometry changes occurring during testing because the specimen ends were not clamped was taken into account (see figure above).

Source: Metals Handbook, Ninth Edition, Volume 8, Mechanical Testing, American Society for Metals, Metals Park OH, 1985, p 173

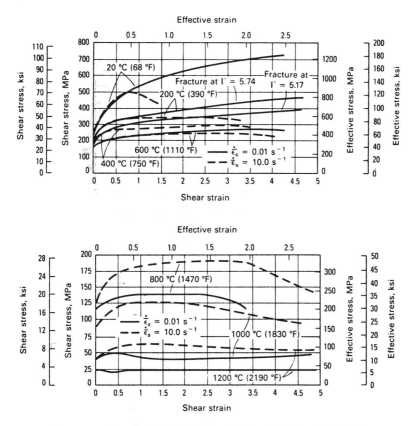

Flow curves from type 304L stainless steel torsion tests. (top) Cold and warm working temperatures. (bottom) Hot working temperatures.

The flow curves for type 304L stainless steel shown in the figures above also indicate the influence of strain rate on flow behavior. At cold and warm working temperatures, strain rate has only a slight effect on flow response. In fact, the 10 s^{-1} curve at a given temperature eventually drops below the curve measured at 0.01 s^{-1}. The lower rate can be considered isothermal and the higher rate adiabatic. Thus, the crossover of flow curves at the two rates is a result of deformation heating and a relatively small strain-rate sensitivity (as shown by the initial portions of the flow curves, in which thermal effects are unimportant). Isothermal flow curves for 10 s^{-1} can be deduced by estimating the associated ΔT values and by constructing σ-T plots. This leads to isothermal high strain rate flow curves that are consistently above the lower strain-rate flow curves.

In contrast to the trends at cold and warm working temperatures, the type 304L flow response in the hot working regime reveals a noticeable strain-rate effect. Under these conditions, the high strain-rate curves are considerably above their low strain-rate counterparts at a given test temperature. Such a response is the result of the high strain-rate sensitivity of most metals at hot working temperatures. The strain-rate sensitivity effect offsets any possible crossover due to deformation heating at the higher strain rates. Flow stresses in this temperature regime are much lower than those at cold and warm working temperatures. Because of this, ΔT values associated with the higher strain rate, which vary with the magnitude of τ and Γ, tend to be smaller at the higher temperatures.

Source: Metals Handbook, Ninth Edition, Volume 8, Mechanical Testing, American Society for Metals, Metals Park OH, 1985, p 162

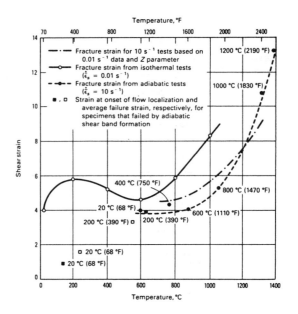

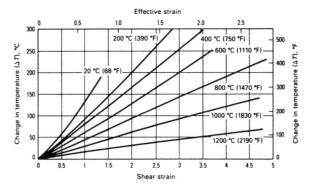

(Top) Fracture-strain data from type 304L austenitic stainless steel torsion tests and (bottom) estimated temperature changes during high rate tests. Low strain-rate ($\bar{\epsilon} = 0.01$ s^{-1}) data are plotted versus the actual test temperature. High strain-rate (10.0 s^{-1}) data are plotted versus temperatures estimated from (bottom) a ΔT-Γ plot and the nominal test temperature, which is shown beside each data point.

The low strain rate data for type 304L shown above illustrate a classical dependence on temperature, i.e., a modest ductility at cold working temperatures, a ductility minimum at warm working temperatures, and large fracture strains at hot working temperatures. The major effect of the higher strain rate is a translation of the lower strain rate data to higher temperatures. For example, the ductility minimum appears to be shifted.

Source: Metals Handbook, Ninth Edition, Volume 8, Mechanical Testing, American Society for Metals, Metals Park OH, 1985, p 167

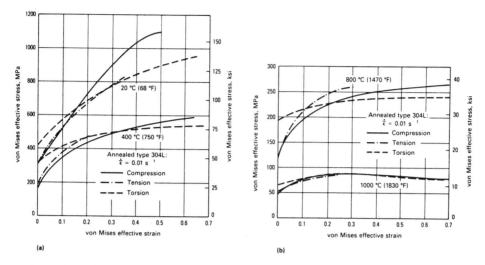

Comparison of effective stress-strain curves determined for type 304L stainless steel in compression, tension, and torsion. (a) Cold and warm working temperatures. (b) Hot working temperatures.

In the figures above, torsion flow stress data for type 304L are compared to compression and tension data in terms of von Mises effective stress and strain. At cold and warm working temperatures, as well as at low hot working temperatures (800 °C, or 1470 °F), the flow curves from the various tests do not coincide. Generally, there is a lower level of strain hardening in torsion. Thus, although the overall stress levels are similar, the actual shapes of the curves are quite different. Even if other definitions of effective stress and strain are employed, the differences between the curves cannot be eliminated.

However, an estimate of the working loads can still be derived from torsion data plotted in von Mises terms. In contrast to the 20, 400, and 800 °C (68, 750, and 1470 °F) behavior, comparison of type 304L torsion data to tension and compression data is quite good at the hot working temperature of 1000 °C (1830 °F) (see b above). This is most likely a result of the absence of textural and strain-hardening effects.

Source: Metals Handbook, Ninth Edition, Volume 8, Mechanical Testing, American Society for Metals, Metals Park OH, 1985, p 164

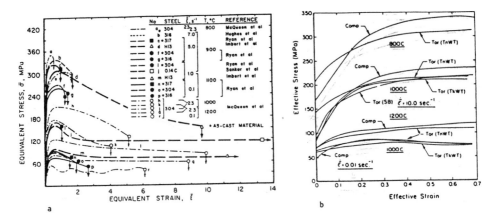

(a) Representative stress-strain curves for 304, 316, and 317 steels with a H13 and C steel for comparison. (b) Compression and torsion curves corrected for deformational heating at high $\dot{\epsilon}$. (TkWT, TnWT, thick-, thin-wall tubes).

Source: Stainless Steels '84, proceedings of the conference sponsored and organized jointly by Chalmers University of Technology and Jernkontoret (Sweden) with the Metals Society (UK), Chalmers University of Technology, Göteborg, 3–4 September 1984, the Institute of Metals, London, 1985, p 51

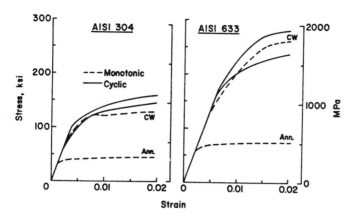

Stress-strain responses in metastable austenitic stainless steels.

In contrast to trends in some steels is the behavior of two austenitic stainless steels (see figure above). In the annealed condition, these steels exhibit pronounced cyclic hardening as a result of a deformation-induced martensitic transformation. Prior cold working is also seen to cause transformation hardening.

Source: Fatigue and Microstructure, papers presented at the 1978 ASM Materials Science Seminar, 14–15 October 1978, St. Louis MO, sponsored by the Materials Science Division of the American Society for Metals, Metals Park OH, 1979, p 443

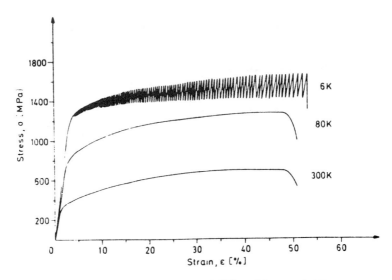

Typical stress-strain curves at different temperatures.

Here it is seen that the alloy has excellent ductility at all temperatures. A slight increase in work hardening is seen by comparing the deformation at 80 K with that at 300 K. No further significant increase in work hardening was observed between 80 and 4 K. Below 40 K the stress-strain curves started to become serrated. However, the flow pattern remained unchanged. No three-stage deformation character was manifest at any temperature, as is generally encountered in some of the austenitic steels alloyed with nitrogen and manganese.

Source: Advances in Cryogenic Engineering Materials, Volume 30, A. F. Clark and R. P. Reed, Eds., Plenum Press, New York, 1984, p 228

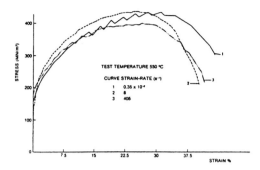

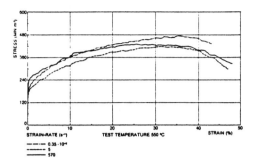

Engineering stress-strain curves for AISI 316L stainless steel, as-received.

Engineering stress-strain curves for AISI 316L thermally aged in sodium at 500 °C for 10,000 h.

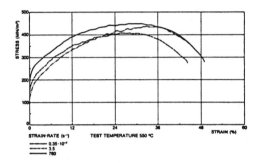

Engineering stress-strain curves for AISI 316L thermally aged in sodium at 500 °C for 30,000 h.

Source: Stainless Steels '84, proceedings of the conference sponsored and organized jointly by Chalmers University of Technology and Jernkontoret (Sweden) with the Metals Society (UK), Chalmers University of Technology, 3–4 September 1984, Göteborg, the Institute of Metals, London, 1985, p 393

11-21. Type 403 Stainless Steel

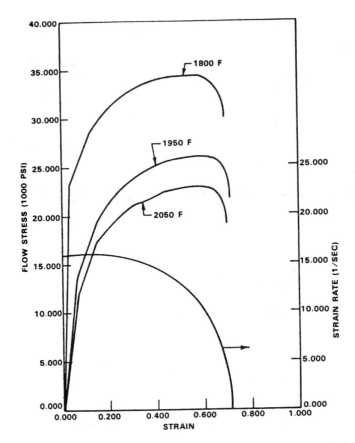

Flow stress versus strain, and strain rate versus strain, for type 403 stainless steel at 1800, 1950, and 2050 °F (tests were conducted in a mechanical press where $\dot{\bar{\epsilon}}$ was not constant).

Source: Metal Forming, Taylan Altan, Soo-Ik Oh and Harold L. Gegel, Eds., American Society for Metals, Metals Park OH, 1983, p 52

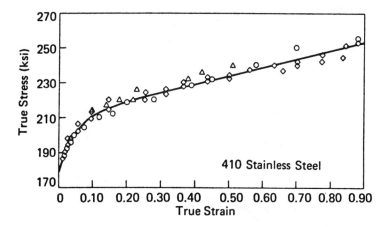

Room-temperature tensile and compressive flow curves.

Source: R. Chait, Factors Influencing the Strength Differential of High Strength Steels, Met. Trans. A, February 1972, American Society for Metals, Metals Park OH, p 367

11-23. Type 410 Stainless Steel

0.750 IN DIAMETER BAR
1800F, 1 HR, OQ, 800F, 4 HR, AC

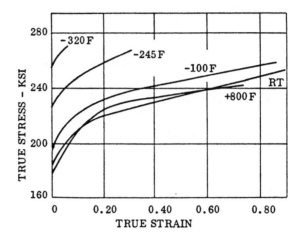

True stress-strain curves for type 410 stainless steel.

Source: Structural Alloys Handbook, Volume 2, Daniel J. Maykuth, Ed., Mechanical Properties Data Center, Battelle
Columbus Laboratories, Columbus OH, 1980, p 23

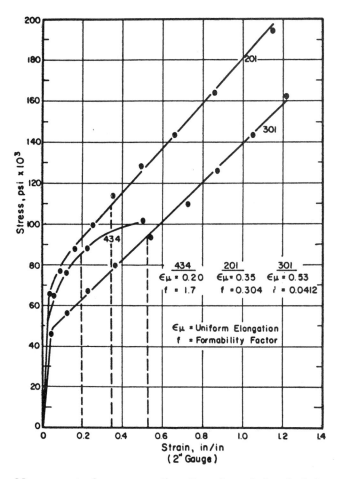

**Measurements along cross sections through cracked and whole
wheel covers indicate thickness variations.**

An analysis of formability is useful to the steel supplier and the fabricator when a failure occurs during production. The figure above shows thickness along cross sections of cracked and uncracked wheel covers. A sample from the coil of type 301 stainless used for that press formed part was evaluated by the true stress-true strain tensile test. The formability factor was found to be 0.125 (upper thin area at break). Various processing changes were made in several coils of strip, and the formability factors computed for these coils were compared with performance during press forming. It was found that type 301 could be consistently formed if the formability factor was 0.035 or less.

After the maximum allowable formability factor for that part had been established as 0.035, quality control could be maintained to ensure that steel shipped would meet requirements. Steel supplied for a part that is less difficult to form might not require as low a formability factor, and a steel with a lower rate of work hardening (to reduce die wear) might be more to the customer's advantage.

Source: Source Book on Cold Forming, American Society for Metals, Metals Park OH, 1975, p 126

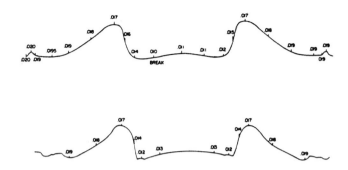

CROSS SECTION SHOWING THICKNESS VARIATIONS IN CRACKED AND UNCRACKED WHEEL COVER

True stress-true strain curves show that work-hardening characteristics differ appreciably among stainless steels.

Tensile strength of stainless grades can be substantially increased by cold working. The figure above compares the work-hardening characteristics of a ferritic grade (434) and two austenitic grades (201 and 301). The higher work-hardening characteristics of the austenitic grades result in significantly improved formability with respect to deep drawing. The high work-hardening rate effectively retards localized reduction, or necking, and permits significant stretching during forming operations. The lower work-hardening characteristics and lower ductility of the ferritic grades are detrimental to deep drawing operations.

Source: Source Book on Cold Forming, American Society for Metals, Metals Park OH, 1975, p 125

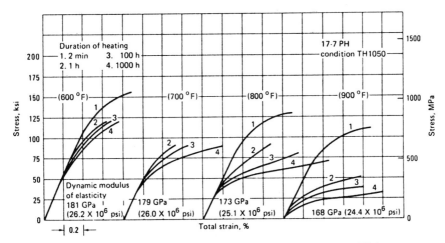

Isochronous stress-strain curves for 17-7 PH sheet 1.3 mm (0.050 in.) thick, TH1050 condition. Room-temperature properties were: tensile strength, 1290 MPa (187 ksi); yield strength at 0.2% offset, 1225 MPa (178 ksi); dynamic modulus of elasticity, 200 GPa (29 × 10⁶ psi). Total strain was adjusted to the indicated modulus values.

Isochronous stress-strain curves such as those shown in the figure above are useful in selection of design stresses for permissible total deformations during short and long periods of time. Because these data are taken from creep curves, extension due to thermal expansion is not included.

Source: Metals Handbook, Ninth Edition, Volume 3, Properties and Selection: Stainless Steels, Tool Materials and Special-Purpose Metals, American Society for Metals, Metals Park OH, 1980, p 192

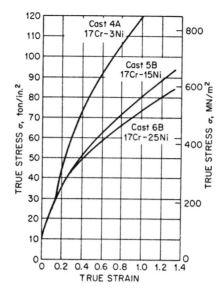

Effect of nickel content on the true stress-true strain curve of 17% Cr steels solution-treated at 1150 °C (2102 °F) and water-quenched.

The figure above shows the effect of nickel content on the true stress-true strain curve of 17% Cr steels containing less than 0.1% carbon. At strain values above about 15%, nickel reduces flow stress. Increasing nickel content from 8 to 10% increases the maximum uniform strain and total strain, i.e., ductility. A further increase in nickel content decreases ductility. Fracture stress decreases with increasing nickel. Holmes and Gladman suggest that these effects are consistent with the effect of nickel in increasing stacking-fault energy of the austenite, thereby reducing the tendency to deform by transformation to martensite, by mechanical twinning, and by movement of dissociated dislocations.

Source: Donald Peckner and I. M. Bernstein, Handbook of Stainless Steels, McGraw-Hill Book Co., New York, 1977, p 12–25

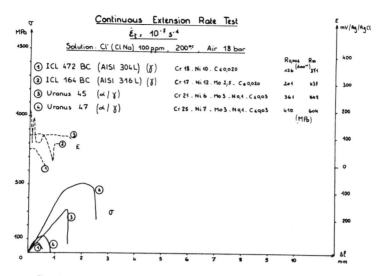

Continuous extension rate tests for austenitic (γ) and duplex alloys (α/γ) in 100 ppm Cl⁻.

Most of the results of S.C.C. of duplex steels in boiling 44% $MgCl_2$ solution have already been published. They clearly indicate the ability of duplex steels to resist in hot chloride solutions in regard with austenitic stainless steels. The same difference has been observed in the 100 ppm Cl⁻ solution at 200 °C as indicated in the figure above. Constant extension rate tests, done at a very low strain rate ($10^{-7}s^{-1}$) highlight the better resistance of duplex alloys. It appears quite clearly that austenitic steels undergo much faster cracking than duplex alloys URANUS 45 and URANUS 47: maximum stresses are much lower (100 MPa or less for austenitic steels, more than 400 MPa for URANUS 47). Likewise, elongations at rupture do not exceed 3% for the austenitic alloys but can reach 12% for duplex alloys. Furthermore, the table below summarizes the results of constant load tests and likewise indicates the better S.C.C. resistance of duplex alloys.

ALLOY	APPLIED STRESS MPa	TIME TO RUPTURE (hours)
AISI 304L	79	60-385-87-41
AISI 316L	151	111-123
URANUS 45	270	> 1000
URANUS 47	307	> 1000

Table   – Constant load tests results at $\sigma = \frac{3}{4} R_{0.002}$
in a solution Cl(NaCl) 100 ppm, 200°C, 18 bars.
($R_{0.002}$ = yield stress at 0.2 % strain)

Source: Duplex Stainless Steel, R. A. Lula, Ed., American Society for Metals, Metals Park OH, 1983, p 537

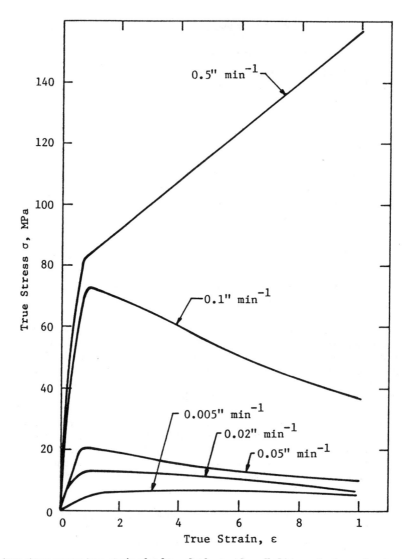

A true stress versus true strain plot for a duplex grade pulled at constant crosshead speeds up to a total strain of one. The data were computed from the load versus elongation plot assuming neck-free elongation and constancy of volume. Strain hardening is seen only at the crosshead speed of 0.5 in. per min.

Source: Duplex Stainless Steels, R. A. Lula, Ed., American Society for Metals, Metals Park OH, 1983, p 10

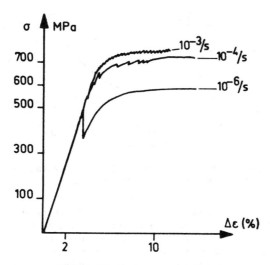

Effect of strain rate on twinning.

As the nucleation of twins depends on the ability of screw dislocations to dissociate in the bcc lattice, an important effect of strain rate can be expected. The graph above gives results for a duplex alloy at room temperature and confirms that the higher the strain rate, the higher the frequency of twinning, at 295 K.

Source: Duplex Stainless Steels, R. A. Lula, Ed., American Society for Metals, Metals Park OH, 1983, p 545

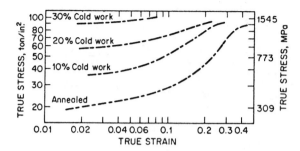

True stress-true strain curves for a 17.5Cr-7Ni stainless steel with varying degrees of prior cold working.

As has been shown, the $\gamma \rightarrow \alpha'$ transformation contributes significantly to work hardening. Since lower temperatures favor transformation, difficult forming operations are sometimes performed at subzero temperatures. The role of the transformation also explains why cold-worked stainless steel does not subsequently work harden as much as annealed material. True stress-true strain curves for a 17.5Cr-7Ni alloy are shown in the curves above.

Source: Donald Peckner and I. M. Bernstein, Handbook of Stainless Steels, McGraw-Hill Book Co., New York, 1977, p 4–36

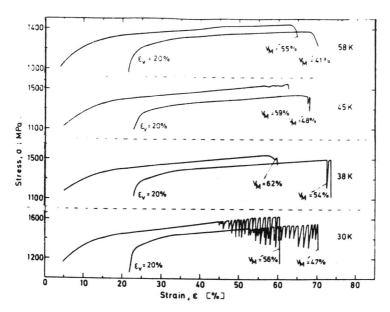

Stress-strain curves with and without prestrain at various temperatures.

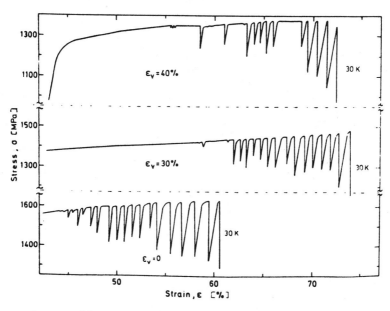

Increase of fracture strain due to different degrees of prestrain (ϵ_v).

Source: Advances in Cryogenic Engineering Materials, Volume 30, A. F. Clark and R. P. Reed, Eds., Plenum Press, New York, 1984, p 230

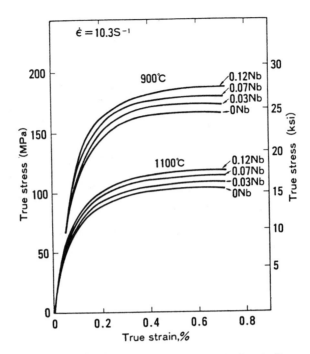

Effect of niobium content on the strength of austenite.

Hot Deformation Strength of Niobium Steel. Recently, automation, using process computer control, is being promoted in plate rolling. This allows such important considerations as the determination of optimum pass schedules and the estimation of roll force needed for the control of plate thickness to be addressed. It is known that an increase in niobium content will result in an increase in resistance to deformation, as shown in the figure above. In addition, there exists a difference in recrystallization and recovery between niobium steel and a conventional steel by applying a true strain of 0.7 at 900 °C and 1000 °C, redeforming after holding a time, and obtaining a softening degree from flow stress ratio. From this, it is clear that the degree of softening with niobium steel of only 20% implies an increase in resistance to deformation with the interval between plate mill passes about 10 s at 900 °C. On the basis of this knowledge, studies have been conducted to estimate optimum rolling schedules for niobium steel, with improved productivity of controlled rolling and achievement of high and consistent quality.

Source: Niobium, Harry Stuart, Ed., proceedings of the International Symposium, 8–11 November 1981, San Francisco CA, AIME, New York, 1984, p 843

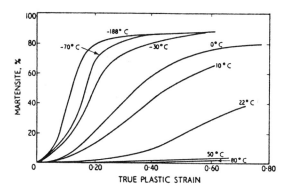

Formation of martensite by plastic tensile strain at various deformation temperatures.

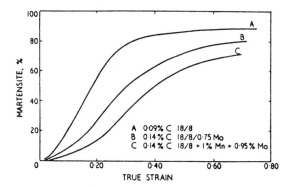

Transformation curves at 0 °C for three different austenitic steels.

Strain Tempering. Considerable increases in strength can be obtained by the straining of martensite in either the untempered or the tempered condition, although so far the process has not been exploited commercially.

It appears that the rate of strain hardening of martensite increases with carbon content, the strengthening effect being principally due to a simple work-hardening mechanism. Whereas the yield strength of low-carbon martensite is increased by subsequent tempering, that of high-carbon martensite decreases even at comparatively low tempering temperatures.

Source: Metallurgical Developments in High-Alloy Steels, proceedings of a joint conference on high-alloy steels held by the British Iron and Steel Research Association and the Iron and Steel Institute, 2–4 June 1964, London, Iron and Steel Institute, 1964, p 28

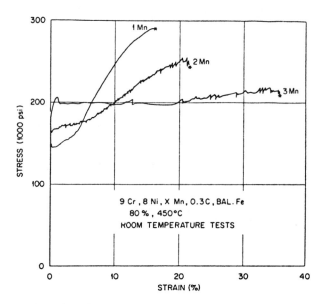

Effect of varying manganese contents on room-temperature
engineering stress-strain curves of 0.3% C alloys after 80%
reduction in thickness at 450 °C. Crosshead speed: 0.04 in.
per min.

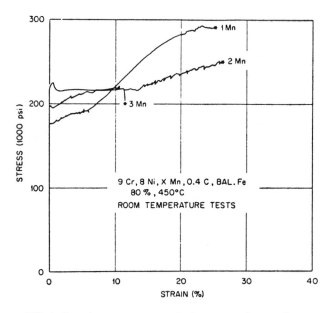

Effect of varying manganese contents on room-temperature
engineering stress-strain curves of 0.4% C alloys after 80%
reduction in thickness at 450 °C. Crosshead speed: 0.04 in.
per min.

Source: Dieter Fahr, Stress and Strain-Induced Formation of Martensite and Its Effects on Strength and Ductility of
Metastable Austenitic Stainless Steels, Met. Trans. A, July 1971, American Society for Metals, Metals Park OH, p
1885–1886

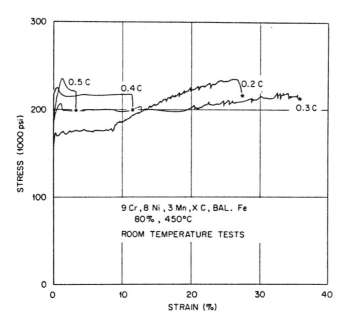

Effect of varying carbon contents on room-temperature engineering stress-strain curves of 3% Mn alloys after 80% reduction in thickness at 450 °C. Crosshead speed: 0.04 in. per min.

Source: Dieter Fahr, Stress and Strain-Induced Formation of Martensite and Its Effects on Strength and Ductility of Metastable Austenitic Stainless Steels, Met. Trans. A, July 1971, American Society for Metals, Metals Park OH, p 1887

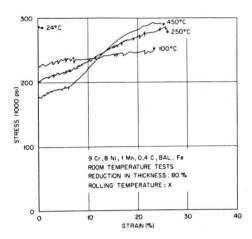

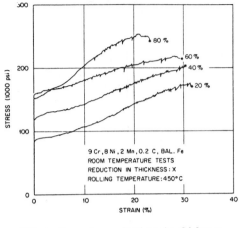

Effect of different rolling temperatures (different austenite stabilities) on the room-temperature engineering stress-strain curves.

Effect of varying reductions in thickness (and rolling times) at 450 °C on the room-temperature engineering stress-strain curves of a relatively unstable alloy. Crosshead speed: 0.04 in. per min.

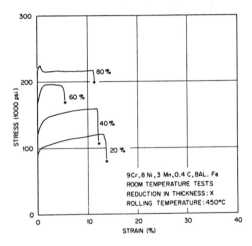

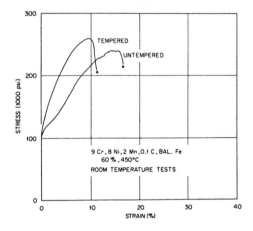

Effect of varying reductions in thickness (and rolling times) at 450 °C on the room-temperature engineering stress-strain curves of a relatively stable alloy. Crosshead speed: 0.04 in. per min.

Effect of annealing (80 min at 450 °C) on the room-temperature engineering stress-strain curve of a partially transformed (M_s > RT) alloy (Alloy 6811-13). Crosshead speed: 0.04 in. per min.

Source: Dieter Fahr, Stress and Strain-Induced Formation of Martensite and Its Effects on Strength and Ductility of Metastable Austenitic Stainless Steels, Met. Trans. A, July 1971, American Society for Metals, Metals Park OH, p 1889–1890

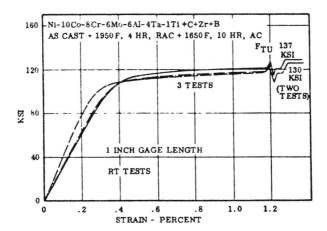

Stress-strain curves for heat treated alloy.

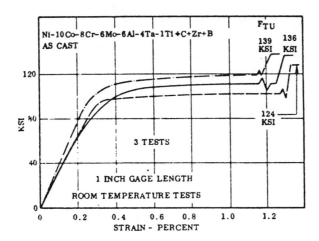

Stress-strain curves for as-cast alloy.

Source: Aerospace Structural Metals Handbook, Volume 5, Mechanical Properties Data Center, Battelle Columbus Laboratories, Columbus OH, December 1978, p 14

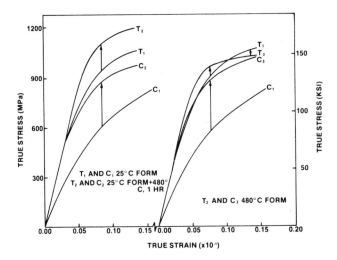

True stress-true strain curves for longitudinal specimens, showing large increases in both tensile and compressive flow curves (left) which occurred as a result of heat treating for 1 h at 480 °C after forming at 25 °C, and an increase in only the compressive flow curve (right) after forming at 480 °C.

Source: Deformation, Processing, and Structure, George Krauss, Ed., papers presented at the ASM Materials Science Seminar, 23 October 1982, St. Louis MO, sponsored by the Seminar Committee of the Materials Science Division of the American Society for Metals, Metals Park OH, 1984, p 466

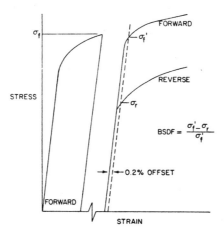

Schematic illustration showing the method used to measure the Bauschinger effect via a parameter referred to as the Bauschinger Strength-Differential Factor (BSDF).

For this study a new Bauschinger parameter, the Bauschinger Strength-Differential Factor (BSDF), shown schematically in the figure above, was formulated:

$$\text{BSDF} = \frac{\sigma_f' - \sigma_r}{\sigma_f'}$$

A distinct advantage of the BSDF parameter is that a measure of the Bauschinger effect can be obtained by testing without knowing the prestrain history, even though the Bauschinger effect is very much a function of prestrain history. Thus, the parameter can be used in tests that incorporate elevated-temperature prestraining and/or heat treatments given after prestrain deformation and prior to re-straining. Furthermore, the BSDF can be used for measuring the extent to which the Bauschinger effect has been developed in formed parts where the prior stress-strain behavior during forming is not readily available.

Source: Deformation, Processing, and Structure, George Krauss, Ed., papers presented at the ASM Materials Science Seminar, 23 October 1982, St. Louis MO, sponsored by the Seminar Committee of the Materials Science Division of the American Society for Metals, Metals Park OH, 1984, p 463

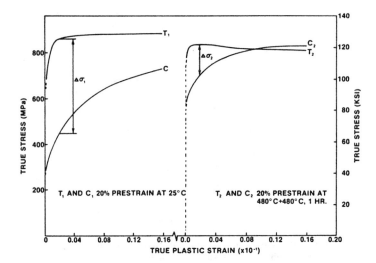

True stress-true plastic strain curves for specimens uniaxially prestrained 20% in tension, showing the large reduction in strength differential ($\Delta\sigma_1$ vs $\Delta\sigma_2$) and elimination of permanent softening with exposure at 480 °C. The material was solution heat treated for 1 h at 950 °C and water quenched prior to testing.

Examination of the values clearly shows that 21-6-9 does exhibit a large Bauschinger effect and that the effect can be reduced by prestraining at elevated temperature (480 °C) and/or by heat treating at 480 °C after prestraining and prior to re-straining. Examples of the flow curves that gave rise to the reduction in the BSDF are shown in the figure above.

Source: Deformation, Processing, and Structure, George Krauss, Ed., papers presented at the ASM Materials Science Seminar, 23 October 1982, St. Louis MO, sponsored by the Seminar Committee of the Materials Science Division of the American Society for Metals, Metals Park OH, 1984, p 464

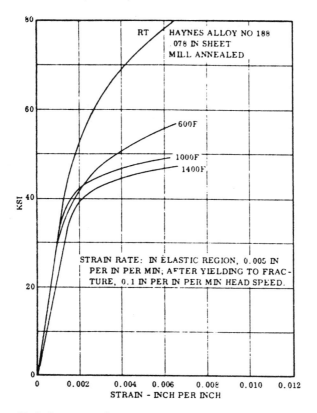

Typical stress-strain curves for sheet tested in the transverse direction at room and elevated temperatures.

Source: Aerospace Structural Metals Handbook, Volume 5, Mechanical Properties Data Center, Battelle Columbus Laboratories, Columbus OH, December 1978, p 12

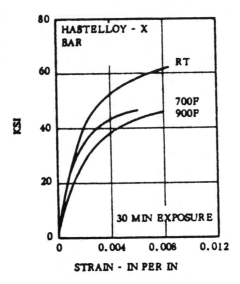

Typical stress-strain curves at elevated temperature for bar in compression.

Source: Aerospace Structural Metals Handbook, Volume 4, Mechanical Properties Data Center, Battelle Columbus Laboratories, Columbus OH, 1981, p 8

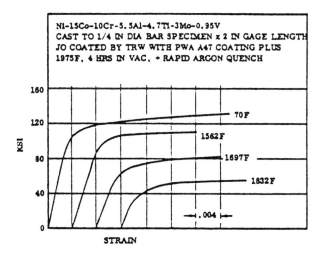

Ni-15Co-10Cr-5.5Al-4.7Ti-3Mo-0.95V
CAST TO 1/4 IN DIA BAR SPECIMEN x 2 IN GAGE LENGTH
JO COATED BY TRW WITH PWA A47 COATING PLUS
1975F, 4 HRS IN VAC. + RAPID ARGON QUENCH

Stress-strain curves for JO coated alloy at room and elevated temperatures.

Source: Aerospace Structural Metals Handbook, Volume 5, Mechanical Properties Data Center, Battelle Columbus
Laboratories, Columbus OH, December 1978, p 25

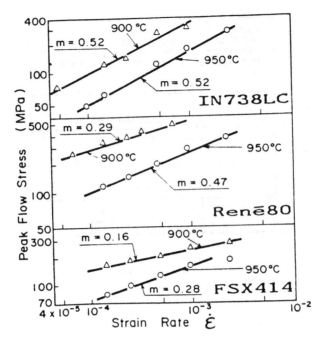

Tensile properties at high temperature, correlation between logσ and log$\dot\varepsilon$ of quench rolled ribbons.

The tensile strengths at high temperatures as a function of strain rate are shown in the figure above. The strain-rate sensitivity, m, is also indicated. The m value for IN 738LC tested at 900 °C and 950 °C is calculated to be 0.52. In the René 80 ribbons, the m value is somewhat less but increases when the testing temperature is raised from 900 °C to 950 °C. On the other hand, the m value of the FSX 414 ribbon is low compared with those of the IN 738LC and the René 80 ribbons.

As m values above 0.3 are considered a prerequisite for superplasticity, it may well be that the IN 738LC and René 80 ribbons would exhibit this phenomenon. It is thought that the lower m values found in the FSX 414 result from the large $Cr_{23}C_6$ carbides formed at triple points that could suppress the grain boundary sliding needed for superplastic properties. It may be possible to produce higher m values in the FSX 414 ribbons by controlling the carbide morphology through heat treatment or chemistry modification.

Source: Superalloys 1984, Maurice Gell, Charles S. Kortovich, Roger H. Bricknell, William B. Kent and John F. Radavich, Eds., proceedings of the Fifth International Symposium on Superalloys, 7–11 October 1984, Champion PA, sponsored by the High Temperature Alloys Committee of the Metallurgical Society of AIME, Warrendale PA, 1984, p 485

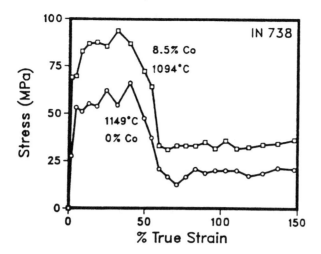

Plot of the resistance to rolling of IN 738.

Results of a preliminary study on a small heat (14 kg) IN 738 showed that roughly a 40% reduction in the peak rolling resistance (a function of hot working flow stress) resulted from the 55 °C increase in working temperature allowed by the removal of the cobalt from the standard alloy containing 8.5% cobalt (see curves above). These stresses are calculated from the current drawn by a constant-speed rolling mill and the strains are determined by the reduction per pass.

Source: Superalloys 1984, Maurice Gell, Charles S. Kortovich, Roger H. Bricknell, William B. Kent and John F. Radavich, Eds., proceedings of the Fifth International Symposium on Superalloys, 7–11 October 1984, Champion PA, sponsored by the High Temperature Alloys Committee of the Metallurgical Society of AIME, Warrendale PA, 1984, p 465

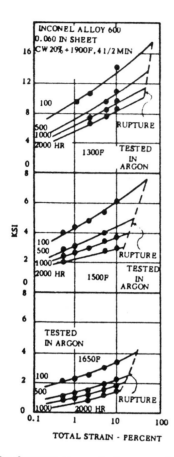

Isochronous stress-strain curves for sheet annealed at 1900 °F and tested in argon at temperatures from 1300 to 1650 °F.

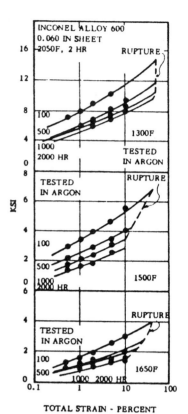

Isochronous stress-strain curves for sheet annealed at 2050 °F and tested in argon at temperatures from 1300 to 1650 °F.

Source: Aerospace Structural Metals Handbook, Volume 4, Mechanical Properties Data Center, Battelle Columbus Laboratories, Columbus OH, 1981, p 14

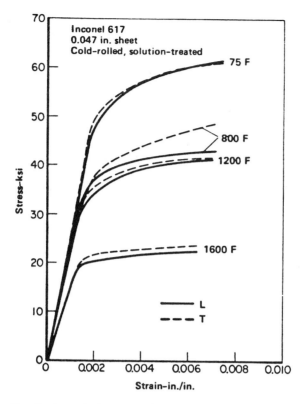

Tensile stress-strain curves at room temperature and elevated temperatures.

Source: Aerospace Structural Metals Handbook, Volume 5, Mechanical Properties Data Center, Battelle Columbus Laboratories, Columbus OH, December 1978, p 8

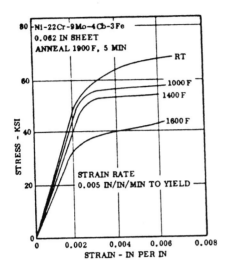

Stress-strain curves for 0.062-in. sheet at room and elevated temperatures.

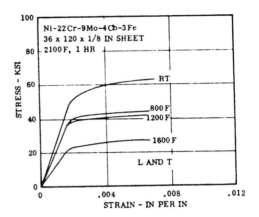

Stress-strain curves in tension for 1/8-in. sheet at room and elevated temperatures.

Source: Aerospace Structural Metals Handbook, Volume 4, Mechanical Properties Data Center, Battelle Columbus Laboratories, Columbus OH, 1981, p 12

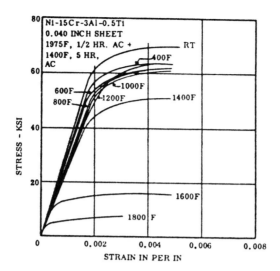

Temperature stress-strain curves at various temperatures.

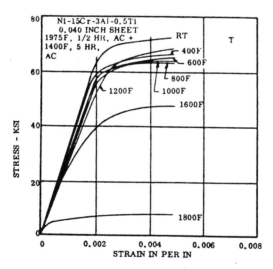

Compressive stress-strain curves at temperature.

Source: Aerospace Structural Metals Handbook, Volume 4, Mechanical Properties Data Center, Battelle Columbus Laboratories, Columbus OH, 1981, p 3

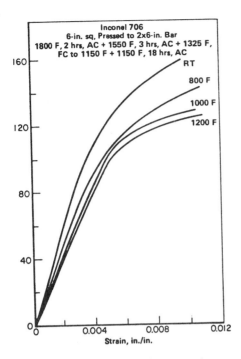

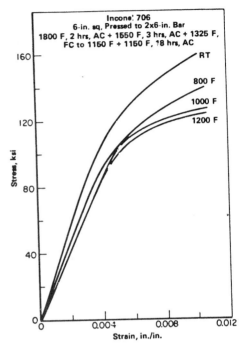

Transverse tensile stress-strain curves at room temperature to 1200 °F.

Longitudinal tensile stress-strain curves at room temperature to 1200 °F.

Source: Aerospace Structural Metals Handbook, Volume 4, Mechanical Properties Data Center, Battelle Columbus Laboratories, Columbus OH, 1981, p 10

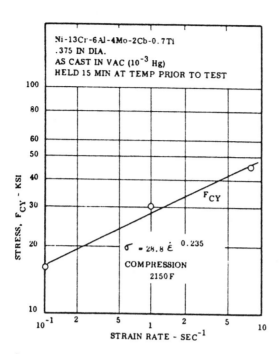

Stress versus strain rate in compression for vacuum cast alloy at 2150 °F.

Source: Aerospace Structural Metals Handbook, Volume 4, Mechanical Properties Data Center, Battelle Columbus Laboratories, Columbus OH, 1981, p 16

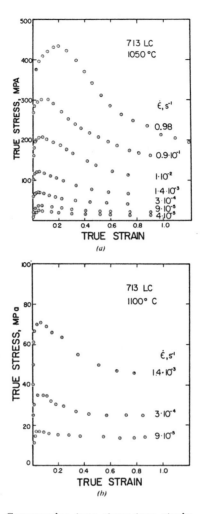

Compressive true stress-true strain curves for 713LC compacts deformed at different strain rates at (*a*) **1050 and** (*b*) **1100 °C.**

To characterize the flow behavior of the as-hipped material, and the microstructures produced during forming, a series of specimens was tested under various conditions of constant true strain rate and temperature to a strain of 0.8 and then quenched for metallographic examinations. This was done at 1050 and 1100 °C and at strain rates between 10^{-5} s^{-1} and 1 s^{-1}.

Source: Production to Near Net Shape Source Book, C. J. Van Tyne and B. Avitzur, Eds., American Society for Metals, Metals Park OH, 1983, p 344

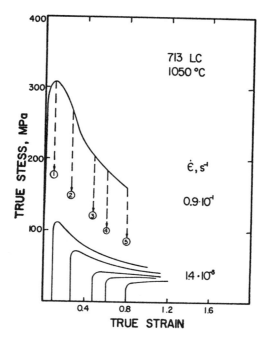

Effects of prestrain, at a strain rate of 9×10^{-2} s^{-1}, on the flow curve at a lower strain rate of 1.4×10^{-3} s^{-1} (1050 °C).

The effects of the amount of prior deformation at a fast strain rate on flow behavior at a slower strain rate are demonstrated in the graph above. With increasing prestrain, a transition in flow behavior can be observed from one where the compacts flow soften during straining to one where the flow strength rises continuously instead. In all cases, however, the stresses developed at high strains all converge to the same level.

Source: Production to Near Net Shape Source Book, C. J. Van Tyne and B. Avitzur, Eds., American Society for Metals, Metals Park OH, 1983, p 345

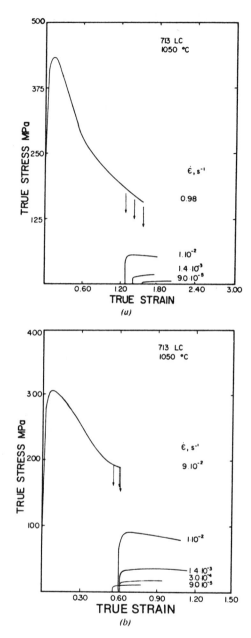

Effects of prestrain at (*a*) **0.98 s⁻¹** and (*b*) **9 × 10⁻² s⁻¹** on the flow curves at different strain rates and 1050 °C.

Source: Production to Near Net Shape Source Book, C. J. Van Tyne and B. Avitzur, Eds., American Society for Metals, Metals Park OH, 1983, p 347

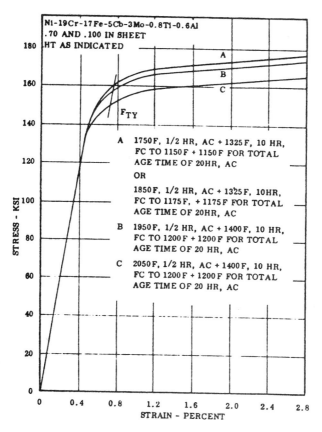

Stress-strain curves for sheet at four heat treatments.

Source: Aerospace Structural Metals Handbook, Volume 4, Mechanical Properties Data Center, Battelle Columbus Laboratories, Columbus OH, 1981, p 17

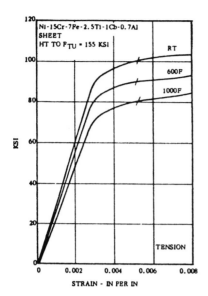

Stress-strain curves at room and elevated temperatures for sheet in tension.

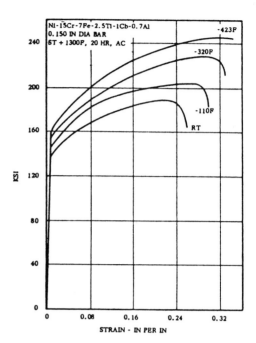

Stress-strain curves for bar at low temperatures.

Source: Aerospace Structural Metals Handbook, Volume 4, Mechanical Properties Data Center, Battelle Columbus Laboratories, Columbus OH, 1981, p 9

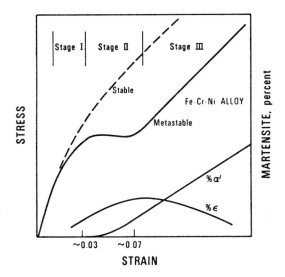

Typical stress-strain curves of stable and metastable austenitic Fe-Cr-Ni alloys at low temperatures and associated relative amounts of strain-induced ϵ and α' that form in the metastable alloy.

Stage 1 (see figure above) of metastable austenite stress-strain behavior represents the microstrain and early macrostrain behavior that includes the 0.2% offset yield strength. The formation of α' is not thought to occur in this range; perhaps stacking-fault clusters, or the ϵ phase, or both contribute to low-temperature deformation in this range. The stacking-fault energy is reduced at low temperatures and apparently becomes low enough to promote ϵ martensite. With X-ray analysis to detect ϵ formation during early deformation and creep at low temperatures, Mirzagev, Goykhanberg, Shteynberg and Rushchin (1973) related the log of volume fraction of ϵ over the volume fraction of $\gamma + \alpha'$ to the log of the plastic deformation for an Fe-18Cr-13Ni-0.02C steel; other X-ray analysis measurements on similar steels have not confirmed this ϵ-strain correspondence (Guntner and Reed, 1962; Reed and Guntner, 1964).

Source: Materials at Low Temperatures, Richard P. Reed and Alan F. Clark, Eds., American Society for Metals, Metals Park OH, 1983, p 310

12-22. Mar-M200 Superalloy

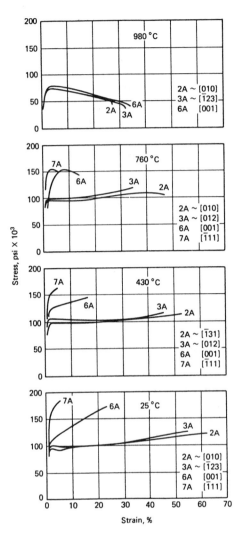

Tensile stress-strain curves for single crystals of Mar-M200. The directions shown were parallel to the initial tensile axis in the tension test. (Adapted from B. H. Kear and B. J. Piearcey, Trans. Met. Soc. AIME, Vol 239, p 1210, 1967)

The creep behavior of as-cast Mar-M200 crystals is shown above. Note that orientations near [100] generally have the best properties. The common orientation in single crystals formed by directional solidification is [100], so that this high-strength orientation can be utilized.

Source: Charlie R. Brooks, Heat Treatment, Structure and Properties of Nonferrous Alloys, American Society for Metals, Metals Park OH, 1982, p 205

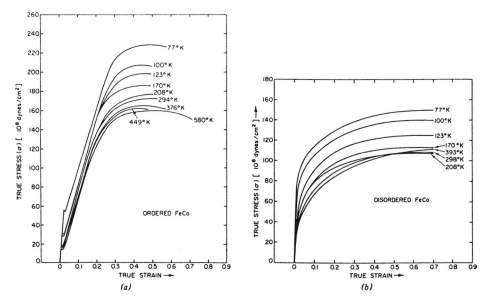

(a) Stress-strain curves for ordered FeCo alloy as a function of temperature. (b) Stress-strain curves for disordered FeCo alloy as a function of temperature.

Experimental Results. The curves above show the stress-strain curves corresponding to the ordered and disordered FeCo alloys, respectively. In all of the following figures the normal stresses are denoted by σ and the shear stresses (one half of the normal stresses) are denoted by τ. The stress-strain curves for the ordered alloy could be separated into three distinct stages: Stage I consisting of a sharp yield point, Stage II characterized by a linear work-hardening rate similar to that of fcc single crystals, and Stage III comprised of the parabolic stress-strain region. Figure (a) further shows that Stage I increases very sharply with a decrease in temperature for temperatures below 100 K, while it is relatively independent of temperature for temperatures greater than 200 K. Such a sharp increase in the yield stress at low temperature has been observed in other ordered alloys. Stage II, on the other hand, appeared quite insensitive to the temperature and Figure (a) shows that all of the curves in Stage II are bundled together. Furthermore, the work-hardening rate, $\theta_{II} = \partial\sigma/\partial\epsilon$, appears to be quite independent of temperature for the range of temperatures investigated. Stage III, however, is quite sensitive to temperature, and correspondingly all the curves in this stage spread apart with decreasing temperature. The disordered alloys, on the other hand, do not show multiple stages of deformation. Instead, the stress-strain curves appear parabolic from the onset of plastic deformation. An intriguing aspect associated with the deformation of the two alloys is that the Stage III deformation in the ordered alloy closely resembles the deformation behavior of the disordered alloy. In particular, both have a similar temperature dependence. Because of the absence of any distinct yield point in the disordered alloys, the yield stress will be defined as the flow stress at 0.2% offset. Figure (b) also shows that the stress-strain curves cross one another in the high temperature regime and high strain regions.

Source: Sheng-Ti Fong, K. Sadananda and M. J. Marcinkowski, Effect of Strain, Temperature, and Atomic Order on Slip Deformation in FeCo, Met. Trans. A, May 1974, American Society for Metals, Metals Park OH, p 1240

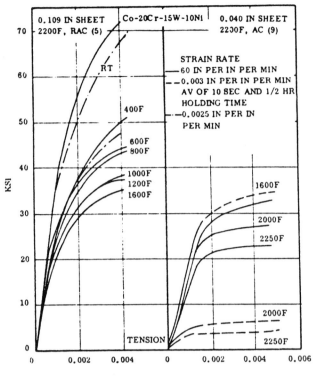

Stress-strain curves for 0.109- and 0.040-in. sheet at room and elevated temperatures at several strain rates.

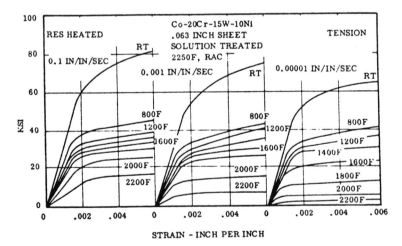

Stress-strain curves in tension for 0.063-in. sheet at room temperatures at several strain rates.

Source: Aerospace Structural Metals Handbook, Volume 5, Mechanical Properties Data Center, Battelle Columbus Laboratories, Columbus OH, December 1978, p 10

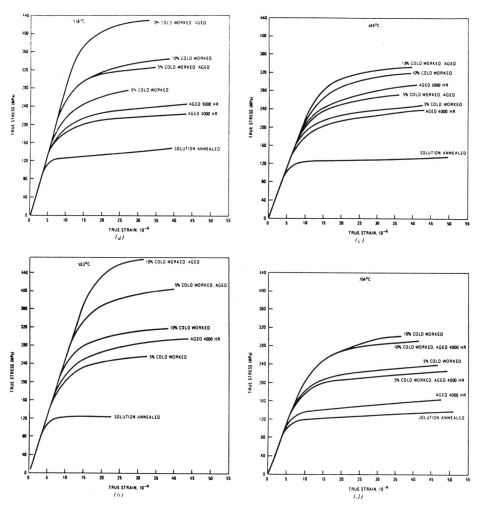

True stress-strain curves of specimens in various cold-worked and aged conditions: (*a*) 538 °C, (*b*) 593 °C, (*c*) 649 °C, (*d*) 704 °C, and (*e*) 760 °C.

The Chemical Composition of Alloy 800H Heat Number HH5556A

Element	Wt Pct	Element	Wt Pct
Fe	44.0	Cu	0.52
Ni	33.0	Ti	0.42
Cr	20.3	Al	0.39
Mn	0.90	C	0.07

(*continued*)

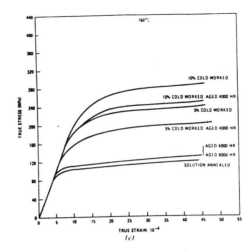

(c)

The true stress-strain curves that were measured in this work are displayed in the graphs on this and the previous page. Each curve in these graphs is an average obtained from several tests and represents the general level of the stress-strain curve, ignoring any effects due to serrated yielding. The 0.2% offset yield strengths that were measured from these curves are given in the table below. Here the data are separated, according to specimen treatment, into three categories: solution annealed and cold worked, aged, and cold worked and aged.

**Summary of 0.2 Pct Offset Yield Strengths (in MPa)
Measured in This Work**

Specimen	Aging and Testing Temperature, °C				
	538	593	649	704	760
Solution annealed	131	124	124	124	110
5 pct c.w.	276	248	234	220	227
10 pct c.w.	331	303	310	296	276
Aged 4000 h	214	282	220++	145++	124
Aged 8000 h	227	–	276+	–	124
5 pct c.w.; aged 4000 h	317	393++	262+	207	193
	(358)	(407)	(331)	(241)	(241)
10 pct c.w.; aged 4000 h	420	462	324+	276+	234
	(413)	(462)	(407)	(317)	(289)

Note: The entries marked with a dagger correspond to specimens which showed γ' formation in the matrix, while the entries marked with a double dagger showed γ' formation in the grain boundaries. The entries enclosed in parentheses correspond to the yield strengths predicted by the Hall-Petch relation.

Source: R. E. Villagrana, J. L. Kaae, J. R. Ellis and P. K. Gantzel, The Effect of Aging and Cold Working on the High-Temperature Low-Cycle Fatigue Behavior of Alloy 800H: Part 1—The Effect of Hardening Processes on the Initial Stress-Strain Curve, Met. Trans. A, July 1978, American Society for Metals, Metals Park OH, p 932–933

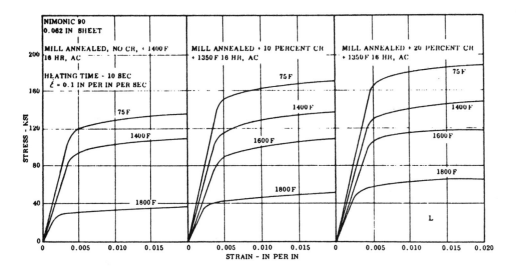

Stress-strain curves determined at various temperatures at a rapid strain rate after rapid heating of sheet cold rolled 0, 10, and 20 percent between mill annealing and aging.

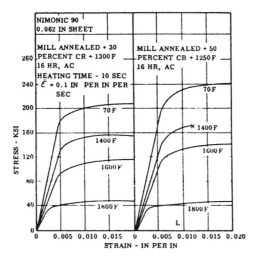

Stress-strain curves determined at various temperatures at a rapid strain rate after rapid heating of sheet cold rolled 30 and 50 percent between mill annealing and aging.

Source: Aerospace Structural Metals Handbook, Volume 5, Mechanical Properties Data Center, Battelle Columbus Laboratories, Columbus OH, December 1978, p 7

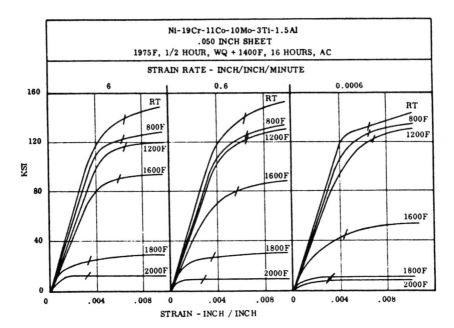

Compressive stress-strain curves at room and elevated temperatures at several strain rates.

Source: Aerospace Structural Metals Handbook, Volume 5, Mechanical Properties Data Center, Battelle Columbus Laboratories, Columbus OH, December 1978, p 34

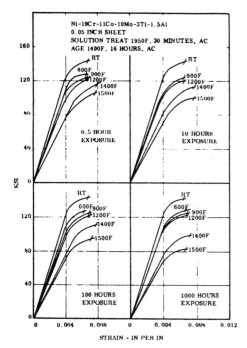

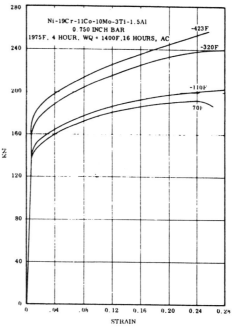

Stress-strain curves at room and elevated temperatures for various exposure times for sheet.

Stress-strain curves of bar at room and low temperatures.

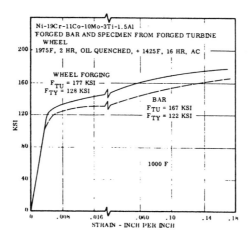

Stress-strain curves at 1000 °F for bar and specimen from turbine wheel forging.

Source: Aerospace Structural Metals Handbook, Volume 5, Mechanical Properties Data Center, Battelle Columbus Laboratories, Columbus OH, December 1978, p 22

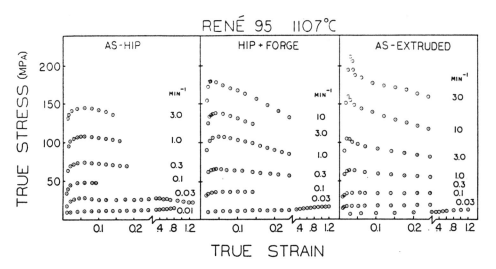

Tensile flow curves for René 95 deformed at different constant true strain rates at 1107 °C.

Experimental Results. The flow curves for the various material conditions are shown in the figure above. The flow curves do not show the final elongations of the tests, as there were limits on the speed and amount of data storage that prevented handling of data from the entire flow curve. Nonetheless, many observations can be made. At the higher strain rates, the elastic portion is followed by rapid work hardening to a peak stress and then by flow softening. The rate of flow softening decreases as strain rate decreases. As the strain rate decreases, a point is reached at which flow softening is not observed, and as strain rate is reduced still further gradual flow hardening occurs throughout the flow curve up to the strains analyzed. Differences exist between the flow curves of the different material conditions. For example, at $\dot{\varepsilon} = 3.0$ min.$^{-1}$, the extruded material work hardens rapidly to a peak stress, then work softens, initially at a high rate and then subsequently at a decreasing rate. In contrast, the HIP and HIP-plus-forge materials work harden more slowly to a peak stress and then very gradually work soften. At $\dot{\varepsilon} = 0.03$ min.$^{-1}$ the HIP material gradually flow softens while both the HIP plus forge and extruded materials gradually work harden.

Source: Superalloys '84, Maurice Gell, Charles S. Kortovich, Roger H. Bricknell, William B. Kent and John F. Radavich, Eds., proceedings of the Fifth International Symposium on Superalloys, 7–11 October 1984, Champion PA, sponsored by the High Temperature Alloys Committee of the Metallurgical Society of AIME, Warrendale PA, 1984, p 279

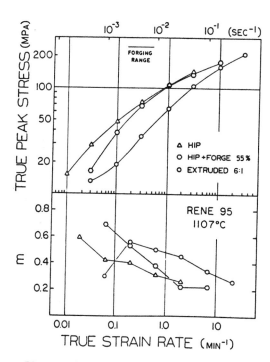

The peak flow stresses as a function of strain rate for the three forms of René 95 deformed at 1107 °C are shown at the top. Below is the strain rate sensitivity $m = \partial\ln\sigma/\partial\ln\dot{\varepsilon}$, determined from the $\sigma - \dot{\varepsilon}$ plots.

From each flow curve the 0.2% yield stress and the peak stress were obtained. The strain rate dependence of the peak flow stress is shown above. The relationship between the logarithms of stress (σ) and strain rate ($\dot{\varepsilon}$) for superplastic materials is sigmoidal. The sigmoidal curve is divided into three regions: the lowest strain rates, region I, in which the slope of the curve $m = \partial\ln\sigma/\partial\ln\dot{\varepsilon}$ is low; an intermediate range of strain rates, region II, in which the slope m is high (the superplastic regime); and the higher strain rates, region III, where m is again low. Most of the data obtained in this work falls within regions II and III. For the extruded material, however, there appears to be a transition in behavior underway between regions I and II at the lowest strain rates tested.

Source: Superalloys '84, Maurice Gell, Charles S. Kortovich, Roger H. Bricknell, William B. Kent and John F. Radavich, Eds., proceedings of the Fifth International Symposium on Superalloys, 7–11 October 1984, Champion PA, sponsored by the High Temperature Alloys Committee of the Metallurgical Society of AIME, Warrendale PA, 1984, p 279

12-31. Udimet 700

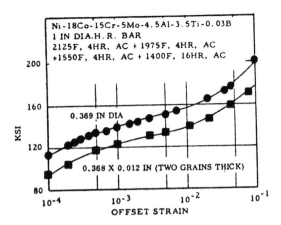

Stress-strain curves at room temperature for round bar and sheet machined from round bar to a thickness of two grains.

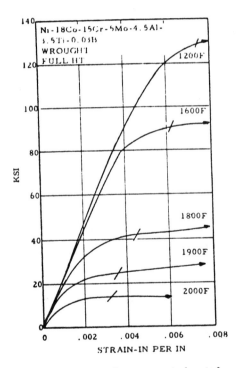

Typical stress-strain curves at elevated temperatures for wrought alloy.

Source: Aerospace Structural Metals Handbook, Volume 5, Mechanical Properties Data Center, Battelle Columbus Laboratories, Columbus OH, December 1978, p 8

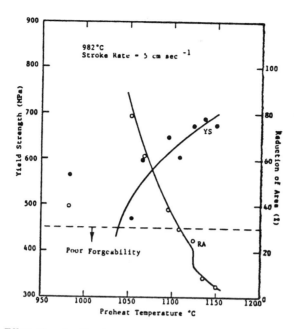

Effect of preheating temperature on the yield strength and ductility of U720. Sample preheated at temperature for 10 min., cooled to 982 °C, then pulled at 5 cm s⁻¹.

Source: Superalloys '84, Maurice Gell, Charles S. Kortovich, Roger H. Bricknell, William B. Kent and John F. Radavich, Eds., proceedings of the Fifth International Symposium on Superalloys, 7–11 October 1984, Champion PA, sponsored by the High Temperature Alloys Committee of the Metallurgical Society of AIME, Warrendale PA, 1984, p 576

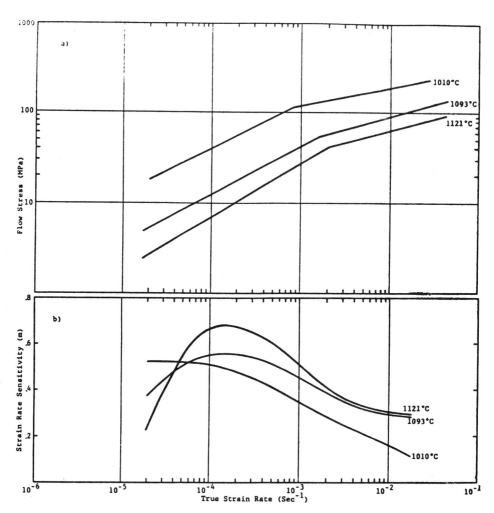

(a) Strain-rate dependence of flow stress for U720 at different temperatures. (b) Strain-rate dependence of strain-rate sensitivity.

Source: Superalloys '84, Maurice Gell, Charles S. Kortovich, Roger H. Bricknell, William B. Kent and John F. Radavich, Eds., proceedings of the Fifth International Symposium on Superalloys, 7–11 October 1984, Champion PA, sponsored by the High Temperature Alloys Committee of the Metallurgical Society of AIME, Warrendale PA, 1984, p 279

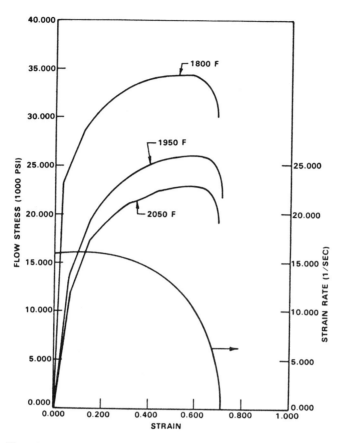

Flow stress versus strain, and strain rate versus strain, for Waspaloy at 1950, 2050 and 2100 °F (tests were conducted in a mechanical press where ἐ was not constant).

Source: Metal Forming: Fundamentals and Applications, Taylan Altan, Soo-Ik Oh and Harold L. Gegel, Eds., American Society for Metals, Metals Park OH, 1983, p 52

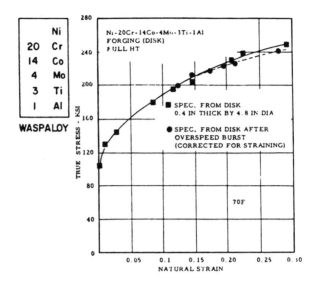

True stress-true strain curve of bar from turbine disk specimen and from fragment of overspeeded (burst) disk.

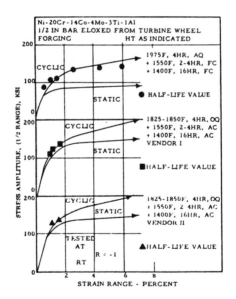

Static and cyclic stress-strain curves at room temperature for specimens from turbine wheel forgings.

Source: Aerospace Structural Metals Handbook, Volume 5, Mechanical Properties Data Center, Battelle Columbus Laboratories, Columbus OH, December 1978, p 6

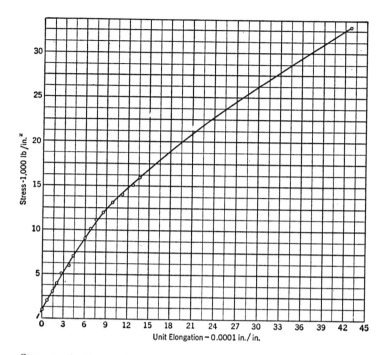

Stress-strain diagram for cast iron: example of behavior of a brittle material.

Source: Donald S. Clark, Engineering Materials and Processes, International Textbook Co., Scranton PA, 1962, p 36

13-2. Gray Iron: Stress-Strain Relationship

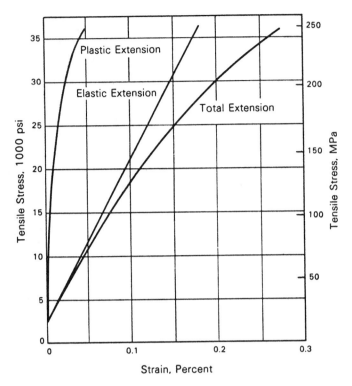

The total strain in gray iron consists of a plastic portion and an elastic portion.

The relation between stress and strain in gray iron is not constant but is a curve of continually decreasing slope. The relation is different for different grades of iron and is influenced by heat treatment. Although gray iron is considered a brittle metal, it can be permanently deformed a measureable amount. Annealed grades can exhibit one-half percent of plastic strain before breaking. A portion of the total strain is elastic and the balance is plastic. This is illustrated in the graph above. The difference between plastic and elastic strain can be demonstrated by loading and unloading tensile specimens equipped with strain gages.

Source: Iron Castings Handbook, Charles F. Walton, Ed., Iron Castings Society, Inc., 1981, p 228

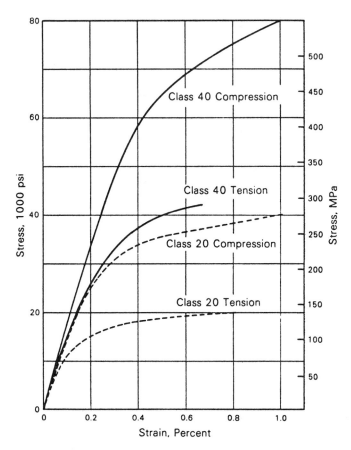

A comparison of stress-strain curves in tension and compression for a class 20 and a class 40 gray iron.

The difference between the tension and compression curves is proportionately larger for lower strength gray irons than for higher strengths in the graph above. The higher effective modulus of elasticity in compression has an important and beneficial effect on the strength of sections that are loaded as beams.

Source: Iron Castings Handbook, Charles F. Walton, Ed., Iron Castings Society, Inc., 1981, p 235

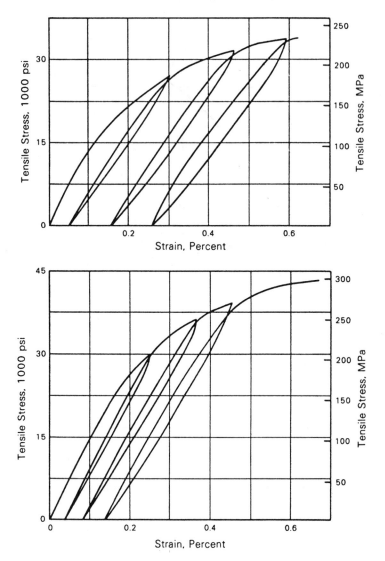

A stress-strain curve in tension for (top) a Class 30 and (bottom) a Class 40 gray iron in which the load was removed to show the permanent deformation.

Source: Iron Castings Handbook, Charles F. Walton, Ed., Iron Castings Society, Inc., 1981, p 229

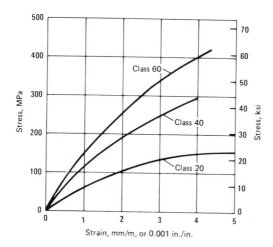

Typical stress-strain curves for three classes of gray iron in tension.

Modulus of Elasticity. Typical stress-strain curves for gray iron are shown in the figure above. Gray iron does not obey Hooke's law, and the modulus in tension is usually determined arbitrarily as the slope of the line connecting the origin of the stress-strain curve with the point corresponding to one-fourth of the tensile strength (secant modulus). Some engineers use the slope of the stress-strain curve near the origin (tangent modulus). The secant modulus is a conservative value suitable for most engineering work; design loads are seldom as high as one-fourth of the tensile strength, and the deviation of the stress-strain curve from linearity is usually less than 0.01% at these loads. However, in the design of certain types of machinery, such as precision equipment, where design stresses are very low, use of the tangent modulus may represent the actual situation more accurately.

Source: Metals Handbook, Ninth Edition, Volume 1, Properties and Selection: Irons and Steels, American Society for Metals, Metals Park OH, 1978, p 20

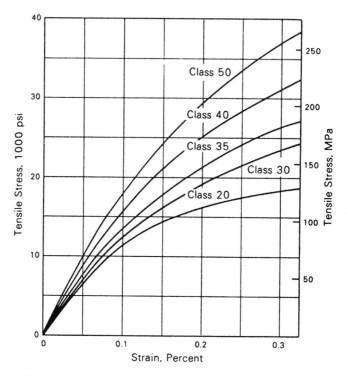

Typical stress-strain curves in tension for gray irons by class.

The tensile properties of gray iron, tensile strength, yield strength, ductility, and modulus of elasticity, can be readily established by a conventional test, as specified by the ASTM. Although yield strength and ductility may be measured, they are seldom determined or specified. The modulus of elasticity of gray iron is not constant as in the case of steel but varies with the class of iron and type of loading. Also, because of the curvature in the stress-strain relation of gray iron, the slope of the curve, the modulus of elasticity, is not a single number. As a result, two approximations, the secant modulus and tangent modulus, are frequently used.

Tensile properties are measured with standard uniaxial tensile testing procedures that provide stress-strain data. Typical curves for different gray irons by class are presented in the graph above. The differing relationships between stress and strain are established by variations in microstructure resulting from changes in section size, cooling rate, and chemical composition.

Source: Iron Castings Handbook, Charles F. Walton, Ed., Iron Castings Society, Inc., 1981, p 211

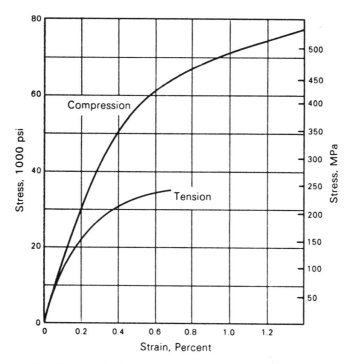

The deviation in the stress-strain relation between tension and compression loading for a class 35 gray iron.

The stress-strain relation for gray iron in compression is similar to that in tension at low stress levels, but deviates at higher loads because the modulus of elasticity in compression is more nearly constant with increasing stress. Thus, the tangent modulus in tension can be properly used as the elastic modulus in compression for engineering calculations. The compressive stress at which 0.1% permanent strain is produced, the proof stress, is approximately twice the proof stress in tension. A comparison of the stress-strain relations in tension and compression for a typical class 35 gray iron is shown in the graph above.

Source: Iron Castings Handbook, Charles F. Walton, Iron Castings Society, Inc., 1981, p 234

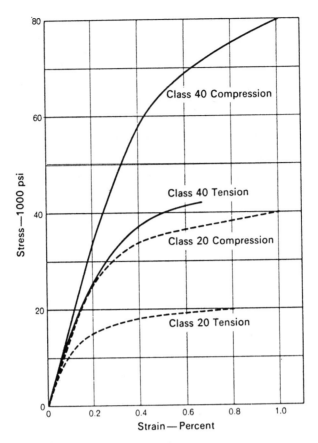

A comparison of stress-strain curves in tension and compression for a class 20 and a class 40 gray iron.

Source: Source Book on Industrial Alloy and Engineering Data, American Society for Metals, Metals Park OH, 1978, p 156

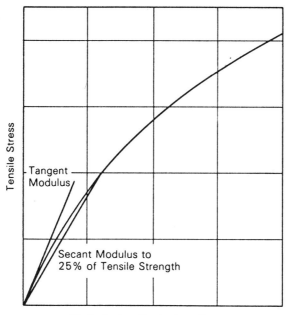

Strain, Inches Per Inch or Percent

Two methods for establishing the modulus of elasticity for gray iron. The secant modulus to 25% of the tensile strength is the more commonly used.

Because of the curvature in the stress-strain relation for gray iron, the slope of the curve cannot be accurately expressed as a single value for the modulus of elasticity. Two approximate values for the modulus are of practical use in gray iron casting design (see graph above). The more commonly quoted value is the secant modulus, which represents the slope of a straight line from the origin at no load to the stress-strain curve at 25% of the ultimate tensile strength. This provides a modulus value that is conservative for most engineering work because design loads are seldom as high as one-fourth of the ultimate strength, and the deviation from a linear relation is less than 0.01% at this load. The tangent modulus is based on the slope of the stress-strain curve at no load. This is useful in the design of precision machinery where applied stresses are low. The tangent modulus can also be determined by measuring the resonant frequency of vibration of an appropriately sized rod with a frequency metering system.

Source: Iron Castings Handbook, Charles F. Walton, Ed., Iron Castings Society, Inc., 1981, p 230

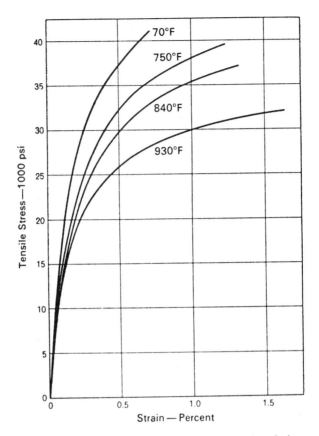

The influence of temperature on the stress-strain relation in tension for a class 40 gray iron.

Source: Source Book on Industrial Alloy and Engineering Data, American Society for Metals, Metals Park OH, 1978, p 159

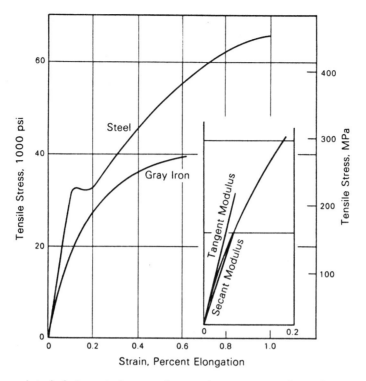

A typical stress-strain curve for gray iron as compared to mild steel. The gray iron curve is enlarged in the insert to show the tangent and secant modulus lines.

The stress-strain relation for gray iron is a curve of gradually decreasing slope, as shown in the graph above. Because of the curvature, there are two useful values for the modulus of elasticity—the secant modulus and the tangent modulus. The secant modulus is the more commonly listed and used. It is the slope of a secant to the stress-strain curve from the origin to a point at 25% of the tensile-strength. Thus, the secant modulus is a conservative value for most designs. The tangent modulus is the slope of a line that is tangent to the stress-strain curve at its origin. Its value is a little higher than the secant modulus and is appropriate for use in the design of precision equipment where the working stress and resulting strain values are low.

Source: Iron Castings Handbook, Charles F. Walton, Ed., Iron Castings Society, Inc., 1981, p 168

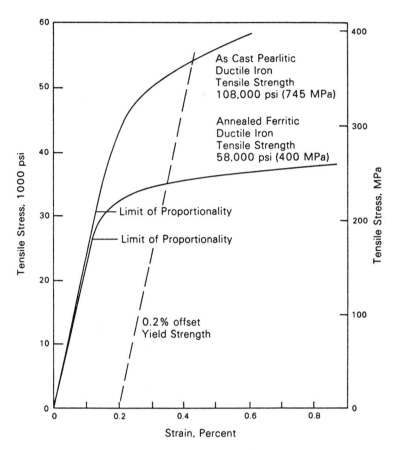

The stress-strain behavior of two types of ductile iron.

Typical stress-strain curves for ductile iron using the 0.2% offset method to establish the yield strength are shown in the graph above. Tensile test samples for ductile iron are machined from sample castings called "Y" blocks, or keel bars, because of their shape. These test castings are designed to provide a sound test bar and are made in several sizes so that they may be related in cooling rate to the controlling section of commercial castings. The test casting size, test bar dimension, and testing procedure are established in ASTM specifications.

Test bar coupons can be attached to the casting or a test bar can be cut out of the casting, but these procedures must be done properly to avoid obtaining erroneous data. Attached coupons can affect the solidification of the casting and result in unsoundness in the test bar or in the casting.

Source: Iron Castings Handbook, Charles F. Walton, Ed., Iron Castings Society, Inc., 1981, p 335

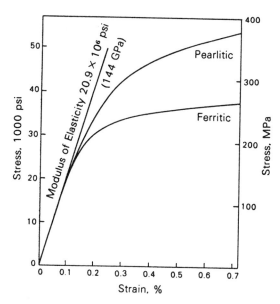

Typical stress-strain curves for ferritic and pearlitic compacted graphite irons. The pearlitic iron had a tensile strength of 59,500 psi (410 MPa) and 1.0% elongation. The ferritic iron had a tensile strength of 46,400 psi (320 MPa) and 3.5% elongation.

Compacted graphite irons provide similar tensile and yield strengths to ferritic ductile and malleable irons, although the ductility is less. The strength is equal to or greater than that of the higher-strength gray irons and with significant ductility. Typical stress-strain curves in tension for ferritic and pearlitic compacted graphite irons are shown in the graph above.

Source: Iron Castings Handbook, Charles F. Walton, Ed., Iron Castings Society, Inc., 1981, p 382

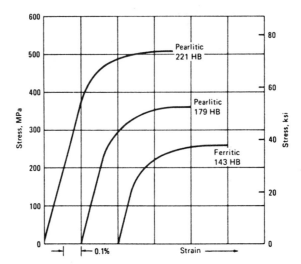

Stress-strain curves for three malleable irons.

Modulus of elasticity in tension is about 170 GPa (25 × 10^6 psi). The figure above shows typical stress-strain curves for ferritic and pearlitic malleable irons. The modulus in compression ranges from 150 to 170 GPa (22 × 10^6 to 25 × 10^6 psi); in torsion, from 65 to 75 GPa (9.5 × 10^6 to 11 × 10^6 psi).

Source: Metals Handbook, Ninth Edition, Volume 1, Properties and Selection: Irons and Steels, American Society for Metals, Metals Park OH, 1978, p 65

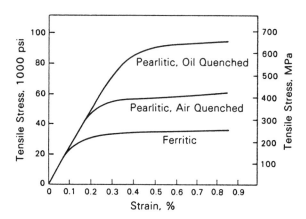

Typical stress-strain curves obtained from machined cast-to-shape test bars for three types of malleable iron.

The typical stress-strain curves for various types of malleable iron shown in the graph above were obtained from machined tensile test bars. The modulus of elasticity generally decreases with an increasing amount of graphite and less compact graphite nodule shape. The matrix structure has little influence except in the case of pearlitic irons in which the amount of graphite is reduced, thereby improving the modulus. Various moduli from 22.8 to 24.6 $\times$ 10^6 psi (157.2 to 169.6 GPa) are reported for ferritic irons and 22.4 to 25.8 $\times$ 10^6 psi (154.4 to 177.9 GPa) for various pearlitic irons. The modulus of elasticity may also be accurately measured dynamically in an Elastomat or in bending.

Source: Iron Castings Handbook, Charles F. Walton, Ed., Iron Castings Society, Inc., 1981, p 304

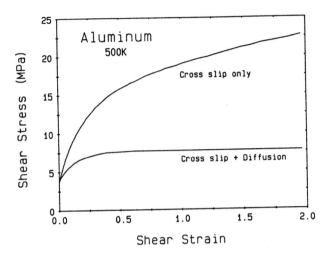

Stress-strain curves for pure Al at 500 K. The effective diffusivity was set equal to zero for one curve. Steady state is not achieved without diffusion, but the strain-hardening rate still decreases substantially even without diffusion.

The above calculations were made using known physical properties for Al including both lattice and core self-diffusion coefficients and some assumed parameters for obstacle cutting and cross slip. In an effort to determine the relative contributions of diffusion and cross slip to plastic flow, we have computed stress-strain curves for Al at 500 K, with and without the effects of diffusion. These results are shown in the figure above. We see, as expected, that diffusion plays a very important role in recovery. Without diffusion, steady-state flow is not reached in these calculations.

Source: Flow and Fracture at Elevated Temperatures, Rishi Raj, Ed., American Society for Metals, Metals Park OH, 1985, p 58

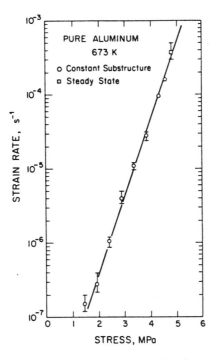

Constant-substructure-creep properties of pure Al tested at 673 K. The straight line indicates an exponential relation between the creep rate and the stress, suggesting rectangular obstacles.

As discussed below, deformation brings about a change in the structure parameter $\hat{\sigma}$, which in turn changes the rate of plastic flow. Thus, to evaluate the form of the kinetic law it is necessary to measure the strain rate–stress relation instantaneously, at a fixed structure. One way to do this is to measure the rate of creep following rapid changes in the applied stress. Assuming that the structure does not change when the stress is changed, the form of the kinetic law can be evaluated. The figure above shows the results of constant-structure-creep experiments for pure Al. In these experiments the samples were first crept to steady state under an applied stress of 4.8 MPa. Then the stress was quickly reduced to a lower value and a new reduced creep rate was measured. The results indicate that the constant-structure-creep properties of Al are described by an exponential kinetic law. A complication arises from the effects of anelasticity, which dominates the creep rate just after the stress change. It is necessary to allow the anelastic creep response to be completed before the "constant structure" creep rate can be measured.

Source: Flow and Fracture at Elevated Temperatures, Rishi Raj, Ed., American Society for Metals, Metals Park OH, 1985, p 31

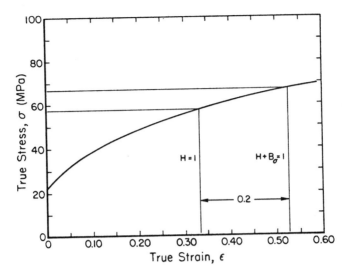

The initial part of a true stress-strain curve for aluminum at $\dot{\epsilon}$ = 5×10^{-4} s^{-1}. Note that the condition H = 1 (maximum load) is separated from the condition H + B$_\sigma$ = 1 (onset of flow localization) by a true strain of 0.2.

As in the case of the copper, agreement at large strains requires the use of the *effective* stress and *effective* stress gradient in the neck regions. Because B$_\sigma$ is no longer negligible at these homologous temperatures, the instability condition $\gamma + B_\sigma = 1$ is not satisfied until about 0.2 strain beyond the point of maximum load ($\gamma = 1$) (see figure above). Only *after* this condition is met does the rate of flow localization attain a practically significant level.

Source: S. L. Semiatin and J. J. Jonas, Formability and Workability of Metals, American Society for Metals, Metals Park OH, 1984, p 185

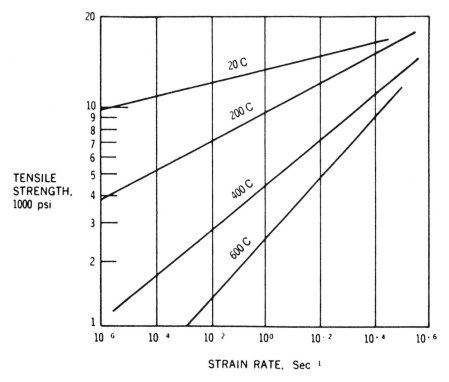

The relationship between tensile strength and strain rate for high-purity aluminum, and the dependence of tensile strength on strain rate in high-purity aluminum for a variety of cold- and hot-working temperatures.

The effect of increasing strain rate on the increase in flow stress can be mitigated to a certain extent by the effects of deformation heating, a phenomenon that occurs at both hot- and cold-working temperatures. This is a result of the fact that more than 90% of the deformation work (area under the stress-strain curve) is converted into heat and only 10% is retained in the metal (in the form of dislocations or subgrains).

Source: Forging Handbook, T. G. Byrer, S. L. Semiatin and Donald C. Vollmer, Eds., Forging Industry Association of America, Cleveland OH, 1985, p 90

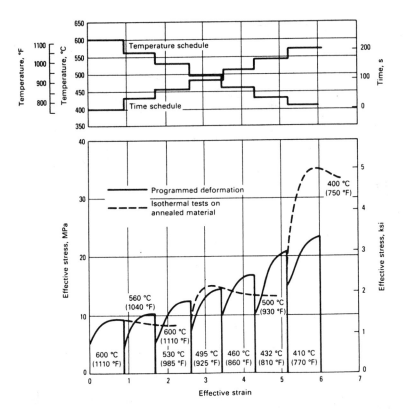

Deformation-temperature-time schedule and resulting flow behavior of super-purity aluminum deformed in torsion at an effective strain rate of 2.3 s⁻¹.

The use of the torsion test to study the effects of history during actual deformation processes is illustrated in the figure above. This figure shows the type of flow stress behavior that might be expected for high-purity aluminum during processes such as rolling, in which the temperature decreases continuously. As the temperature decreases, the flow stress increases, but not as much as would be expected based on isothermal measurements. This phenomenon is particularly evident once the temperature has dropped below 450 °C (840 °F) and is a result of the retention of a soft high-temperature substructure.

Source: Metals Handbook, Ninth Edition, Volume 8, Mechanical Testing, American Society for Metals, Metals Park OH, 1985, p 178

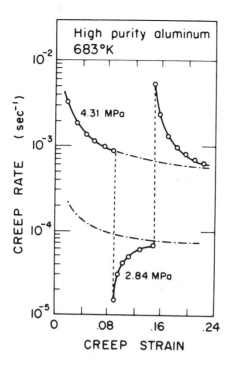

High purity aluminum 683°K

Creep transients associated with stress changes in pure Al. (Data of Horiuchi and Otsuka, replotted by Sherby, Klundt, and Miller.)

Constant Structure Creep. The distinction between constant-structure and variable-structure flow kinetics was also observed by Sherby, Trozera, and Dorn in their now-classic transient-creep experiments on Al and Al-Mg alloys. The results of an experiment of this kind conducted by Horiuchi and Otsuka and analyzed by Sherby, Klundt, and Miller are shown in the figure above. The two dot-dash curves show the monotonic creep response of pure Al at two different stresses. The data points show the creep response when the stress is changed twice during the experiment. When the stress reduction is made, the creep rate falls well below the monotonic creep rate at the reduced stress. This indicates that a "harder" structure is produced by first creeping the sample at a high stress. The difference in strain rate at the reduced stress might be regarded as an irreversible difference because the stress and temperature in this comparison are the same. Eventually the creep rate tends toward the monotonic curve. When the stress is increased back to the initial value, the observed creep rate again overshoots the monotonic value. Here we find that the structure produced by creep at the lower stress is softer than that created by monotonic testing at the higher stress. Again the difference in strain rate is an irreversible difference.

Source: Flow and Fracture at Elevated Temperatures, Rishi Raj, Ed., American Society for Metals, Metals Park OH, 1985, p 25

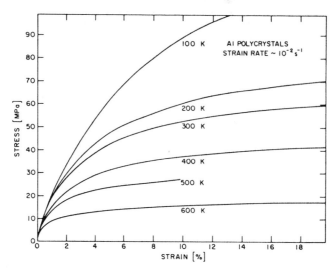

Nominal tensile stress-strain curves for aluminum (99.99%, grain size d ≃ 0.2 mm). After Kocks et al.

In the figure above is a set of stress-strain curves, as a function of temperature, that is typical at least for pure fcc materials, both as polycrystals and as single crystals in multiple-slip orientations. A noteworthy feature of these curves is that they coincide for small strains. This observation, together with more detailed considerations, led Mecking and Kocks to identify an initial "athermal hardening" component Θ_h of the strain-hardening rate $\Theta = d\sigma/d\epsilon$ (much as a "stage II" of hardening had been identified previously in single crystals). The remaining component is then defined as the rate of dynamic recovery, Θ_r:

$$\Theta = \Theta_h - \Theta_r(\sigma, T, \dot{\epsilon})$$

As is indicated in this equation, all the dependence on stress, temperature, and strain rate is associated with the dynamic-recovery term. (A slight stress dependence of Θ_h is not ruled out but is assumed to be negligible with respect to that of Θ_r.) In fact, the athermal hardening rate is insensitive even to material; in tension it is on the order of 1/50 of Young's modulus.

Source: Deformation, Processing, and Structure, George Krauss, Ed., papers presented at the ASM Materials Science Seminar, 23 October 1982, St. Louis MO, sponsored by the Seminar Committee of the Materials Science Division of the American Society for Metals, Metals Park OH, 1984, p 91

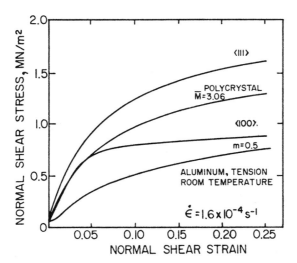

Engineering stress-strain curves for 99.99% Al for three single crystals of different orientations and polycrystal with grain size of 0.2 mm.

The figure above shows stress-strain curves for aluminum mono- and polycrystals (0.2 mm grain size, giving 15% surface grains). Three single crystals were tested. The lowest curve corresponds to a tensile axis in the central region of the stereographic triangle. This curve is characteristic of all orientations inside the triangle. For the specific orientation, $M = 0.5$ and one can clearly see an easy glide plateau, corresponding to slip in the primary system. The two other single crystals are oriented in such a way that the tensile axis coincides with $\langle 100 \rangle$ and $\langle 111 \rangle$.

Source: Marc André Meyers and Krishan Kumar Chawla, Mechanical Metallurgy: Principles and Applications, Prentice-Hall, Inc., Englewood Cliffs NJ, 1984, p 341

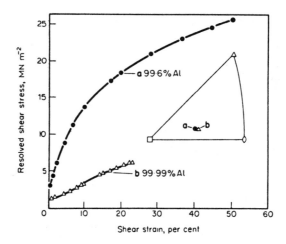

Influence of purity on resolved shear stress–shear strain curves of single crystals. a, 99.6% Al; b, 99.99% Al.

Impurities can, however, influence in a pronounced way the extent of stage 1, but the mode of dispersion of the impurity is important. In general, impurities that form a dispersion of a second phase, even when present in low concentration, tend to reduce and finally eliminate stage 1 hardening. The less-pure aluminum used in the earlier single crystal experiments did not exhibit stage 1 because the impurities present (primarily silicon and iron) formed a fine dispersion of other phases. These small inclusions encourage localized slip on other than the primary slip plane, and this eliminates stage 1 hardening. This effect is illustrated in the figure above where stress-strain curves for two crystals of different-purity aluminum, but of similar orientation, are plotted. In the curve from the purer crystal some stage 1 hardening is visible, but it is completely absent in the less pure metal which exhibits a parabolic stress-strain curve markedly different from that of the purer metal.

Source: R. W. K. Honeycombe, The Plastic Deformation of Metals, Second Edition, American Society for Metals, Metals Park OH, 1984, p 82

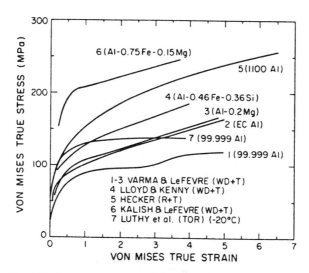

Von Mises effective stress-strain curves for aluminums of different purities and for several dilute aluminum alloys. WD + T denotes wire drawing followed by tension; R + T, rolling plus tension; and TOR, torsion.

A careful review of the large-strain literature shows the important effect of purity on solute hardening. The figure above presents flow curves for aluminums of different purities and for several dilute aluminum alloys. These curves clearly show that high-purity aluminum tends to saturate at low stress levels (note that the torsion curve of Luthy et al. is for a temperature of −20 °C). Low-purity or alloyed aluminum exhibits continued hardening to large strains.

Source: Deformation, Processing, and Structure, George Krauss, Ed., papers presented at the ASM Materials Science Seminar, 23 October 1982, St. Louis MO, sponsored by the Seminar Committee of the Materials Science Division of the American Society for Metals, Metals Park OH, 1984, p 25

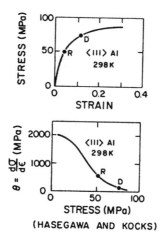

(HASEGAWA AND KOCKS)

Stress-strain relation for a ⟨111⟩ single crystal of Al tested at room temperature. The points D and R refer to hypothetical states after deformation and recovery, respectively.

The need for a second structure parameter can also be demonstrated by recovery experiments. For this discussion, we consider the stress-strain relation for pure Al shown in the figure above. As is illustrated, this relation can be expressed either in a conventional way or as the strain-hardening rate, θ, versus stress, σ. Suppose a sample is deformed to the point D on the stress-strain curve. At this point the structure might be characterized by a single parameter $\hat{\sigma} = 3\mu b \rho^{1/2}$. If the sample were then recovered by annealing, the flow strength would naturally decrease, say to point R. For the case of a single structure parameter, the structural state after recovery (at R) should be identical to that state that existed when the point R was passed during the original deformation, before recovery.

Source: Flow and Fracture at Elevated Temperatures, Rishi Raj, Ed., American Society for Metals, Metals Park OH, 1985, p 35

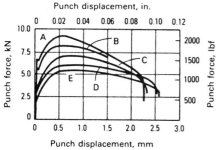

Curve	Punch speed	
	m/s	**ft/s**
A............	19.5	64
B............	12.4	41
C............	23.8×10^{-2}	7.8×10^{-1}
D............	14.7×10^{-5}	4.8×10^{-4}
E............	14.7×10^{-7}	4.8×10^{-6}

Strain-rate sensitivity of aluminum. LYS: lower yield stress.

Load-displacement curves for commercial-purity aluminum. Curve B corresponds to a test condition for which punching was incomplete.

Typical punch load/displacement curves for tests on commercial-purity aluminum specimens at mean punching speeds of 14.7×10^{-7} to 19.5 m/s (4.8×10^{-6} to 64 ft/s) are shown above at left. As anticipated, there is a marked increase in the punching load with increased speed of punching. For the aluminum tests shown in the same figure, the work done in punching—i.e., the area under the load-displacement curve—increases with punching speed. For other materials, such as a high-strength steel, the opposite effect is observed, because of a marked reduction in punch displacement at fracture that cancels out the increased punch loads.

Although principally a technological test, this technique has been used to study the strain-rate sensitivity of several materials at strain rates up to about 10^4 s^{-1}. As shown above at right, aluminum becomes strongly rate-dependent at strain rates on this order. Such behavior corroborates that reported in tests using the double-notch shear technique. Nevertheless, the validity of and the reason for the apparently greatly increased rate sensitivity exhibited by many materials at these strain rates remain unclear.

Source: Metals Handbook, Ninth Edition, Volume 8, Mechanical Testing, American Society for Metals, Metals Park OH, 1985, p 231

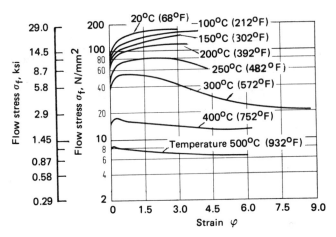

Flow stress of aluminum (technical purity) as a function of strain at different temperatures.

The flow curves in the figure above exhibit a maximum stress at about 220 °C (428 °F). Their shapes can be explained qualitatively by the "climbing" of dislocations, which takes place at elevated temperatures.

Since both recovery and recrystallization take place at a finite temperature-dependent rate, flow stress depends strongly on the strain rate. At a given temperature the effect of the strain rate on flow stress can be approximated by the relation

$$\sigma_f \simeq \sigma_{f,1} \left(\frac{\dot{\varphi}}{\dot{\varphi}_1} \right)^m$$

where $\sigma_{f,1}$ is the flow stress at the strain rate $\dot{\varphi}_1$. For steels typical values of the exponent m range from -0.02 to $+0.05$ at 20–450 °C (68–845 °F) and from 0.1 to 0.2 at temperatures above 880 °C (1616 °F).

Source: Handbook of Metal Forming, Kurt Lange, Ed , McGraw-Hill Book Co., New York, 1985, p 4.4

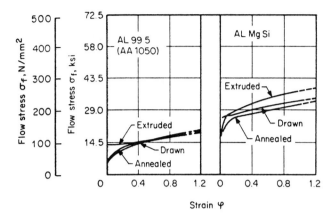

Flow curves of some aluminum alloys.

In the more general case the hardening coefficient cannot be assumed to be constant, that is,

$$n = n(\varphi).$$

Source: Handbook of Metal Forming, Kurt Lange, Ed., McGraw-Hill Book Co., New York, 1985, p 4.13

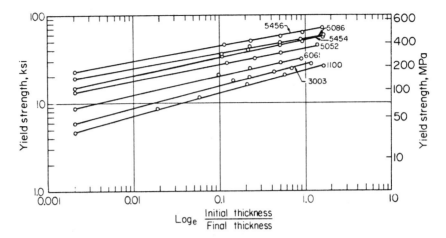

Strain-hardening curves for annealed aluminum alloys.

Non-heat treatable alloys initially in a cold worked or hot worked condition have rates of strain hardening substantially below those of material in the annealed temper. For the cold worked tempers, this difference is caused by the strain necessary to produce the temper. If this strain equals ϵ_0, then the equation for strain hardening becomes:

$$\sigma = k(\epsilon_0 + \epsilon)^n$$

A similar situation exists for products initially in the hot worked condition. The strain hardening resulting from hot working or forming is assumed to be equivalent to that achieved by a certain amount of cold work. From the tensile properties of the hot worked product, the amount of equivalent cold work can be estimated, using the work-hardening curve for the annealed temper. By such procedures, it is usually possible to calculate work-hardening curves for hot worked products that are in reasonable agreement with those for annealed products.

Source: Aluminum: Properties and Physical Metallurgy, John E. Hatch, Ed., American Society for Metals, Metals Park OH, 1984, p 111

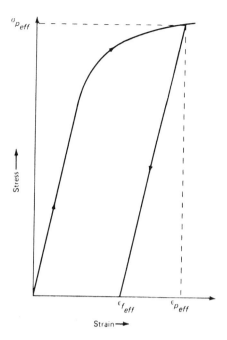

Stress-strain forming path.

Because the elastic modulus of aluminum is lower than that of steel, formed aluminum panels exhibit more elastic recovery, or springback, than formed steel panels. This must be compensated for by increasing overcrown in the draw die and/or incorporating locking beads in the binder, to ensure that all material has been "set" by plastic deformation. When the metal is plastically deformed, there will always be a component of elastic recovery. The amount of recovery depends on total plastic strain, as shown in the figure above. Of the total strain induced by tooling during forming, $\epsilon_{p_{eff}}$, there is elastic recovery back to $\epsilon_{f_{eff}}$ after forming.

Source: Metals Handbook, Ninth Edition, Volume 2, Properties and Selection: Nonferrous Alloys and Pure Metals, American Society for Metals, Metals Park OH, 1979, p 184

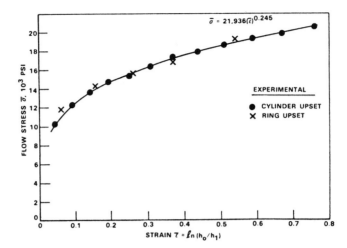

Flow stress-strain ($\bar{\sigma}$-$\bar{\epsilon}$) curve for annealed 1100 aluminum obtained from uniform cylinder and ring-upset tests.

At room temperature the flow stresses of most metals (except that of lead, for example) are only slightly strain-rate dependent. Therefore, any testing machine or press can be used for the compression test, regardless of its ram speed. Adequate lubrication of the platens is usually accomplished by using lubricants such as Teflon, molybdenum disulfide, or high-viscosity oil, and by machining grooves on both flat faces of the compression specimen, to hold the lubricant. A typical load-displacement curve obtained in uniform compression of an aluminum alloy (Al 1100, annealed) at room temperature in a testing machine is shown in the curve above.

Source: Metal Forming: Fundamentals and Applications, Taylan Altan, Soo-Ik Oh and Harold Gegel, Eds., American Society for Metals, Metals Park OH, 1983, p 51

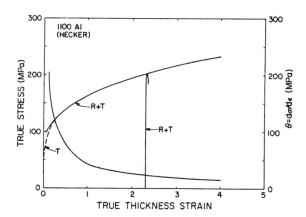

Stress-strain curve and hardening rate (θ) for 1100 aluminum. Dashed curve labeled T represents uniaxial tension on annealed material. The curve at a true thickness strain of 2.3 represents tension following a prestrain of 2.3.

As shown, the flow stress can be increased much beyond the ultimate tensile strength by rolling prestrain. However, the intrinsic hardening rate ($d\sigma/d\epsilon$) decreases. For the prestrain of 2.3 (90% rolling reduction) chosen in the graph above, we demonstrate how the flow stress has been increased much above the level that could be supported by the intrinsic hardening rate. Hence, a tensile specimen should go unstable immediately upon yielding, and little tensile elongation is to be expected.

Source: Deformation, Processing, and Structure, George Krauss, Ed., papers presented at the ASM Materials Science Seminar, 23 October 1982, St. Louis MO, sponsored by the Seminar Committee of the Materials Science Division of the American Society for Metals, Metals Park OH, 1984, p 38

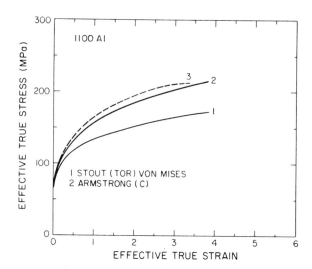

Crystallographic effective stress-strain comparison for 1100 aluminum. Curve 1 represents the von Mises torsion curve, and curve 3 the crystallographic torsion curve (based on the Taylor factors). Curve 2 represents uniaxial compression for comparison.

Source: Deformation, Processing, and Structure, George Krauss, Ed., papers presented at the ASM Materials Science Seminar, 23 October 1982, St. Louis MO, sponsored by the Seminar Committee of the Materials Science Division of the American Society for Metals, Metals Park OH, 1984, p 19

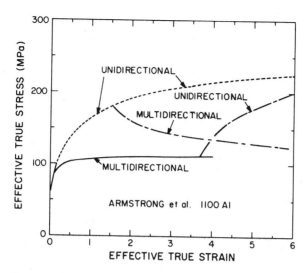

Stress-strain curves for 1100 aluminum in unidirectional (dashed line) versus multidirectional (solid line) compression. The dash-dot curves represent changes in deformation mode from unidirectional to multidirectional and vice versa.

This composite flow curve is compared with the monotonic curve in the graph above. Initial hardening behavior is identical, but the multidirectional curve soon deviates and saturates at low stress levels. This behavior is similar to saturation observed in tension-compression fatigue where there is complete stress reversal. The graph also shows that when the loading is changed from monotonic to multidirectional or vice versa, the flow curves tend toward the current mode of loading. The transition from one type of hardening to the other is very gradual.

Source: Deformation, Processing, and Structure, George Krauss, Ed., papers presented at the ASM Materials Science Seminar, 23 October 1982, St. Louis MO, sponsored by the Seminar Committee of the Materials Science Division of the American Society for Metals, Metals Park OH, 1984, p 13

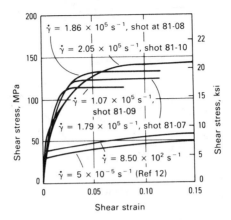

Shear strain

Dynamic stress-strain curves for 1100-O aluminum at a hydrostatic pressure of 2.8 GPa (406 ksi).

Shot No.	Impact angle, θ, degrees	Specimen thickness, h		Projectile velocity, V_o	
		mm	in.	mm/μs	in./μs
81-07 ..	26.6	0.273	0.0107	0.0582	0.00229
81-08 ..	26.6	0.276	0.0109	0.0624	0.00246
81-09 ..	14	0.212	0.0083	0.0319	0.00126
81-10 ..	26.6	0.246	0.0096	0.0623	0.00245

This relative independence of hydrostatic pressure is demonstrated by the stress-strain curves shown in the figure above for four shots conducted at the same nominal hydrostatic pressure of 2.8 GPa (406 ksi). Furthermore, systematic variation of the impact angle θ and the projectile velocity V_o in 27 tests designed to study the relative importance of strain rate and pressure on the flow stress leads to the conclusion that, in these experiments, no measurable change occurred in flow stress for changes in hydrostatic pressure up to approximately 1.0 GPa (145 ksi).

Source: Metals Handbook, Ninth Edition, Volume 8, Mechancial Testing, American Society for Metals, Metals Park OH, 1985, p 237

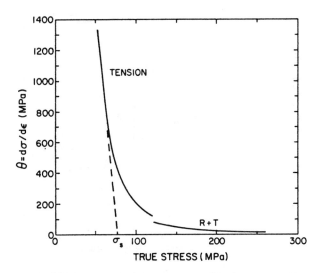

Strain-hardening rate (θ) as a function of stress.

As shown above, hardening does not saturate but persists to high stresses at low rates (θ). Apparently, some process intervenes at large strains (high stresses) and prevents a steady-state balance. This may be the result of deformation mode and texture development or of microstructural effects such as different deformation mechanisms, deformation banding, or shear banding. A clear answer does not exist at present because no systematic studies have been conducted combining measurement of hardening, texture evolution, and substructural evolution with analytical predictions.

Source: Deformation, Processing, and Structure, George Krauss, Ed., papers presented at the ASM Materials Science Seminar, 23 October 1982, St. Louis MO, sponsored by the Seminar Committee of the Materials Science Division of the American Society for Metals, Metals Park OH, 1984, p 4

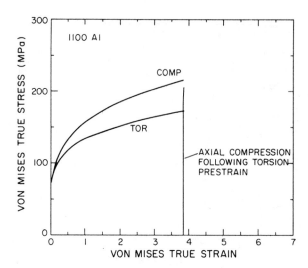

Results of path-change experiment on 1100 aluminum. The additional curve represents axial compression following torsional prestraining.

Torsional prestraining was conducted on a solid round rod. The rod was then drilled out and bored to provide a thin-wall tube 5.1 mm long, 4.8 mm in diameter, and 0.78 mm in wall thickness. The thin-wall tube was then tested in compression. The resulting stress-strain curve for a von Mises prestrain of 3.9 is shown in the graph above. Again, the flow curve for axisymmetric flow (this time in compression) was much higher than that for torsion. The specimen buckled at a stress very close to that observed for monotonic compression at the same von Mises strain level.

Source: Deformation, Processing, and Structure, George Krauss, Ed., papers presented at the ASM Materials Science Seminar, 23 October 1982, St. Louis MO, sponsored by the Seminar Committee of the Materials Science Division of the American Society for Metals, Metals Park OH, 1984, p 13

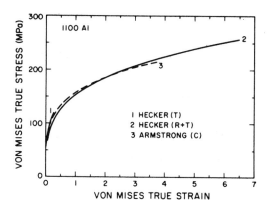

Comparison of stress-strain curves as determined by tension, rolling + tension, and compression of annealed 1100 aluminum.

Source: Deformation, Processing, and Structure, George Krauss, Ed., papers presented at the ASM Materials Science Seminar, 23 October 1982, St. Louis MO, sponsored by the Seminar Committee of the Materials Science Division of the American Society for Metals, Metals Park OH, 1984, p 3

14-25. 1100 Aluminum

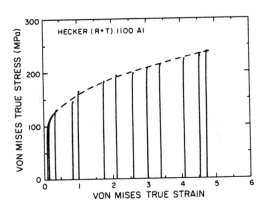

Construction of a flow curve from rolling prestrain followed by uniaxial tension (R + T). The rolling thickness reduction is converted to an effective von Mises strain.

The resulting stress-strain curve can be represented as an effective flow curve by converting the rolling prestrain to a von Mises effective strain. The validity of such a curve depends, of course, on the independence of hardening from deformation mode, which has been demonstrated not to be the case. Hence, flow curves such as the one in the figure above must be recognized as being complicated composites of two deformation modes.

Source: Deformation, Processing, and Structure, George Krauss, Ed., papers presented at the ASM Materials Science Seminar, 23 October 1982, St. Louis MO, sponsored by the Seminar Committee of the Materials Science Division of the American Society for Metals, Metals Park OH, 1984, p 10

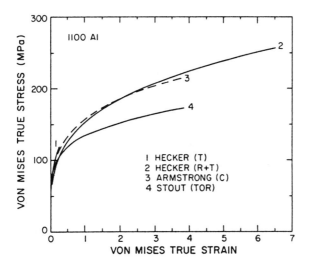

Comparison of torsion with other deformation modes in annealed 1100 aluminum. Torsion was performed on rod specimens using the method of Nadai to reduce torque-twist to stress-strain. All comparisons were made on the basis of von Mises effective stress and strain.

Source: Deformation, Processing, and Structure, George Krauss, Ed., papers presented at the ASM Materials Science Seminar, 23 October 1982, St. Louis MO, sponsored by the Seminar Committee of the Materials Science Division of the American Society for Metals, Metals Park OH, 1984, p 5

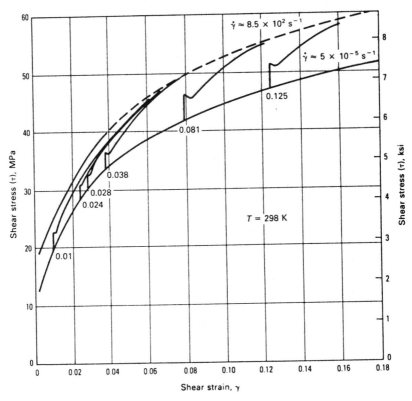

**Behavior of 1100-O aluminum under static, dynamic, and incremental strain rate
loading in shear. Strain rate changes from 5×10^{-5} to 850 s^{-1} in all incremental rate
tests.**

In incremental strain rate tests, the Kolsky bar provides an important advantage: the trans-
mitted signal furnishes a measure not of the total stress in the specimen, but of the excess
stress $\Delta\tau_s$ imposed by the stress pulse above the existing stress as a result of loading at the
quasi-static strain rate. Thus, rather than evaluation of a small difference between two large
numbers, measurement of the stress increment $\Delta\tau_s$ can be made directly from oscilloscope
records. The results of incremental tests on aluminum are given in the figure above. The
static portion of the curve is obtained from the output of the x-y recorder. The dynamic
incremental results are obtained separately from the oscillogram.

Source: Metals Handbook, Ninth Edition, Volume 8, Mechanical Testing, American Society for Metals, Metals Park
OH, 1985, p 226

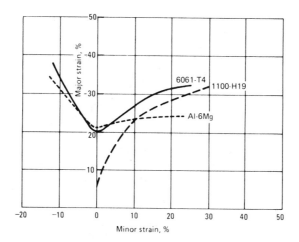

Forming-limit diagrams for aluminum 1100-H19 and alloys 6061-T4 and Al-6Mg.

The diagrams for fully hardened aluminum (1100-H19), a lightly rolled Al-6Mg alloy, and a heat treated Cu-Mg-Si alloy (6001-T4) are presented above.

Source: Metals Handbook, Ninth Edition, Volume 2, Properties and Selection: Nonferrous Alloys and Pure Metals, American Society for Metals, Metals Park OH, 1979, p 183

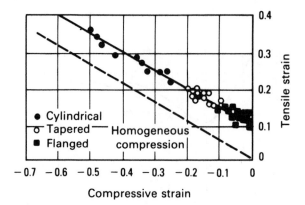

Fracture loci in cylindrical, tapered, and flanged upset test specimens of aluminum alloy 2024-T351 deformed at room temperature.

The use of flanged and tapered specimens expands the range of strains that can be obtained in the upset test. As shown in the figure above, tests with tapered and flanged specimens on aluminum alloy 2024-T351 have permitted the accumulation of data close to the $\epsilon_\theta = 0$ position. In this case, the data fit a straight line of constant slope.

Source: Metals Handbook, Ninth Edition, Volume 8, Mechanical Testing, American Society for Metals, Metals Park OH, 1985, p 580

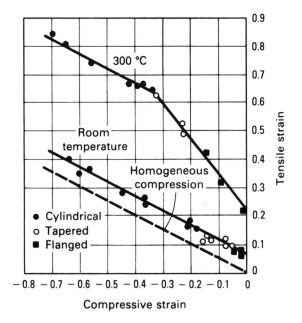

Fracture loci for aluminum alloy 2024 tested at room temperature and 300 °C (570 °F) at $\dot{\epsilon} = 0.1$ s^{-1}.

For some materials the fracture locus lines are straight and parallel with a slope of $-1/2$ (most steels, for example). For other materials (see above) the fracture lines have two slopes. In this case a slope of unity is found at small strains, using the tapered and flanged compression specimen.

Source: Metals Handbook, Ninth Edition, Volume 8, Mechanical Testing, American Society for Metals, Metals Park OH, 1985, p 583

14-31. Aluminum Alloy 3003-O: Effects of Strain Rate and Temperature

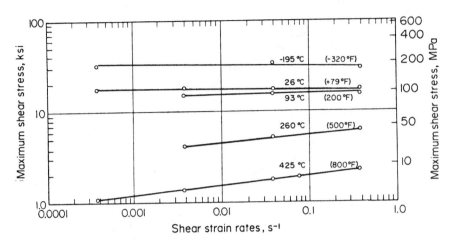

Effects of strain rate and temperature on the maximum shear stress for 3003-O.

Studies have shown that the yield strength of aluminum alloys increases as strain rates increase. These effects are not great at room temperature; rate changes of several orders of magnitude are required to produce an appreciable increase in yield strength. At elevated temperatures, the relative increase in yield strength is much larger. This is illustrated in the figure above where the maximum shear stress in torsion is plotted against strain rate for the aluminum-manganese alloy 3003 at various temperatures. Although the yield strength of aluminum increases as strain rates increase, this should not be interpreted as a substantial, or necessarily permanent, increase in strain hardening. Except for shock loading, the effects are small and may be offset by recovery phenomena resulting from the heat generated by rapid plastic deformation.

Source: Aluminum: Properties and Physical Metallurgy, John E. Hatch, Ed., American Society for Metals, Metals Park OH, 1984, p 112

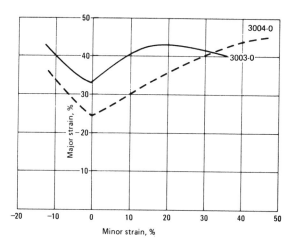

Forming-limit diagrams for two annealed Al-Mn alloys, 3003-O and 3004-O.

Source: Metals Handbook, Ninth Edition, Volume 2, Properties and Selection: Nonferrous Alloys and Pure Metals, American Society for Metals, Metals Park OH, 1979, p 182

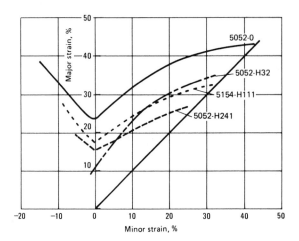

Forming-limit diagrams for four Al-Mg alloys (5052-O, 5154-H111, 5052-H241, and 5052-H32) with different levels of work hardening.

Source: Metals Handbook, Ninth Edition, Volume 2, Properties and Selection: Nonferrous Alloys and Pure Metals, American Society for Metals, Metals Park OH, 1979, p 183

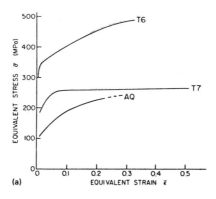

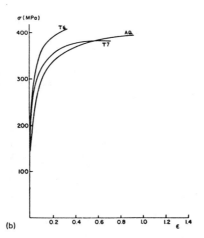

Stress-strain curves for (a) 6010 Al-Mg-Si alloy from hydraulic bulge tests (from Campbell *et al.*) and (b) Al-3.6% Cu alloy from uniaxial compression tests (from Martin). AQ, as-quenched; T6, peak-aged; T7, over-aged.

In precipitation-hardening alloys, obstacles can be classified either as "hard" obstacles or as "soft" obstacles. The "hard" obstacles, by virtue of their size, structure, and different elastic moduli, remain plastically undeformed until large strains. Both Al-Mg-Si alloys and Al-Cu alloys in over-aged condition are examples of alloys containing "hard" obstacles. This results in high initial rates of work hardening in over-aged alloys. This can be seen by comparing the stress-strain curves for the Al-Mg-Si and Al-Cu alloys in the over-aged condition with those in as-quenched and peak-aged conditions (see graphs above).

Source: Yield, Flow and Fracture of Polycrystals, T. N. Baker, Ed., Applied Science Publishers Ltd., Essex, England, 1983, p 279

14-35. Aluminum-Magnesium Alloys

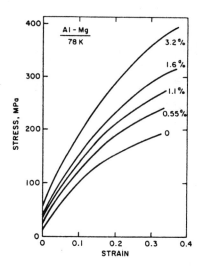

True tensile stress-strain curves for aluminum-magnesium alloys (d ≃ 0.3 mm), tested at 78 K at a strain rate of 2×10^{-3} s⁻¹.

There are some sets of stress-strain curves for solution-hardened alloys that appear to diverge monotonically, right from the beginning. The graph above shows such a series for Al-Mg alloys at 78 K; the behavior is qualitatively similar at temperatures up to about 500 K. Alloys of Cu-Zn, Cu-Al and Al-Mg-Mn exhibit similar behavior, as do Fe-C alloys, at least in some observations, and single crystals of Nb-W and Nb-Mo. In some cases, dynamic strain-aging is observed in the same regime, but in many it is not; at least one may say that the phenomenon occurs over a much wider range of temperature than jerky flow (as shown above).

Source: Deformation, Processing, and Structure, George Krauss, Ed., papers presented at the ASM Materials Science Seminar, 23 October 1982, St. Louis MO, sponsored by the Seminar Committee of the Materials Science Division of the American Society for Metals, Metals Park OH, 1984, p 100

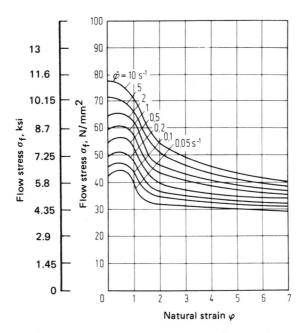

Flow-stress curves for AlMgSi1 (AA 6082) at 450 °C (842 °F).

The figure above shows, for example, the stress-strain curves of AlMgSi1 (AA 6082) at 450 °C (842 °F) in which the flow stress σ_f at $\phi > 1$ falls back to a more or less constant value. Stress-strain curves at other temperatures show a similar behavior.

Source: Handbook of Metal Forming, Kurt Lange, Ed., McGraw-Hill Book Co., New York, 1985, p 16.11

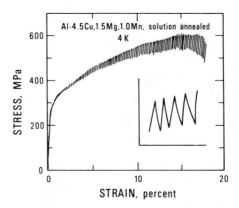

Serrated stress-strain curve obtained at very low temperature.

Tensile stress-strain curves having the form shown in the figure above are described as serrated, having many load drops, or exhibiting discontinuous yielding. Such curves are a low-temperature phenomenon, except for some materials that undergo strain aging during testing at ambient temperatures. The discontinuous yielding process consists of an initiation event (which is not visible in the stress-strain curve), a sudden drop in the load accompanied by sudden specimen extension, an elastic increase in the load, possibly some plastic deformation, another initiation event, and another sudden drop in the load. Adiabatic deformation has been advanced as a probable explanation for the serrations (Basinski, 1957, 1960). A nucleating deformation, perhaps at a stress concentration within the specimen and aided by a thermal fluctuation, occurs that produces enough heat to drive the sample into an unstable condition. The heat produced by the deformation raises the temperature of the sample (adiabatic as opposed to isothermal conditions) and lowers the flow stress enough for the deformation to continue at the applied stress level, which produces more heat, continuing the process. The low specific heat and thermal conductivity of impure metals at low temperatures make such a process quite plausible, although the details of the nucleating deformation are not well understood.

Source: Materials at Low Temperatures, Richard P. Reed and Alan F. Clark, Eds., American Society for Metals, Metals Park OH, 1983, p 248

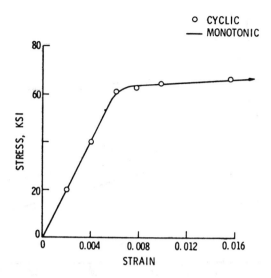

Monotonic and cyclic stress-strain curves for 7075-T73 aluminum alloy.

Monotonic and cyclic stress-strain curves for 7075-T73 aluminum alloy tested in laboratory air (20 to 50% relative humidity) are shown in the curve above. Note that the material is cyclically stable. Data points for the cyclic stress-strain curve were obtained from companion-specimen results of controlled-strain-amplitude tests performed at a constant total strain rate of $\dot\epsilon = 2.4 \times 10^{-3}$ s^{-1} ($\dot\epsilon = f \times \epsilon$ = frequency × strain amplitude). A saline environment (or, for that matter, even relative humidity) has a more pronounced effect on the long-life fatigue behavior of aluminum than on short lives.

Source: Fatigue and Microstructure, papers presented at the 1978 ASM Materials Science Seminar, 14–15 October 1978, St. Louis MO, sponsored by the Materials Science Division of the American Society for Metals, Metals Park OH, 1979, p 432

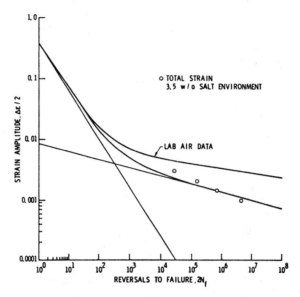

Strain versus life curves for 7075-T73 aluminum alloy tested in laboratory air and 3.5 wt% NaCl. The values of the slope, b, of the respective elastic strain-life lines are −0.15 (3.5% NaCl) and −0.11 (air).

Source: Fatigue and Microstructure, papers presented at the 1978 ASM Materials Science Seminar, 14–15 October 1978, St. Louis MO, sponsored by the Materials Science Division of the American Society for Metals, Metals Park OH, 1979, p 433

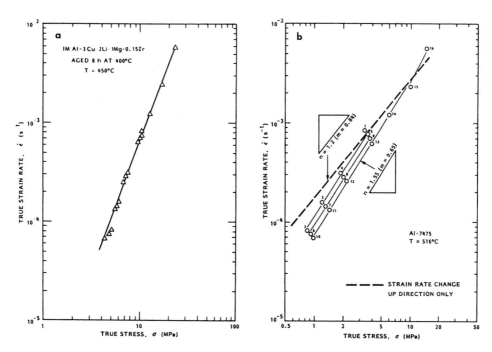

Log true strain rate versus log stress for (a) Al-3Cu-2Li-1Mg-0.15Zr alloy and (b) Al-7475 alloy.

Source: Superplastic Forming, Suphal P. Agrawal, Ed., proceedings of a symposium, 22 March 1984, sponsored by the Los Angeles Chapter of the American Society for Metals, Metals Park OH, 1985, p 47

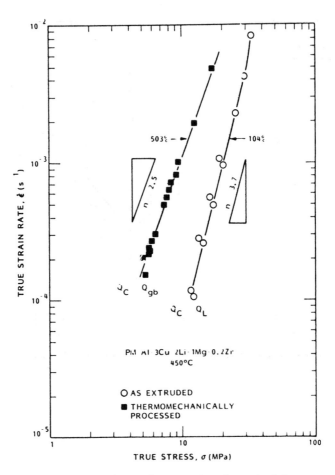

A summary of the changes in properties of as-extruded versus thermomechanically processed PM Al-3Cu-2Li-1Mg-0.2Zr alloy.

In the figure above, the transition in properties for the case of the PM Al-3Cu-2Li-1Mg-0.2Zr alloy, from the consolidated to the thermomechanically processed condition, is summarized. At a single temperature of 450 °C, the strain-rate sensitivity changes from m = 0.28 to m = 0.4, the elongations-to-failure increase from about 100 to 500%, the activation energy for plastic flow changes from lattice diffusion to grain boundary diffusion, and the strength of the alloy is significantly reduced, making forming operations easy.

Source: Superplastic Forming, Suphal P. Agrawal, Ed., proceedings of a symposium, 22 March 1984, sponsored by the Los Angeles Chapter of the American Society for Metals, Metals Park OH, 1985, p 54

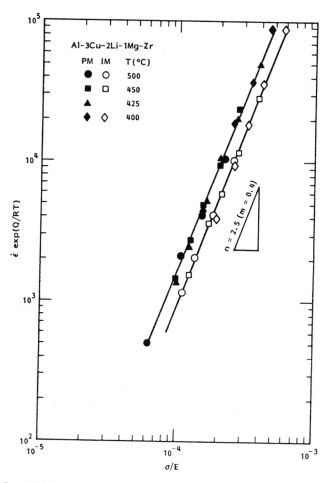

Log diffusion-compensated strain rate versus modulus-compensated stress for the PM and IM Al-3Cu-2Li-1Mg-0.2Zr alloy after thermomechanical processing.

The measured value of activation energy can be used to write a phenomenological equation for superplastic flow in the PM and IM alloys. Specifically, the data can be represented by:

$$\dot{\epsilon} = A'(\sigma/E)^n D_{gb}$$

where A' is 1.5×10^{17} for the PM alloy and 7.8×10^{16} for the IM alloy, $n = 4.0$, and $D_{gb} = D_o \exp(-Q_{gb}/RT)$ (D_o is assumed to be equal to 10^{-4} m^2 s^{-1} and Q is the measured value of 93 kJ/mole). In the figure above, the data are plotted in an $\dot{\epsilon}/D$ versus σ/E format and the lines through the data represent the equation above.

Source: Superplastic Forming, Suphal P. Agrawal, Ed., proceedings of a symposium, 22 March 1984, sponsored by the Los Angeles Chapter of the American Society for Metals, Metals Park OH, 1985, p 53

14-43. Aluminum Alloys PM Al-3Cu-2Li-1Mg-0.2Zr and PM Al-2.5Li-1.5Cu-1Mg-0.2Zr

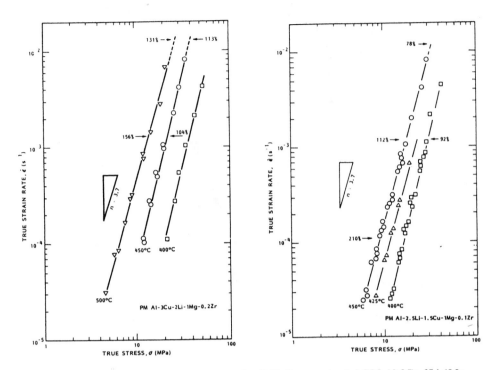

Log true strain rate versus log true stress for (left) the as-extruded PM Al-3Cu-2Li-1Mg-0.2Zr alloy and (right) the as-extruded PM Al-2.5Li-1.5Cu-1Mg-0.1Zr alloy. Selected values of elongation-to-failure at various strain rates and temperatures are shown.

Source: Superplastic Forming, Suphal P. Agrawal, Ed., proceedings of a symposium, 22 March 1984, sponsored by the Los Angeles Chapter of the American Society for Metals, Metals Park OH, 1985, p 50

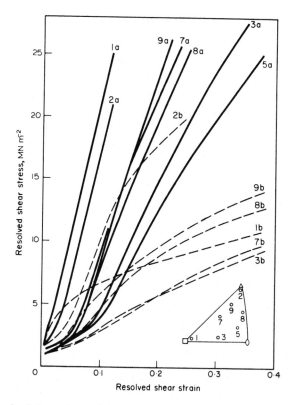

Resolved shear stress–shear strain curves of aluminum crystals at (a) 77K and (b) room temperature. (After Staubwasser, 1957.) Note that aluminum does not exhibit a well-defined stage 2 hardening at room temperature, stage 1 merging into stage 3.

Source: R. W. K. Honeycombe, The Plastic Deformation of Metals, Second Edition, American Society for Metals, Metals Park OH, 1984, p 92

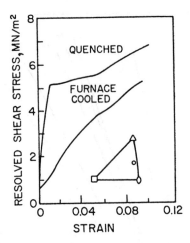

Stress-versus-strain curves for aluminum monocrystals. The crystallographic orientation is shown in the stereographic triangle.

There are also alterations in the plastic portion of the stress-versus-strain curve, seen in the graph above. The initial work-hardening rate of the quenched aluminum is lower. At greater strains, however, the two work-hardening rates become fairly similar. Hence, the effect of quenching disappears at higher strains. This is thought to be due to the elimination of the excess concentrations during plastic deformation; at the same time, excess vacancies are generated by dislocation motion, so that the concentrations in the quenched and annealed materials become the same.

Source: Marc André Meyers and Krishan Kumar Chawla, Mechanical Metallurgy: Principles and Applications, Prentice-Hall, Inc., Englewood Cliffs NJ, 1984, p 221

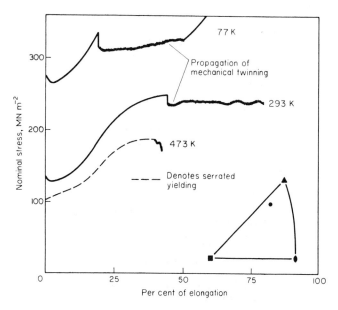

Stress-strain curve of polycrystalline aluminum at 77 K.

It is interesting to explore how far the stress-strain behavior of face-centered cubic crystals is reflected in that of polycrystalline aggregates. In single crystals, stage 1 hardening represents essentially uninterrupted slip on the primary system; this stage would not be expected to be extensive in polycrystalline aggregates because the grain-boundary restraints lead to multiple slip, early in the stress-strain curve. On the other hand, stage 2 linear hardening in single crystals seems to be closely paralleled by the linear hardening in polycrystalline stress-strain curves, and stage 3 parabolic hardening is comparable in both single and polycrystalline specimens (see above).

Source: R. W. K. Honeycombe, The Plastic Deformation of Metals, Second Edition, American Society for Metals, Metals Park OH, 1984, p 214

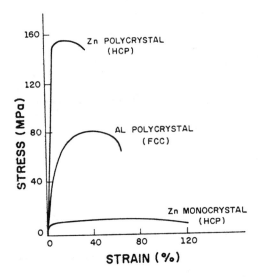

Comparison of stress-strain curves (engineering) for single and polycrystals.

The plastic deformation and the consequent work hardening results in an increase in the dislocation density. An annealed metal, for example, will have about 10^6 to 10^8 dislocations per cm^2, while a plastically cold-worked metal may contain up to 10^{12} dislocations per cm^2. The relationship between the flow stress and the dislocation density is the same as that observed for single crystals, that is,

$$\tau = \tau_0 + \alpha G b \sqrt{\rho}$$

A similar relationship exists between the flow stress, τ, and the mean grain diameter or the cell diameter. This linear relationship between τ and the square root of the grain size (D) is known as the Hall-Petch relationship. A comparison of the stress-strain curves of a few polycrystalline metals is shown in the figure above. One notes that stages 1 and 2 are absent in the polycrystalline metals.

Source: Marc André Meyers and Krishan Kumar Chawla, Mechanical Metallurgy: Principles and Applications, Prentice-Hall, Inc., Englewood Cliffs NJ, 1984, p 374

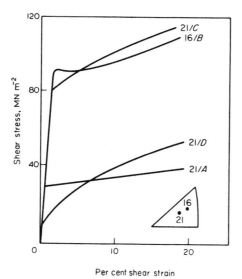

Shear stress–shear strain curves of Al-4.5 wt% Cu crystals. 21/A As air cooled; 16/B aged 2 days at 403 K; 21/C aged 27$\frac{1}{2}$ h at 463 K; 21/D over-aged at 623 K and slowly cooled (Greetham and Honeycombe).

The figure above shows a typical set of shear stress–shear strain curves at 77K for similarly oriented crystals in the four conditions. The most striking feature of these curves is the combined effect of solute and aging on the yield stress. Pure aluminum crystals have a τ_0 of about 1 MNm^{-2}, whereas the figure above shows that a supersaturated solution of 4.5 wt% copper in aluminum raises this figure to around 30 MNm^{-2}. On allowing aging to take place, τ_0 is further raised to over 80 MNm^{-2}. This illustrates very effectively the progressive role first of solute atoms, then aggregates of solute atoms, and finally precipitates in the strengthening of a relatively weak metal.

Source: R. W. K. Honeycombe, The Plastic Deformation of Metals, Second Edition, American Society for Metals, Metals Park OH, 1984, p 188

14-49. Aluminum Crystals: Effect of Quenching

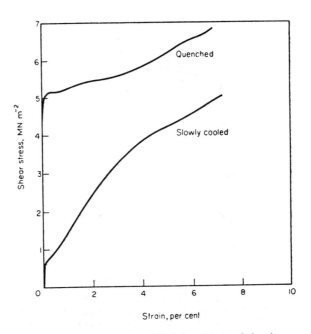

Effect of quenching on the critical shear stress of aluminum crystals. (After Maddin and Cottrell, 1955.)

Source: R. W. K. Honeycombe, The Plastic Deformation of Metals, Second Edition, American Society for Metals, Metals Park OH, 1984, p 269

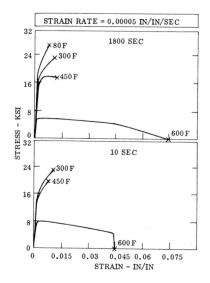

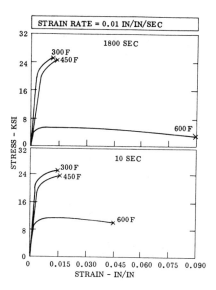

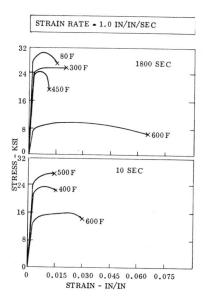

Effect of strain rate and temperature on stress-strain curves for 356-T6 cast aluminum alloy.

Source: Structural Alloys Handbook, Volume 2, Daniel J. Maykuth, Ed., Mechanical Properties Data Center, Battelle Columbus Laboratories, Columbus OH, 1980, p 71

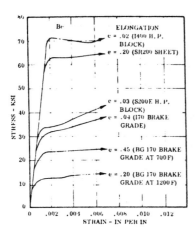

Tension stress-strain curves for various grades of beryllium.

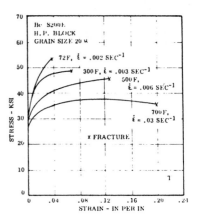

Tension stress-strain curves at several temperatures and strain rates for S200 E block.

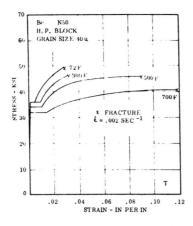

Tension stress-strain curves for N-50 block at $\dot{\epsilon} = 0.002$ s^{-1}.

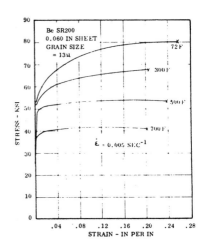

Tension stress-strain curves at several temperatures for SR200 sheet.

Source: Aerospace Structural Metals Handbook, Volume 5, Mechanical Properties Data Center, Battelle Columbus Laboratories, Columbus OH, 1978, p 12

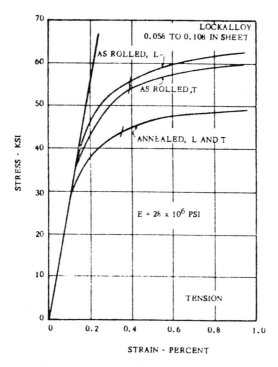

Stress-strain curves for sheet in tension.

Source: Aerospace Structural Metals Handbook, Volume 5, Mechanical Properties Data Center, Battelle Columbus Laboratories, Columbus OH, 1978, p 4

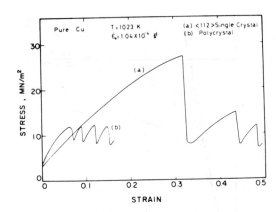

Flow curves for copper deformed in tension at 1023 K and 10^{-4} s^{-1}: (a) $\langle 112 \rangle$ single crystal; (b) polycrystalline specimen (Kikuchi). The single-crystal curve is plotted in terms of the average *normal* rather than resolved *shear* stresses and strains (Schmid factor m = 0.408).

As can be seen in the curves above, the initiation stress in a typical single crystal oriented for multiple slip can be two or more times as high (depending on orientation) as that in a polycrystal of the same material and purity, tested at the same temperature and strain rate.

Source: Deformation, Processing, and Structure, George Krauss, Ed., papers presented at the ASM Materials Science Seminar, 23 October 1982, St. Louis MO, sponsored by the Seminar Committee of the Materials Science Division of the American Society for Metals, Metals Park OH, 1984, p 219

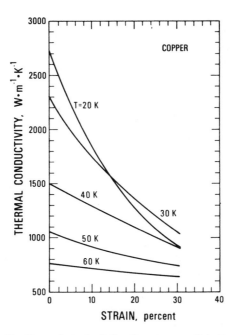

The thermal conductivity of copper as a function of tensile strain at constant temperature. (After Gladun and Holzhauser.)

The elongation limit of copper is near 50%, so considerably more thermal resistivity could have been introduced by additional strain. Thermal resistivity as a function of strain at several temperatures is plotted in the curves above. It appears from these data that as much as $0.005 \ \text{m} \cdot \text{K} \cdot \text{W}^{-1}$ could be introduced by plastic deformation at 4 K.

Source: Materials at Low Temperatures, Richard P. Reed and Alan F. Clark, Eds., American Society for Metals, Metals Park OH, 1983, p 148

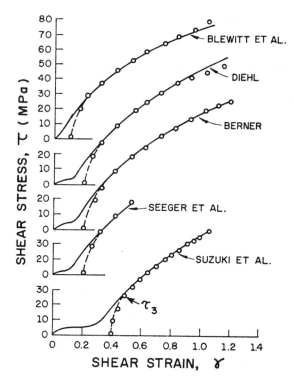

Experimental shear stress–shear strain curves for copper (from different authors as indicated) with Bell's calculated points.

In the figure above, the experimental curves are interpolated through points calculated by Bell. According to Bell, the experimental stress–strain curves follow the following relations:

$$\tau = \theta_{III}\sqrt{\gamma - \gamma_b} = \sqrt{2\theta_{II}\tau_3(\gamma - \gamma_b)}$$

with

$$\theta_{III} = \sqrt{2\theta_{II}\tau_3}$$

The symbols in these relations have familiar meanings. The only new symbol is γ_b, which is the strain at the midpoint of stage II counting it from its linear extrapolation back to the γ axis. Thus, the stage III is the parabola centering on the strain axis and joining stage II at τ_3 without discontinuity in τ or $d\tau/d\gamma$.

Source: Marc André Meyers and Krishan Kumar Chawla, Mechanical Metallurgy: Principles and Applications, Prentice-Hall, Inc., Englewood Cliffs NJ, 1984, p 367

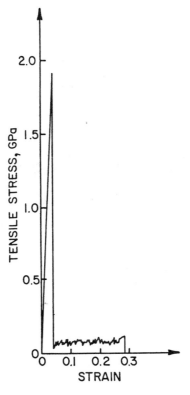

Stress–strain curve for a copper whisker with a fiber direction ⟨100⟩. The whisker diameter is 6.8 μm.

Once the surface sources are activated, a large enough number of dislocations is generated so that they can interact and multiply by other mechanisms; hence, the crystal loses its "whisker" characteristic (above).

There are a variety of whisker fabrication techniques: stress-induced growth, growth by vapor deposition, reaction with a gaseous phase.

Source: Marc André Meyers and Krishan Kumar Chawla, Mechanical Metallurgy: Principles and Applications, Prentice-Hall, Inc., Englewood Cliffs NJ, 1984, p 198

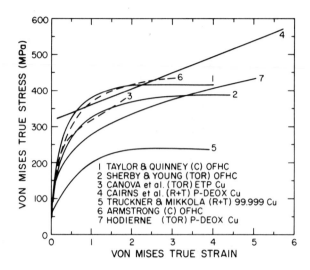

Von Mises effective stress-strain curves for coppers of different purities. C denotes compression. These curves are generally similar to those for other pure metals.

Source: Deformation, Processing, and Structure, George Krauss, Ed., papers presented at the ASM Materials Science Seminar, 23 October 1982, St. Louis MO, sponsored by the Seminar Committee of the Materials Science Division of the American Society for Metals, Metals Park OH, 1984, p 26

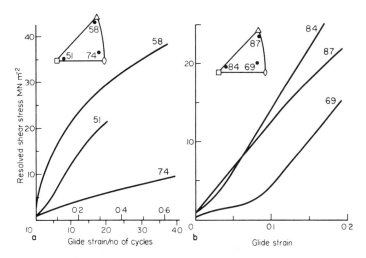

Fatigue-hardening of copper crystals. (a) Resolved shear stress plotted against total glide strain (upper scale) and number of cycles (lower scale). (b) Shear stress–shear strain curves in tensile tests.

Further evidence for hardening during fatigue comes from tensile tests on fatigued crystals. Patterson made a comparison of the hardening of copper crystals in tensile deformation, and as a result of alternate tension and compression between shear (glide) strain limits ±0.004. The hardening during fatigue was determined by measuring the maximum resolved shear stress in each cycle, which was plotted against the cumulative shear strain (above curves).

Source: R. W. K. Honeycombe, The Plastic Deformation of Metals, Second Edition, American Society for Metals, Metals Park OH, 1984, p 495

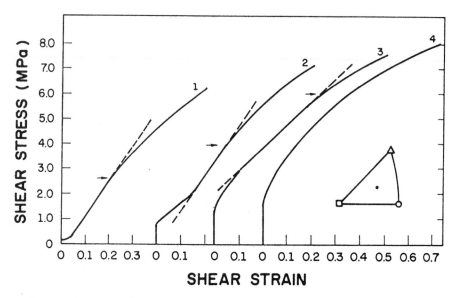

Stress-strain curves of (a) a pure Cu single crystal; (b) Cu + 1/3% SiO$_2$; (c) Cu + 2/3% SiO$_2$; (d) Cu + 1% SiO$_2$.

Basically work hardening occurs because stored groups of dislocations, ρ^G and ρ^S, impede the movement of other mobile dislocations. The stress-strain curves for pure copper single crystal and copper containing single crystal of the same orientation but with varying volume fractions of SiO$_2$ particles—that is, nondeforming particles—are shown above. In pure copper monocrystal, the work hardening is entirely due to ρ^S (figure above). Increasing the volume fraction of SiO$_2$ particles, we manage to reduce λ^G from 20 μm to 2 μm. In the last curve (No. 4) only λ^G controls the work hardening, and the stress-strain curve is parabolic.

Source: Marc André Meyers and Krishan Kumar Chawla, Mechanical Metallurgy: Principles and Applications, Prentice-Hall, Inc., Englewood Cliffs NJ, 1984, p 428

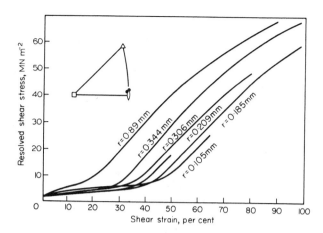

Influence of crystal radius on the extent of stage 1 hardening in copper crystals. (After Suzuki _et al._)

Source: R. W. K. Honeycombe, The Plastic Deformation of Metals, Second Edition, American Society for Metals, Metals Park OH, 1984, p 87

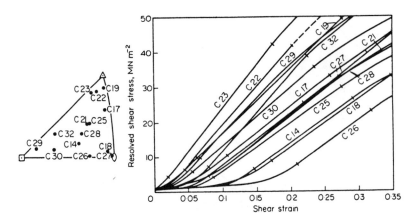

Resolved shear stress–shear strain curves of copper crystals as a function of orientation. (After Diehl.)

The figure above shows stress-strain curves from copper crystals covering a wide range of orientations. In the hardest orientation, stage 1 hardening is absent and the crystal is exhibiting primarily stage 2 hardening. On the other hand, the softest crystal gives about 15 percent shear strain in stage 1. This behavior has been confirmed for aluminum and for silver. The rate of hardening during stage 1 as a function of orientation has been particularly considered by Diehl for copper crystals. He found that θ_1 ($d\tau/d\epsilon$ in stage 1) increased in the same way as the length of stage 1 decreased, with maximum values near [$\bar{1}$11] and [001] and minimum values towards [011].

Source: R. W. K. Honeycombe, The Plastic Deformation of Metals, Second Edition, American Society for Metals, Metals Park OH, 1984, p 84

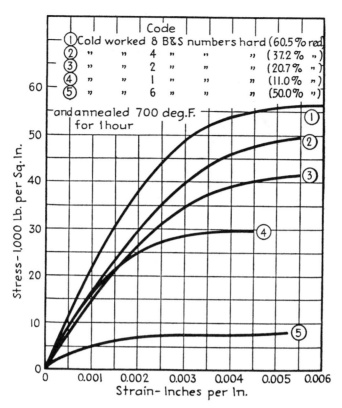

The effect of cold rolling on the stress-strain characteristics of electrolytic (tough-pitch) copper strip having a ready-to-finish grain size of 0.015 mm. (0.040-in.-thick stock); 5,000-lb. capacity hydraulic testing machine and Templin automatic extensometer accurate to 0.00001 in. were used.

Source: R. A. Wilkins and E. S. Bunn, Copper and Copper Base Alloys, McGraw-Hill Book Co., New York, 1943, p 7

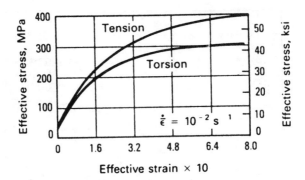

Flow curves determined at room temperature in tension and torsion on oxygen-free high-conductivity copper.

Source: Metals Handbook, Ninth Edition, Volume 8, Mechanical Testing, American Society for Metals, Metals Park OH, 1985, p 164

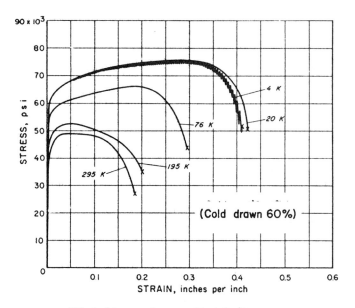

Effect of temperature on stress-strain curves.

Source: Source Book on Copper and Copper Alloys, American Society for Metals, Metals Park OH, 1979, p 34

16-13. Electrolytic Tough-Pitch and OFHC Coppers
(C10100 and C11000)

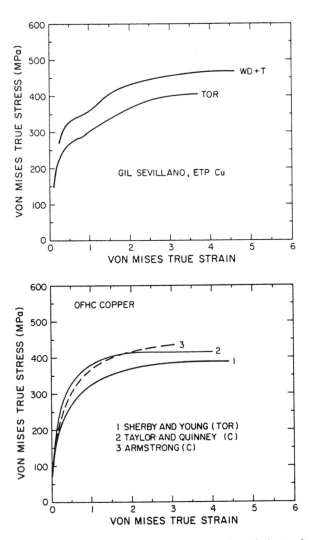

(Top) Comparison of stress-strain curves for electrolytic tough-pitch copper deformed by wire drawing plus tension and by torsion (Gil-Sevillano). (Bottom) Comparison of stress-strain curves for oxygen-free, high-conductivity copper deformed by compression (Taylor and Quinney, and unpublished work by Armstrong) and by torsion (Sherby and Young) of solid rods.

Source: Deformation, Processing, and Structure, George Krauss, Ed., papers presented at the ASM Materials Science Seminar, 23 October 1982, St. Louis MO, sponsored by the Seminar Committee of the Materials Science Division of the American Society for Metals, Metals Park OH, 1984, p 6

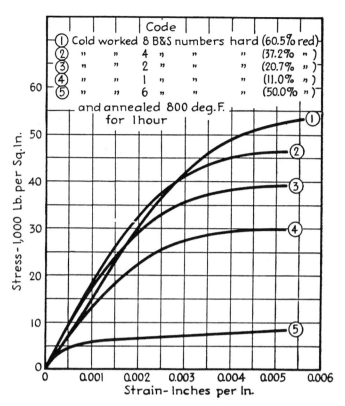

The effect of cold rolling on the stress-strain characteristics of electrolytic (tough-pitch) copper strip having a ready-to-finish grain size of 0.045 mm. (0.040-in.-thick stock); 5,000-lb. capacity hydraulic testing machine and Templin automatic extensometer accurate to 0.00001 in. were used.

Source: R. A. Wilkins and E. S. Bunn, Copper and Copper Base Alloys, McGraw-Hill Book Co., New York, 1943, p 7

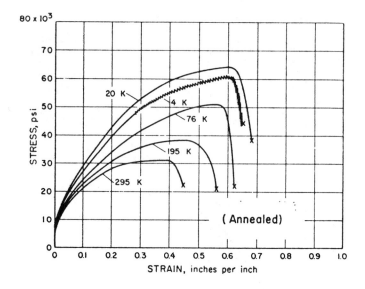

Source: Source Book on Copper and Copper Alloys, American Society for Metals, Metals Park OH, 1979, p 34

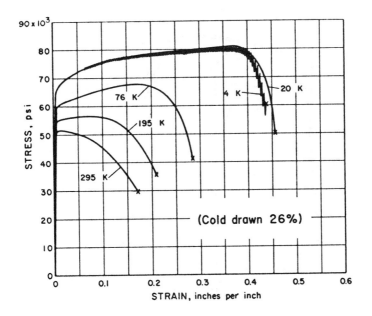

Source: Source Book on Copper and Copper Alloys, American Society for Metals, Metals Park OH, 1979, p 35

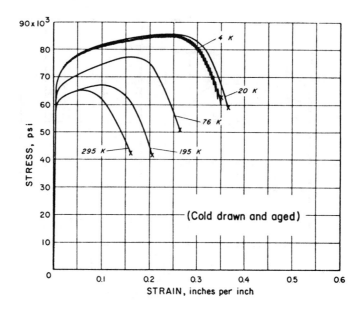

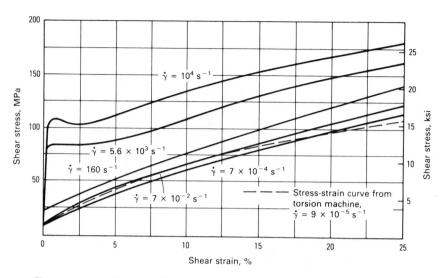

Shear stress-strain curves for copper. Torsion stress-strain results were used as a base.

Source: Metals Handbook, Ninth Edition, Volume 8, Mechanical Testing, American Society for Metals, Metals Park OH, 1985, p 231

16-19. Copper Deformed in Torsion

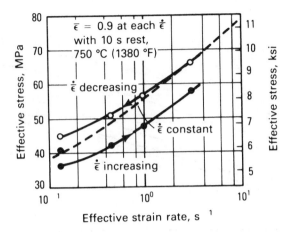

Effect of increasing or decreasing strain rate on the flow stress of copper deformed in torsion at 750 °C (1380 °F).

The effect of strain-rate history on the flow stress of copper is shown in the figure above. Under constant strain-rate conditions, harder substructures are produced at higher strain rates. However, if the strain rate is increased or decreased during torsion testing, the inertia of the acquired substructure prevents changes in flow stress as high as those observed in a series of constant strain-rate tests. Thus, strain-rate sensitivities measured in rate change tests are often lower than those based on constant strain-rate or so-called continuous flow curves in materials such as copper, aluminum, and austenitic stainless steels. In materials that exhibit dynamic strain aging (e.g., carbon steels at cold working temperatures), the relationship between the two rate sensitivity parameters may be reversed depending on the strain-rate regime and the kinetics of strain aging.

Source: Metals Handbook, Ninth Edition, Volume 8, Mechanical Testing, American Society for Metals, Metals Park OH, 1985, p 177

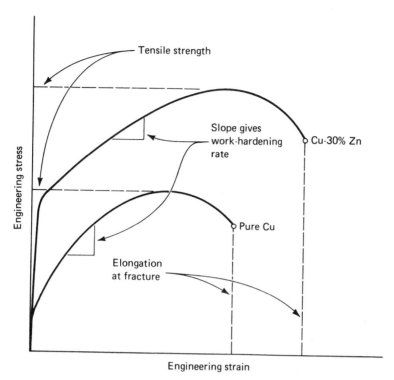

Schematic engineering stress–engineering strain curves for pure Cu and a Cu-30Zn solid solution alloy, showing that the alloy has a higher tensile strength and a lower work-hardening rate, necks at a higher strain, and has a greater fracture strain (elongation at fracture).

Source: Charlie R. Brooks, Heat Treatment, Structure and Properties of Nonferrous Alloys, American Society for Metals, Metals Park OH, 1982, p 291

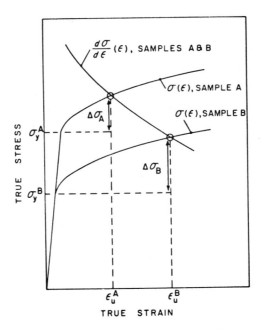

Schematic diagram showing the effect of yield strength on maximum uniform strain for equivalent strain-dependent strain-hardening-rate function.

These two stress-strain curves involve different yield strengths but equivalent strain-dependent strain-hardening-rate functions. In this case an increase in yield strength corresponds to a decrease in uniform strain due to the decrease in the magnitude of $\Delta\sigma$ required to satisfy instability. The behavior presented in the curves above closely approximates the effect of grain size on ductility in alpha brass.

Source: Deformation, Processing, and Structure, George Krauss, Ed., papers presented at the ASM Materials Science Seminar, 23 October 1982, St. Louis MO, sponsored by the Seminar Committee of the Materials Science Division of the American Society for Metals, Metals Park OH, 1984, p 65

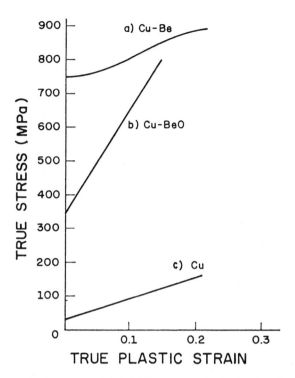

Stress-strain curves at 77 K. (a) For a precipitation hardened Cu-Be single crystal (precipitate volume fraction = 20%). These precipitates are cut by dislocations. (b) For a Cu crystal containing 2.8% of BeO that does not deform plastically. (c) For a pure Cu single crystal.

In the case of hard, incoherent particles that do not deform plastically, the initial yield stress will be controlled by interparticle spacing ($\tau \simeq Gb/x$) and the strain-hardening rate will be much greater than that shown by isolated matrix, as shown by the Cu-BeO curve in the figure above. The dislocation density increases very rapidly in this case.

We define, following Ashby, a crystalline material containing inclusions of a second phase that do not deform together with the matrix as plastically nonhomogeneous. Although the plastic deformation in the matrix may be quite large, the inclusions only deform elastically. We assume throughout that the inclusions adhere strongly to the matrix. These nondeforming inclusions change the slip distribution in the matrix. A homogeneous and monophase single crystal (free of inclusions), appropriately oriented and uniformly stressed, will deform principally on one slip system.

Source: Marc André Meyers and Krishan Kumar Chawla, Mechanical Metallurgy: Principles and Applications, Prentice-Hall, Inc., Englewood Cliffs NJ, 1984, p 414

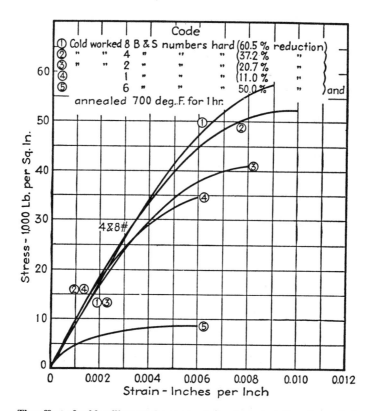

The effect of cold rolling on the stress-strain characteristics of arsenical tough-pitch copper strip (0.040-in. thick) having a ready-to-finish grain size of 0.020 mm. (99.50% copper, 0.45% arsenic); 5,000-lb. capacity hydraulic testing machine and Templin automatic extensometer accurate to 0.00001 in. were used.

Source: R. A. Wilkins and E. S. Bunn, Copper and Copper Base Alloys, McGraw-Hill Book Co., New York, 1943, p 21

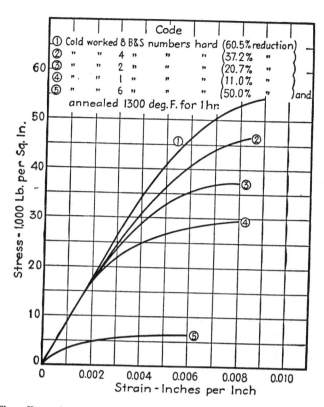

The effect of cold rolling on the stress-strain characteristics of arsenical tough-pitch copper strip (0.040 in. thick) having a ready-to-finish grain size of 0.050 mm. (99.50% copper, 0.45% arsenic); 5,000-lb. capacity hydraulic testing machine and Templin automatic extensometer accurate to 0.00001 in. were used.

Source: R. A. Wilkins and E. S. Bunn, Copper and Copper Base Alloys, McGraw-Hill Book Co., New York, 1943, p 21

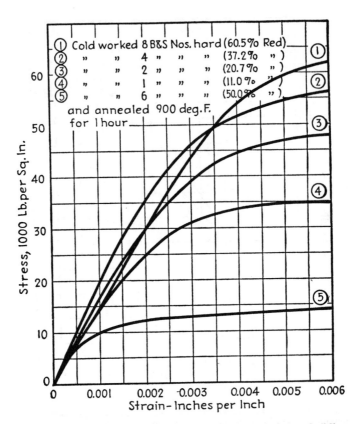

The effect of cold-working on the stress-strain characteristics of gilding-metal strip (0.040-in. thick) having a ready-to-finish grain size of 0.015 mm.; 5,000-lb. capacity hydraulic testing machine and Templin automatic extensometer accurate to 0.00001 in. were used (94.59% copper).

Source: R. A. Wilkins and E. S. Bunn, Copper and Copper Base Alloys, McGraw-Hill Book Co., New York, 1943, p 33

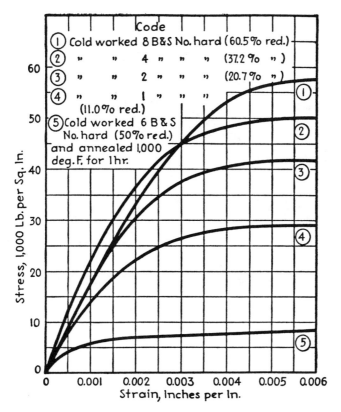

The effect of cold working on the stress-strain characteristics of gilding-metal strip (0.040 in. thick) having a ready-to-finish grain size of 0.070 mm.; 5,000-lb. capacity hydraulic testing machine and Templin automatic extensometer accurate to 0.00001 in. were used (94.59% copper).

Source: R. A. Wilkins and E. S. Bunn, Copper and Copper Base Alloys, McGraw-Hill Book Co., New York, 1943, p 33

16-27. Commercial Bronze (C22000)

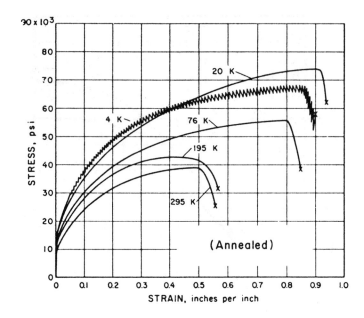

Source: Source Book on Copper and Copper Alloys, American Society for Metals, Metals Park OH, 1979, p 34

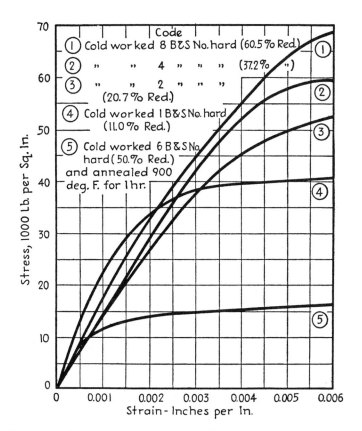

The effect of cold working on the stress-strain characteristics of commercial bronze (Government-gilding) strip (0.040 in. thick) having a ready-to-finish grain size of 0.015 mm.; 5,000-lb. capacity hydraulic testing machine and Templin automatic extensometer accurate to 0.00001 in. were used (89.74% copper).

Source: R. A. Wilkins and E. S. Bunn, Copper and Copper Base Alloys, McGraw-Hill Book Co., New York, 1943, p 37

16-29. Commercial Bronze (C22000)

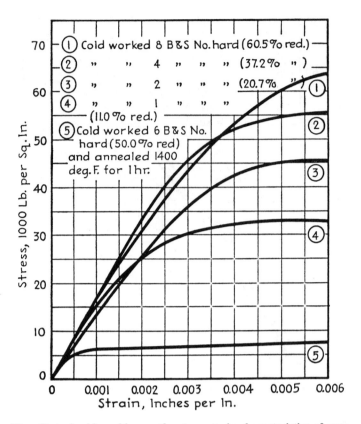

The effect of cold working on the stress-strain characteristics of commercial bronze (Government-gilding) strip (0.040 in. thick) having a ready-to-finish grain size of 0.070 mm. (89.74% copper); 5,000-lb. capacity hydraulic testing machine and Templin automatic extensometer accurate to 0.00001 in. were used.

Source: R. A. Wilkins and E. S. Bunn, Copper and Copper Base Alloys, McGraw-Hill Book Co., New York, 1943, p 38

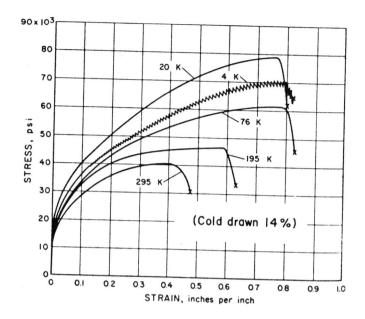

Source: Source Book on Copper and Copper Alloys, American Society for Metals, Metals Park OH, 1979, p 34

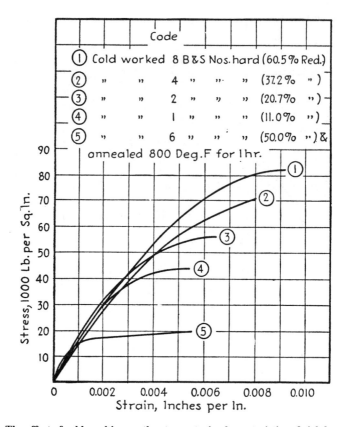

The effect of cold working on the stress-strain characteristics of rich low-brass strip (0.040 in. thick) having a ready-to-finish grain size of 0.015 mm. (85.42% copper); 5,000-lb. capacity hydraulic testing machine and Templin automatic extensometer accurate to 0.00001 in. were used.

Source: R. A. Wilkins and E. S. Bunn, Copper and Copper Base Alloys, McGraw-Hill Book Co., New York, 1943, p 44

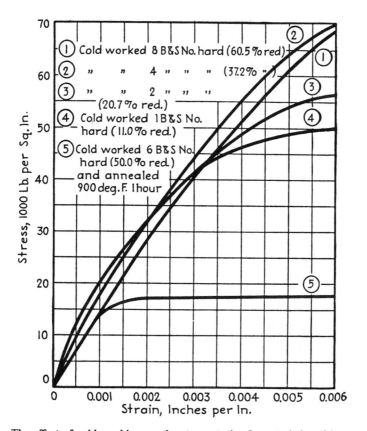

The effect of cold working on the stress-strain characteristics of low-brass (80-20) strip (0.040 in. thick) having a ready-to-finish grain size of 0.020 mm. (80.41% copper); 5,000-lb. capacity hydraulic testing machine and Templin automatic extensometer accurate to 0.00001 in. were used.

Source: R. A. Wilkins and E. S. Bunn, Copper and Copper Base Alloys, McGraw-Hill Book Co., New York, 1943, p 50

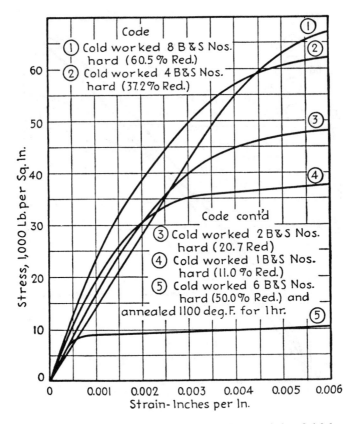

The effect of cold working on the stress-strain characteristics of rich low-brass strip (0.040 in. thick) having a ready-to-finish grain size of 0.070 mm. (85.42% copper); 5,000-lb. capacity hydraulic testing machine and Templin automatic extensometer accurate to 0.00001 in. were used.

Source: R. A. Wilkins and E. S. Bunn, Copper and Copper Base Alloys, McGraw-Hill Book Co., New York, 1943, p 44

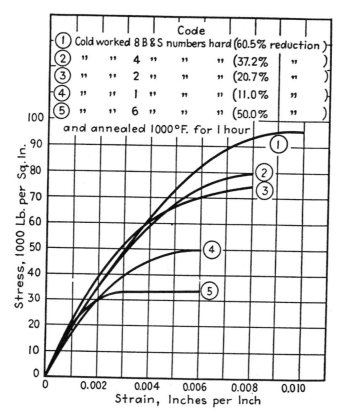

The effect of cold rolling on the stress-strain characteristics of pen-metal strip (0.040 in. thick) having a ready-to-finish grain size of 0.015 mm.; 5,000 lb. capacity hydraulic testing machine and Templin automatic extensometer accurate to 0.00001 in. were used (83.32% copper, 1.32% tin, balance zinc).

Source: R. A. Wilkins and E. S. Bunn, Copper and Copper Base Alloys, McGraw-Hill Book Co., New York, 1943, p 143

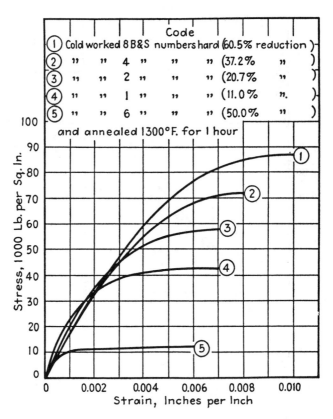

The effect of cold rolling on the stress-strain characteristics of pen-metal strip (0.040 in. thick) having a ready-to-finish grain size of 0.080 mm.; 5,000-lb. capacity hydraulic testing machine and Templin automatic extensometer accurate to 0.00001 in. were used (83.32% copper, 1.32% tin, balance zinc).

Source: R. A. Wilkins and E. S. Bunn, Copper and Copper Base Alloys, McGraw-Hill Book Co., New York, 1943, p 143

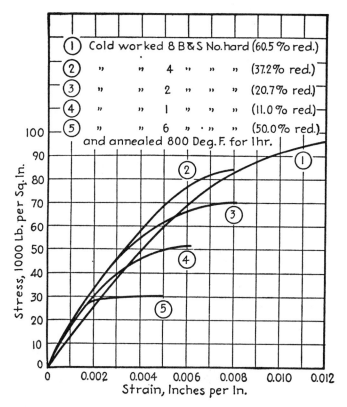

The effect of cold rolling on the stress-strain characteristics of special spring-brass strip (0.040 in. thick) having a ready-to-finish grain size of 0.015 mm. (74.69% copper); 5,000-lb. capacity hydraulic testing machine and Templin automatic extensometer accurate to 0.00001 in. were used.

Source: R. A. Wilkins and E. S. Bunn, Copper and Copper Base Alloys, McGraw-Hill Book Co., New York, 1943, p 57

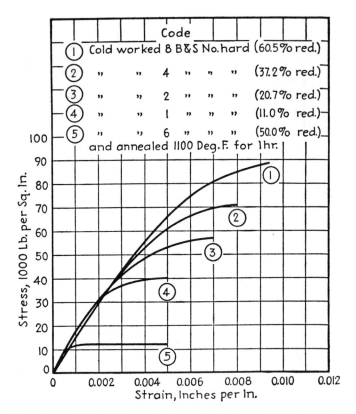

The effect of cold rolling on the stress-strain characteristics of special spring-brass strip (0.040 in. thick) having a ready-to-finish grain size of 0.095 mm. (74.69% copper); 5,000-lb. capacity hydraulic testing machine and Templin automatic extensometer accurate to 0.00001 in. were used.

Source: R. A. Wilkins and E. S. Bunn, Copper and Copper Base Alloys, McGraw-Hill Book Co., New York, 1943, p 57

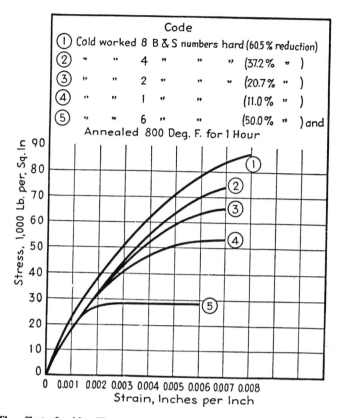

Code
① Cold worked 8 B & S numbers hard (60.5 % reduction)
② " " 4 " " " (37.2 % ")
③ " " 2 " " " (20.7 % ")
④ " " 1 " " (11.0 % ")
⑤ " " 6 " " (50.0 % ") and
Annealed 800 Deg. F. for 1 Hour

The effect of cold rolling on the stress-strain characteristics of Lancashire brass strip (0.040 in. thick) having a ready-to-finish grain size of 0.015 mm.; 5,000-lb. capacity hydraulic testing machine and Templin automatic extensometer accurate to 0.00001 in. were used (73.53% copper, 2.24% lead, balance zinc).

Source: R. A. Wilkins and E. S. Bunn, Copper and Copper Base Alloys, McGraw-Hill Book Co., New York, 1943, p 96

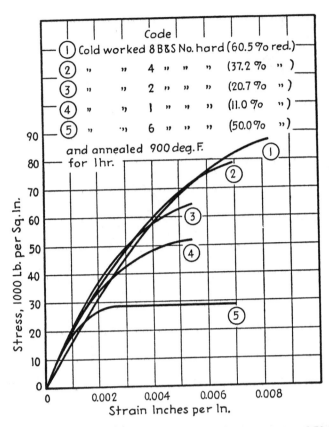

The effect of cold working on the stress-strain characteristics of 70-30 (cartridge brass) strip (0.040 in. thick) having a ready-to-finish grain size of 0.015 mm. (69.83% copper); 5,000-lb. capacity hydraulic testing machine and Templin automatic extensometer accurate to 0.00001 in. were used.

Source: R. A. Wilkins and E. S. Bunn, Copper and Copper Base Alloys, McGraw-Hill Book Co., New York, 1943, p 62

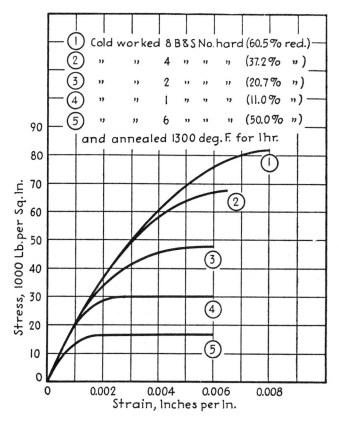

The effect of cold working on the stress-strain characteristics of 70-30 (cartridge brass) strip (0.040 in. thick) having a ready-to-finish grain size of 0.070 mm. (69.83% copper); 5,000-lb. capacity hydraulic testing machine and Templin automatic extensometer accurate to 0.00001 in. were used.

Source: R. A. Wilkins and E. S. Bunn, Copper and Copper Base Alloys, McGraw-Hill Book Co., New York, 1943, p 62

16-41. 70-30 Brass (C26000)

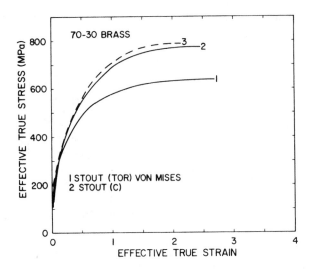

Crystallographic effective stress-strain comparison for 70-30 brass.
Curve 1 represents the von Mises torsion curve, and curve 3 the
crystallographic torsion curve (based on the Taylor factors). Curve
2 represents uniaxial compression for comparison.

Source: Deformation, Processing, and Structure, George Krauss, Ed., papers presented at the ASM Materials Science
Seminar, 23 October 1982, St. Louis MO, sponsored by the Seminar Committee of the Materials Science Division
of the American Society for Metals, Metals Park OH, 1984, p 19

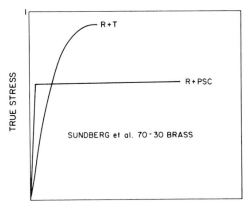

Comparison of stress-strain curves for 70-30 brass determined by rolling plus tension and by rolling plus plane-strain compression. Curves are only schematic because original reference by Sundberg *et al.* contains no units for stress or strain.

The graph above shows a most dramatic effect of a change in deformation mode. Sundberg *et al.* found that rolling plus tension in brass produced a rapidly rising flow curve, whereas rolling plus plane-strain compression resulted in immediate saturation. Unfortunately, Sundberg attached no scale to his plot; but the results are still most interesting. He noted that the plane-strain compression tests exhibited immediate shear-band formation.

Source: Deformation, Processing, and Structure, George Krauss, Ed., papers presented at the ASM Materials Science Seminar, 23 October 1982, St. Louis MO, sponsored by the Seminar Committee of the Materials Science Division of the American Society for Metals, Metals Park OH, 1984, p 11

16-43. 70-30 Brass (C26000)

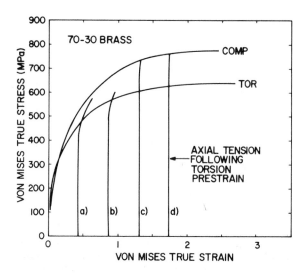

Results of path-change experiments on 70-30 brass. Curves (a) through (d) represent axial tension in thin-wall tubes following torsional prestraining to von Mises strains indicated.

A series of experiments was conducted by prestraining in torsion followed by uniaxial tension. All specimens were thin-wall tubes. Test sections were 25.4 mm long, 12.14 mm in diameter, and 0.589 mm in wall thickness. Specimens were carefully machined, annealed, and electropolished before twisting. After twisting, they were unloaded, re-electropolished and strain gaged for tension testing. The resulting tensile curves are shown in the graph above superimposed on the previous torsion and compression curves. The two curves at smaller prestrains showed little uniform elongation; most of the deformation occurred in a localized neck. Hence, these flow curves are questionable. The two curves for large prestrains definitely show that significant plastic flow in tension following torsional prestraining takes much higher stresses than does continued torsion. In fact, the flow curves are very close to that observed for compression at the same von Mises strain level.

Source: Deformation, Processing, and Structure, George Krauss, Ed., papers presented at the ASM Materials Science Seminar, 23 October 1982, St. Louis MO, sponsored by the Seminar Committee of the Materials Science Division of the American Society for Metals, Metals Park OH, 1984, p 12

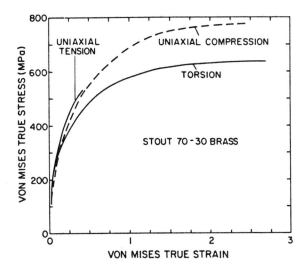

Comparison of stress-strain curves for 70-30 brass for uniaxial
tension, uniaxial compression, and torsion (Stout and Hecker).
Tension and torsion were carried out on identical thin-wall tubes.
Compression was carried out on solid rod, which was remachined
often to avoid barreling.

Source: Deformation, Processing, and Structure, George Krauss, Ed., papers presented at the ASM Materials Science
Seminar, 23 October 1982, St. Louis MO, sponsored by the Seminar Committee of the Materials Science Division
of the American Society for Metals, Metals Park OH, 1984, p 7

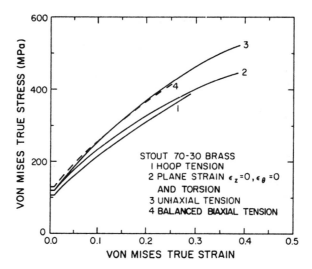

Comparison of stress-strain curves for thin-wall 70-30 brass tubes. Curve 2 represents the results for three different stress states: torsion, plane strain with no length change ($\epsilon_z = 0$), and plane strain with no diameter change ($\epsilon_\theta = 0$). Curve 1 represents uniaxial hoop tension.

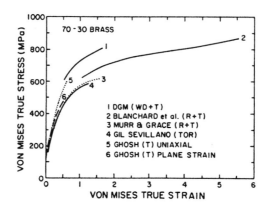

Comparison of stress-strain curves for 70-30 brass for different deformation modes.

Source: Deformation, Processing, and Structure, George Krauss, Ed., papers presented at the ASM Materials Science Seminar, 23 October 1982, St. Louis MO, sponsored by the Seminar Committee of the Materials Science Division of the American Society for Metals, Metals Park OH, 1984, p 8

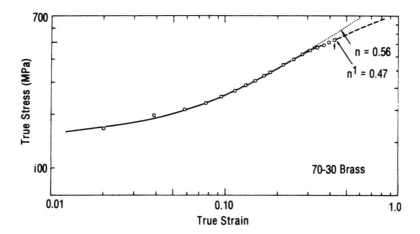

A logarithmic plot of true stress versus true strain from uniaxial tension test of brass exhibits a drop in instantaneous n from 0.56 to 0.47 at large strain.

Biaxial Hardening. In the case of brass it is surprising that the plane-strain limit is considerably less than n. While rate sensitivity is zero for brass, accurate measurement of uniaxial strain hardening shows that the instantaneous slope of log $\bar{\sigma}$ versus log $\bar{\epsilon}$ curve (i.e. instantaneous n) falls to as low as 0.47 before maximum load in tensile test. This is shown in the figure above. Since strain hardening at large strains is of interest in the prediction of necking, $n = 0.47$ may be a better approximation than the best fit value of 0.56. However, this strain dependence and the low plane-strain limit led us to question whether the strain-hardening behavior for brass was invariant with stress state.

Source: Mechanics of Sheet Metal Forming: Material Behavior and Deformation Analysis, Donald P. Koistinen and Neng-Ming Wang, Eds., proceedings of a symposium, 17–18 October 1977, Warren MI, sponsored by General Motors Research Laboratories, Plenum Press, New York, 1978, p 301

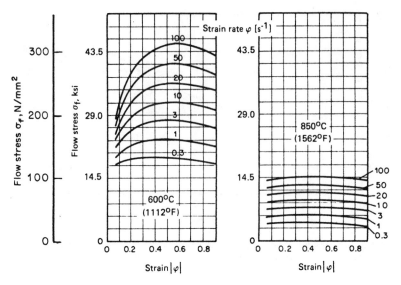

Flow curves of Cu-28Zn.

Source: Handbook of Metal Forming, Kurt Lange, Ed., McGraw-Hill Book Co., New York, 1985, p 4.13

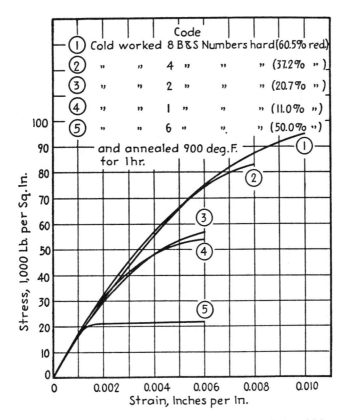

The effect of cold rolling on the stress-strain characteristics of Muntz metal strip (0.040 in. thick) having a ready-to-finish grain size of 0.015 mm. (60.50% copper); 5,000-lb. capacity hydraulic testing machine and Templin automatic extensometer accurate to 0.00001 in. were used.

Source: R. A. Wilkins and E. S. Bunn, Copper and Copper Base Alloys, McGraw-Hill Book Co., New York, 1943, p 82

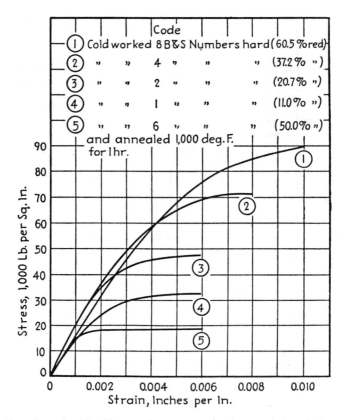

The effect of cold rolling on the stress-strain characteristics of Muntz metal strip (0.040 in. thick) having a ready-to-finish grain size of 0.045 mm. (60.50% copper); 5,000-lb. capacity hydraulic testing machine and Templin automatic extensometer accurate to 0.00001 in. were used.

Source: R. A. Wilkins and E. S. Bunn, Copper and Copper Base Alloys, McGraw-Hill Book Co., New York, 1943, p 82

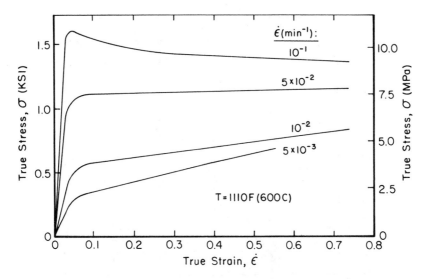

Constant strain rate flow curves for 60/40 brass determined at 600 °C (1110 °F) showing strain hardening due to grain growth at low strain rates and flow softening due to evolution of an equiaxed microstructure at the higher strain rates.

In cases where grain coarsening is taking place, the *current* value of the rate sensitivity m gradually diminishes with strain (see figure above). Thus, the rate sensitivity can decrease from the vicinity of 0.5 to values as low as 0.25.

Source: S. L. Semiatin and J. J. Jonas, Formability and Workability of Metals, American Society for Metals, Metals Park OH, 1984, p 175

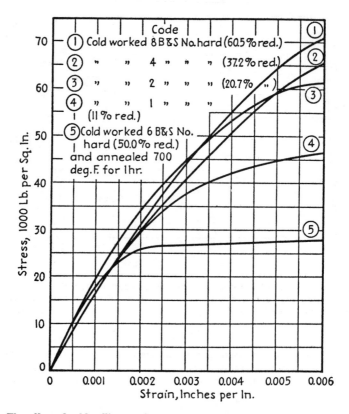

The effect of cold rolling on the stress-strain characteristics of common high-brass strip (0.040 in. thick) having a ready-to-finish grain size of 0.015 mm. (66.49% copper); 5,000-lb. capacity hydraulic testing machine and Templin automatic extensometer accurate to 0.00001 in. were used.

Source: R. A. Wilkins and E. S. Bunn, Copper and Copper Base Alloys, McGraw-Hill Book Co., New York, 1943, p 72

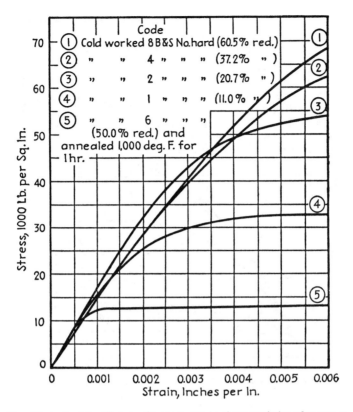

The effect of cold rolling on the stress-strain characteristics of common high-brass strip (0.040 in. thick) having a ready-to-finish grain size of 0.070 mm. (66.49% copper); 5,000-lb. capacity hydraulic testing machine and Templin automatic extensometer accurate to 0.00001 in. were used.

Source: R. A. Wilkins and E. S. Bunn, Copper and Copper Base Alloys, McGraw-Hill Book Co., New York, 1943, p 72

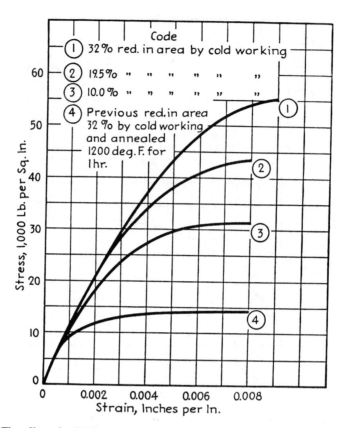

The effect of cold drawing on the stress-strain characteristics of deep-drilling rod (under 1 in. in diameter), previously extruded to a grain size of 0.050 mm. 100,000-lb. capacity hydraulic testing machine and Templin automatic extensometer accurate to 0.00001 in. were used (62.11% copper, 4.00% lead, balance zinc).

Source: R. A. Wilkins and E. S. Bunn, Copper and Copper Base Alloys, McGraw-Hill Book Co., New York, 1943, p 122

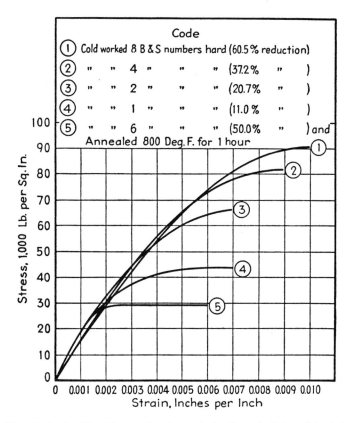

The effect of cold rolling on the stress-strain characteristics of leaded high-brass strip (65.19% copper, 1.09% lead, balance zinc) having a ready-to-finish grain size of 0.015 mm.; 5,000-lb. capacity hydraulic testing machine and Templin automatic extensometer accurate to 0.00001 in. were used (0.040-in. stock).

Source: R. A. Wilkins and E. S. Bunn, Copper and Copper Base Alloys, McGraw-Hill Book Co., New York, 1943, p 100

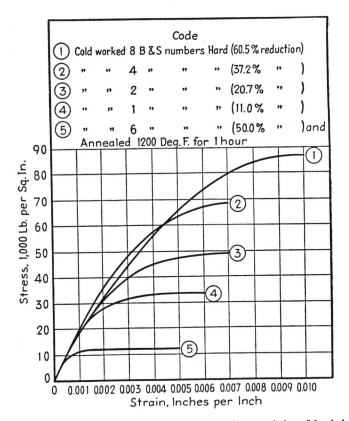

The effect of cold rolling on the stress-strain characteristics of leaded high-brass strip (65.19% copper, 1.09% lead, balance zinc) having a ready-to-finish grain size of 0.080 mm.; 5,000-lb. capacity hydraulic testing machine and Templin automatic extensometer accurate to 0.00001 in. were used (0.040-in. stock).

Source: R. A. Wilkins and E. S. Bunn, Copper and Copper Base Alloys, McGraw-Hill Book Co., New York, 1943, p 100

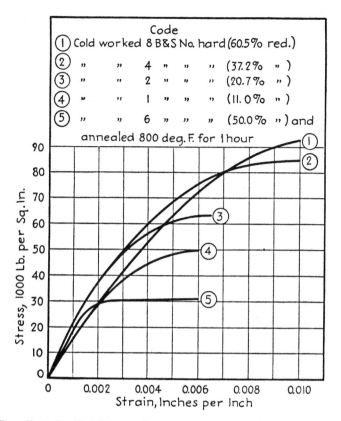

Code
1. Cold worked 8 B&S No. hard (60.5% red.)
2. " " 4 " " " (37.2% ")
3. " " 2 " " " (20.7% ")
4. " " 1 " " " (11.0% ")
5. " " 6 " " " (50.0% ") and
   annealed 800 deg. F. for 1 hour

The effect of cold rolling on the stress-strain characteristics of heavy leaded-brass strip (0.040 in. thick) (63.35% copper, 2.79% lead, balance zinc) having a ready-to-finish grain size of 0.015 mm.; 5,000-lb. capacity hydraulic testing machine and Templin automatic extensometer accurate to 0.00001 in. were used.

Source: R. A. Wilkins and E. S. Bunn, Copper and Copper Base Alloys, McGraw-Hill Book Co., New York, 1943, p 106

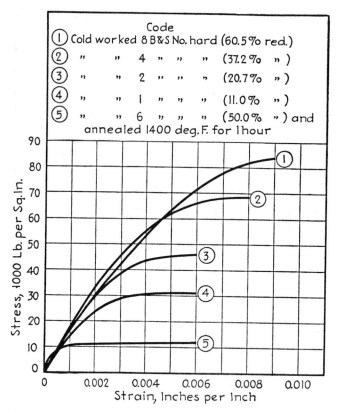

The effect of cold rolling on the stress-strain characteristics of heavy leaded-brass strip (0.040 in. thick) (63.35% copper, 2.79% lead, balance zinc) having a ready-to-finish grain size of 0.080 mm.; 5,000-lb. capacity hydraulic testing machine and Templin automatic extensometer accurate to 0.00001 in. were used.

Source: R. A. Wilkins and E. S. Bunn, Copper and Copper Base Alloys, McGraw-Hill Book Co., New York, 1943, p 106

16-58. Forging Brass (C37700)

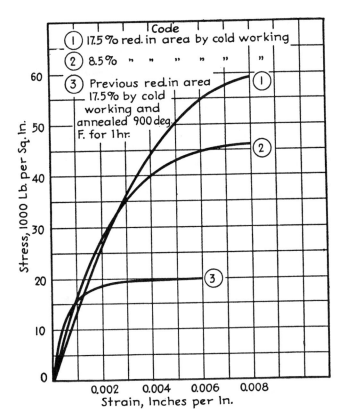

The effect of cold drawing on the stress-strain characteristics of standard-brass forging rod (under 1 in. in diameter), previously extruded to a grain size of 0.010 mm.; 100,000-lb. capacity hydraulic testing machine and Templin automatic extensometer accurate to 0.00001 in. were used (60.05% copper, 2.12% lead, balance zinc).

Source: R. A. Wilkins and E. S. Bunn, Copper and Copper Base Alloys, McGraw-Hill Book Co., New York, 1943, p 124

16-59. Admiralty Metal (C44300)

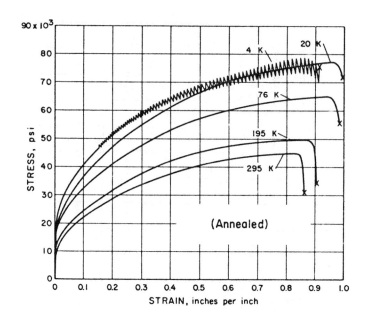

Source: Source Book on Copper and Copper Alloys, American Society for Metals, Metals Park OH, 1979, p 35

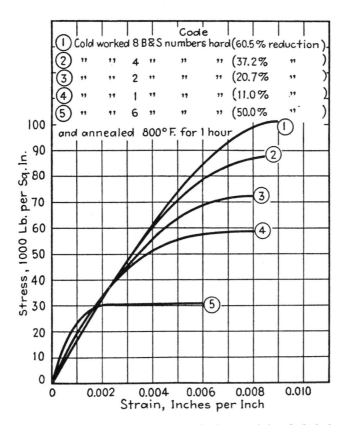

The effect of cold rolling on the stress-strain characteristics of admiralty-metal strip (0.040 in. thick) having a ready-to-finish grain size of 0.015 mm.; 5,000-lb. capacity hydraulic testing machine and Templin automatic extensometer accurate to 0.00001 in. were used (70.37% copper, 1.01% tin, balance zinc).

Source: R. A. Wilkins and E. S. Bunn, Copper and Copper Base Alloys, McGraw-Hill Book Co., New York, 1943, p 147

16-61. Admiralty Metal (C44400)

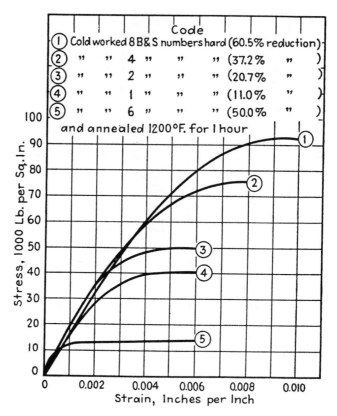

The effect of cold rolling on the stress-strain characteristics of admiralty-metal strip (0.040 in. thick) having a ready-to-finish grain size of 0.080 mm.; 5,000-lb. capacity hydraulic testing machine and Templin automatic extensometer accurate to 0.00001 in. were used (70.37% copper, 1.01% tin, balance zinc).

Source: R. A. Wilkins and E. S. Bunn, Copper and Copper Base Alloys, McGraw-Hill Book Co., New York, 1943, p 147

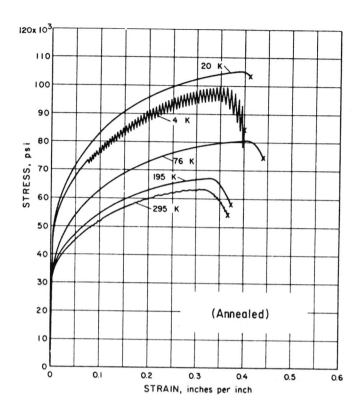

Source: Source Book on Copper and Copper Alloys, American Society for Metals, Metals Park OH 1979, p 35

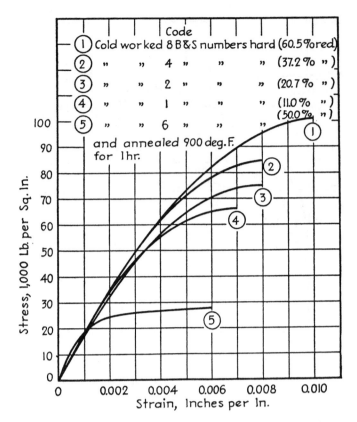

The effect of cold rolling on the stress-strain characteristics of Government naval-brass strip (0.040 in. thick) having a ready-to-finish grain size of 0.015 mm.; 5,000-lb. capacity hydraulic testing machine and Templin automatic extensometer accurate to 0.00001 in. were used (61.51% copper, 0.57% tin, balance zinc).

Source: R. A. Wilkins and E. S. Bunn, Copper and Copper Base Alloys, McGraw-Hill Book Co., New York, 1943, p 155

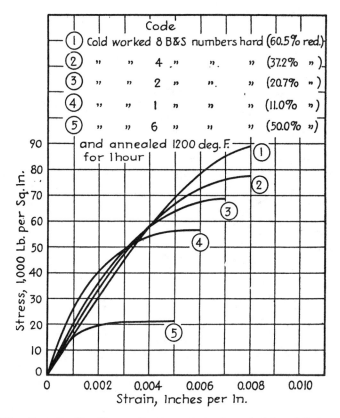

The effect of cold rolling on the stress-strain characteristics of Government naval-brass strip (0.040 in. thick) having a ready-to-finish grain size of 0.080 mm.; 5,000-lb. capacity hydraulic testing machine and Templin automatic extensometer accurate to 0.00001 in. were used (61.51% copper, 0.57% tin, balance zinc).

Source: R. A. Wilkins and E. S. Bunn, Copper and Copper Base Alloys, McGraw-Hill Book Co., New York, 1943, p 155

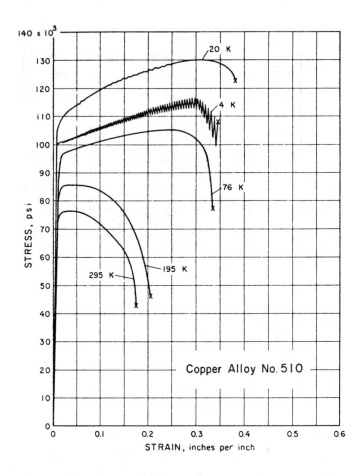

Source: Source Book on Copper and Copper Alloys, American Society for Metals, Metals Park OH, 1979, p 36

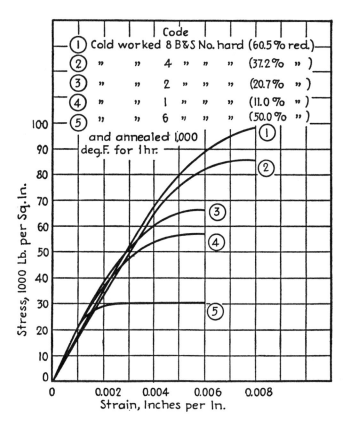

The effect of cold rolling on the stress-strain characteristics of 5 percent grade A phosphor bronze (0.040 in. thick) having a ready-to-finish grain size of 0.015 mm.; 5,000-lb. capacity hydraulic testing machine and Templin automatic extensometer accurate to 0.00001 in. were used (4.09% tin, 0.035% phosphorus, balance copper).

Source: R. A. Wilkins and E. S. Bunn, Copper and Copper Base Alloys, McGraw-Hill Book Co., New York, 1943, p 269

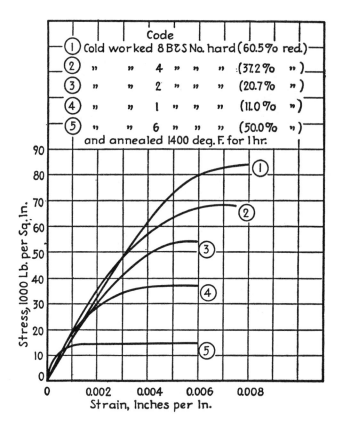

The effect of cold rolling on the stress-strain characteristics of 5 percent grade A phosphor bronze (0.040 in. thick) having a ready-to-finish grain size of 0.070 mm.; 5,000-lb. capacity hydraulic testing machine and Templin automatic extensometer accurate to 0.00001 in. were used (4.09% tin, 0.035% phosphorus, balance copper).

Source: R. A. Wilkins and E. S. Bunn, Copper and Copper Base Alloys, McGraw-Hill Book Co., New York, 1943, p 269

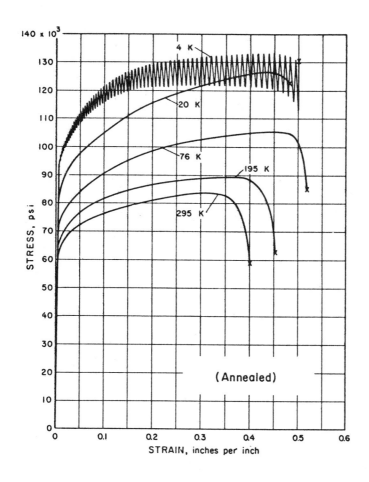

Source: Source Book on Copper and Copper Alloys, American Society for Metals, Metals Park OH, 1979, p 36

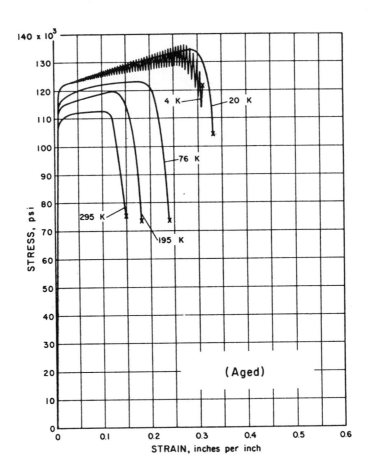

Source: Source Book on Copper and Copper Alloys, American Society for Metals, Metals Park OH, 1979, p 37

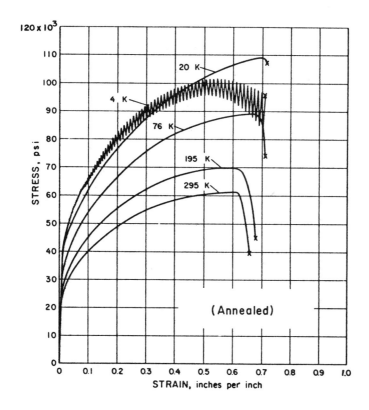

Source: Source Book on Copper and Copper Alloys, American Society for Metals, Metals Park OH, 1979, p 37

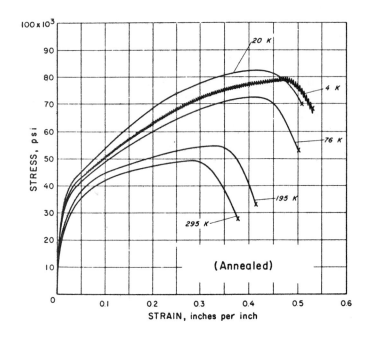

Source: Source Book on Copper and Copper Alloys, American Society for Metals, Metals Park OH, 1979, p 36

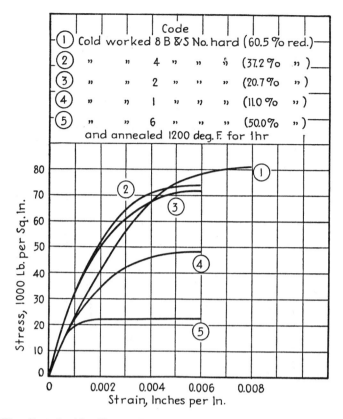

The effect of cold rolling on the stress-strain characteristics of 80-20 cupro-nickel strip (0.040 in. thick) having a ready-to-finish grain size of 0.015 mm.; 5,000-lb. capacity hydraulic testing machine and Templin automatic extensometer accurate to 0.00001 in. were used (78.18% copper, 20.65% nickel, 0.51% manganese).

Source: R. A. Wilkins and E. S. Bunn, Copper and Copper Base Alloys, McGraw-Hill Book Co., New York, 1943, p 237

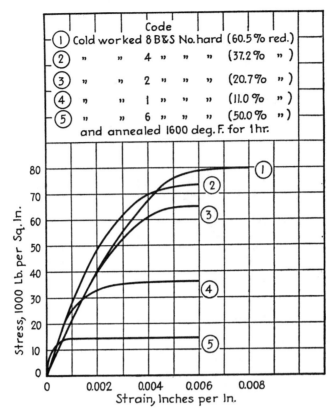

The effect of cold rolling on the stress-strain characteristics of 80-20 cu-pro-nickel strip (0.040 in. thick) having a ready-to-finish grain size of 0.055 mm.; 5,000-lb. capacity hydraulic testing machine and Templin automatic extensometer accurate to 0.00001 in. were used (79.18% copper, 20.65% nickel, 0.51% manganese).

Source: R. A. Wilkins and E. S. Bunn, Copper and Copper Base Alloys, McGraw-Hill Book Co., New York, 1943, p 237

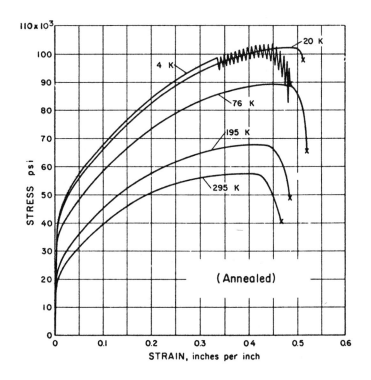

Source: Source Book on Copper and Copper Alloys, American Society for Metals, Metals Park OH, 1979, p 36

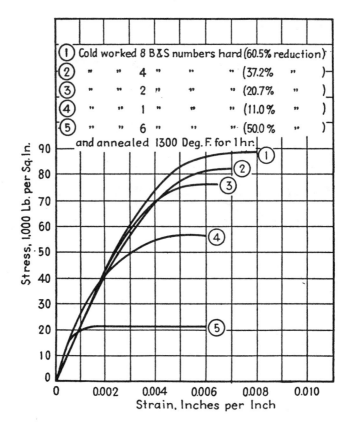

The effect of cold rolling on the stress-strain characteristics of 70-30 cu-pro-nickel strip (0.040 in. thick) having a ready-to-finish grain size of 0.015 mm.; 5,000-lb. capacity hydraulic testing machine and Templin automatic extensometer accurate to 0.00001 in. were used (68.94% copper, 29.61% nickel).

Source: R. A. Wilkins and E. S. Bunn, Copper and Copper Base Alloys, McGraw-Hill Book Co., New York, 1943, p 230

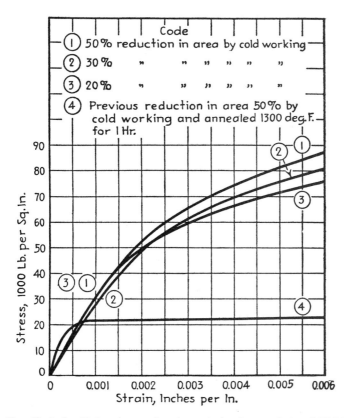

The effect of cold drawing on the stress-strain characteristics of 70-30 cupro-nickel rod (under 1 in. in diameter) having a ready-to-finish grain size of 0.035 mm.; 100,000-lb. capacity hydraulic testing machine and Templin automatic extensometer accurate to 0.00001 in. were used (68.56% copper, 30.48% nickel, 0.39% iron, 0.57% manganese).

Source: R. A. Wilkins and E. S. Bunn, Copper and Copper Base Alloys, McGraw-Hill Book Co., New York, 1943, p 233

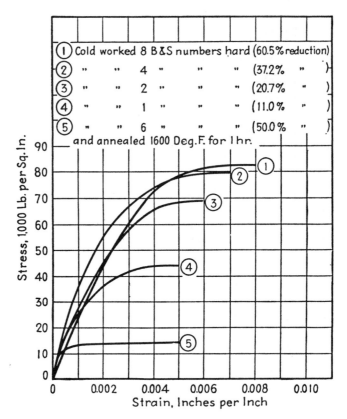

The effect of cold rolling on the stress-strain characteristics of 70-30 cu-pro-nickel strip (0.040 in. thick) having a ready-to-finish grain size of 0.070 mm.; 5,000-lb. capacity hydraulic testing machine and Templin automatic extensometer accurate to 0.00001 in. were used (68.94% copper, 29.61% nickel).

Source: R. A. Wilkins and E. S. Bunn, Copper and Copper Base Alloys, McGraw-Hill Book Co., New York, 1943, p 230

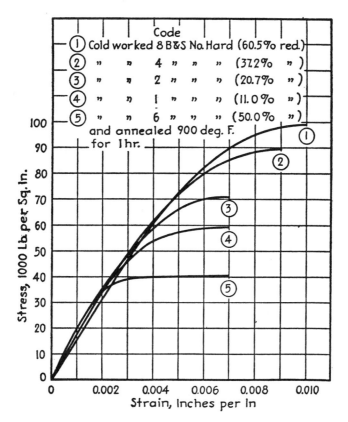

The effect of cold rolling on the stress-strain characteristics of 5% nickel-silver strip (0.040 in. thick) having a ready-to-finish grain size of 0.015 mm.; 5,000-lb. capacity hydraulic testing machine and Templin automatic extensometer accurate to 0.00001 in. were used (63.55% copper, 5.14% nickel, balance zinc).

Source: R. A. Wilkins and E. S. Bunn, Copper and Copper Base Alloys, McGraw-Hill Book Co., New York, 1943, p 220

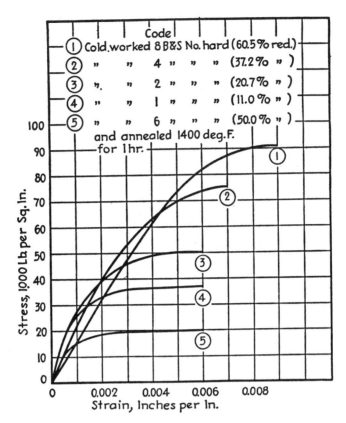

The effect of cold rolling on the stress-strain characteristics of 5% nickel-silver strip (0.040 in. thick) having a ready-to-finish grain size of 0.110 mm.; 5,000-lb. capacity hydraulic testing machine and Templin automatic extensometer accurate to 0.00001 in. were used (63.55% copper, 5.14% nickel, balance zinc).

Source: R. A. Wilkins and E. S. Bunn, Copper and Copper Base Alloys, McGraw-Hill Book Co., New York, 1943, p 220

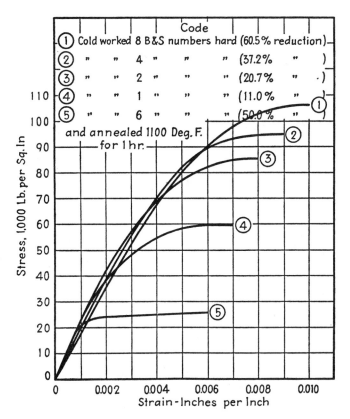

The effect of cold rolling on the stress-strain characteristics of 10% nickel-silver strip (0.040 in. thick) having a ready-to-finish grain size of 0.015 mm.; 5,000-lb. capacity hydraulic testing machine and Templin automatic extensometer accurate to 0.00001 in. were used (66.02% copper, 10.73% nickel, balance zinc).

Source: R. A. Wilkins and E. S. Bunn, Copper and Copper Base Alloys, McGraw-Hill Book Co., New York, 1943, p 215

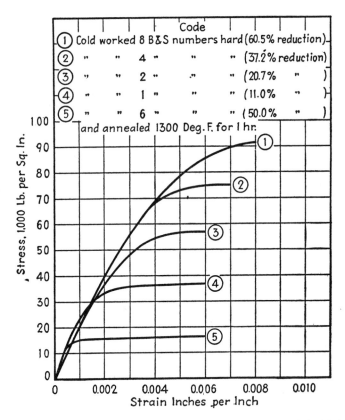

The effect of cold rolling on the stress-strain characteristics of 10% nickel-silver strip (0.040 in. thick) having a ready-to-finish grain size of 0.080 mm.; 5,000-lb. capacity hydraulic testing machine and Templin automatic extensometer accurate to 0.00001 in. were used (66.02% copper, 10.73% nickel, balance zinc).

Source: R. A. Wilkins and E. S. Bunn, Copper and Copper Base Alloys, McGraw-Hill Book Co., New York, 1943, p 215

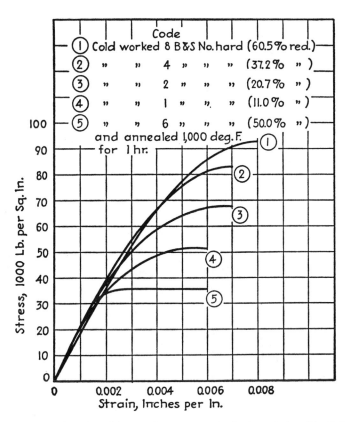

The effect of cold working on the stress-strain characteristics of leaded 12% nickel-silver strip (0.040 in. thick) having a ready-to-finish grain size of 0.015 mm.; 5,000-lb. capacity hydraulic testing machine and Templin automatic extensometer accurate to 0.00001 in. were used (65.49% copper, 12.11% nickel, 1.96% lead, balance zinc).

Source: R. A. Wilkins and E. S. Bunn, Copper and Copper Base Alloys, McGraw-Hill Book Co., New York, 1943, p 225

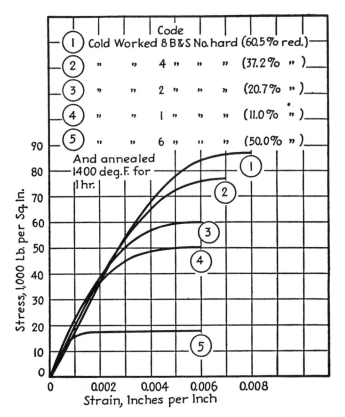

The effect of cold rolling on the stress-strain characteristics of 12% nickel-silver strip (0.040 in. thick) having a ready-to-finish grain size of 0.080 mm.; 5,000-lb. capacity hydraulic testing machine and Templin automatic extensometer accurate to 0.00001 in. were used (66.24% copper, 11.57% nickel, balance zinc).

Source: R. A. Wilkins and E. S. Bunn, Copper and Copper Base Alloys, McGraw-Hill Book Co., New York, 1943, p 212

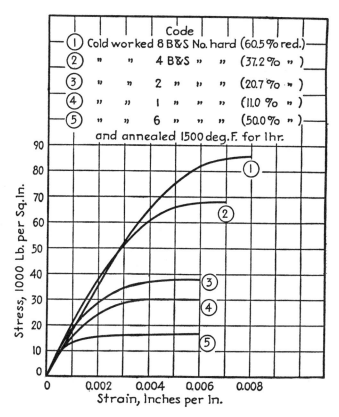

The effect of cold working on the stress-strain characteristics of leaded 12% nickel-silver strip (0.040 in. thick) having a ready-to-finish grain size of 0.060 mm.; 5,000-lb. capacity hydraulic testing machine and Templin automatic extensometer accurate to 0.00001 in. were used (65.49% copper, 12.11% nickel, 1.96% lead, balance zinc).

Source: R. A. Wilkins and E. S. Bunn, Copper and Copper Base Alloys, McGraw-Hill Book Co., New York, 1943, p 225

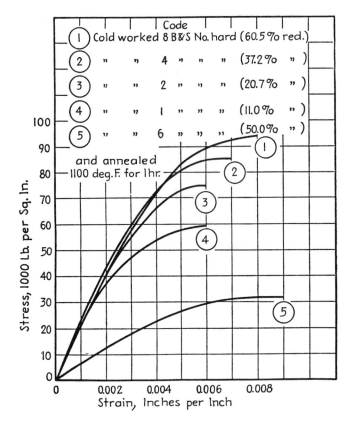

The effect of cold rolling on the stress-strain characteristics of 15% nickel-silver strip (0.040 in. thick) having a ready-to-finish grain size of 0.015 mm.; 5,000-lb. capacity hydraulic testing machine and Templin automatic extensometer accurate to 0.00001 in. were used (66.18% copper, 15.05% nickel, balance zinc).

Source: R. A. Wilkins and E. S. Bunn, Copper and Copper Base Alloys, McGraw-Hill Book Co., New York, 1943, p 208

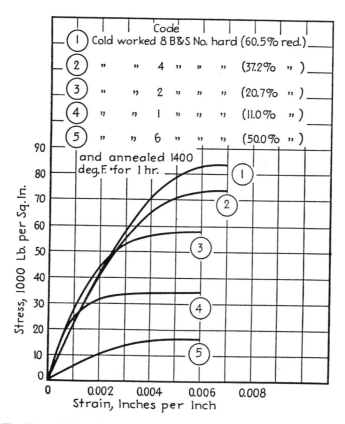

The effect of cold rolling on the stress-strain characteristics of 15% nickel-silver strip (0.040 in. thick) having a ready-to-finish grain size of 0.100 mm.; 5,000-lb. capacity hydraulic testing machine and Templin automatic extensometer accurate to 0.00001 in. were used (66.18% copper, 15.05% nickel, balance zinc).

Source: R. A. Wilkins and E. S. Bunn, Copper and Copper Base Alloys, McGraw-Hill Book Co., New York, 1943, p 208

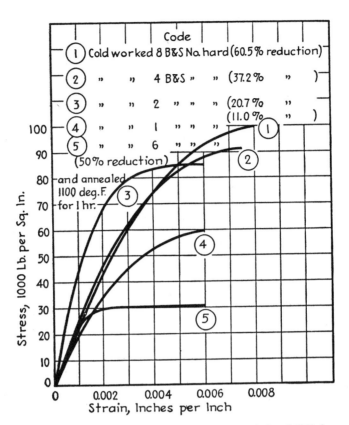

The effect of cold rolling on the stress-strain characteristics of 18% deep-drawing nickel-silver strip (0.040 in. thick) having a ready-to-finish grain size of 0.015 mm.; 5,000-lb. capacity hydraulic testing machine and Templin automatic extensometer accurate to 0.00001 in. were used (66.00% copper, 18.00% nickel, balance zinc).

Source: R. A. Wilkins and E. S. Bunn, Copper and Copper Base Alloys, McGraw-Hill Book Co., New York, 1943, p 200

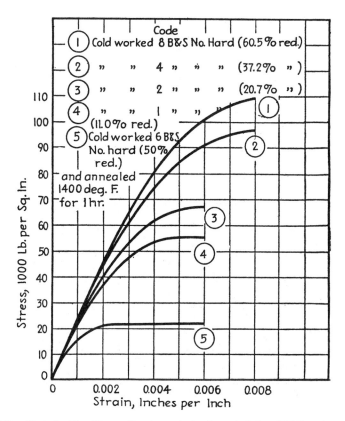

The effect of cold rolling on the stress-strain characteristics of 18% spring-stock nickel-silver strip (0.040 in. thick) having a ready-to-finish grain size of 0.080 mm.; 5,000-lb. capacity hydraulic testing machine and Templin automatic extensometer accurate to 0.00001 in. were used (56.56% copper, 17.77% nickel, balance zinc).

Source: R. A. Wilkins and E. S. Bunn, Copper and Copper Base Alloys, McGraw-Hill Book Co., New York, 1943, p 203

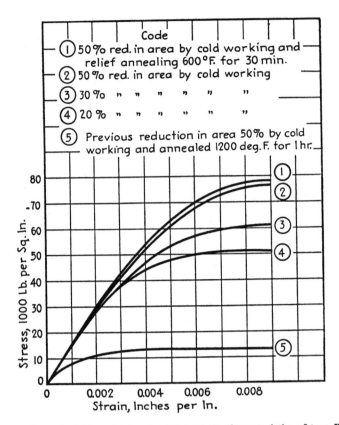

The effect of cold drawing on the stress-strain characteristics of type B silicon-bronze rod (under 1 in. in diameter) having a ready-to-finish grain size of 0.115 mm.; 100,000-lb. capacity hydraulic testing machine and Templin automatic extensometer accurate to 0.00001 in. were used (1.76% silicon, 0.35% manganese, balance copper).

Source: R. A. Wilkins and E. S. Bunn, Copper and Copper Base Alloys, McGraw-Hill Book Co., New York, 1943, p 248

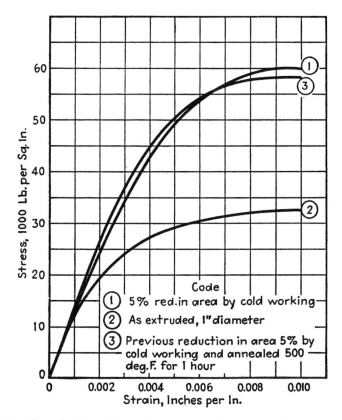

The effect of cold working and annealing on the stress-strain characteristics of a 10% aluminum-bronze rod, previously extruded (rod under 1.00 in. diameter); 100,000-lb. capacity hydraulic testing machine and Templin automatic extensometer accurate to 0.00001 in. were used (88.83% copper, 10.02% aluminum, 0.77% iron, 0.31% manganese).

Source: R. A. Wilkins and E. S. Bunn, Copper and Copper Base Alloys, McGraw-Hill Book Co., New York, 1943, p 262

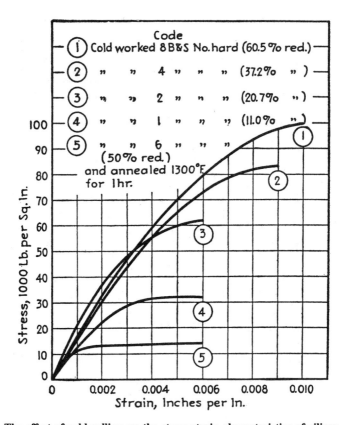

The effect of cold rolling on the stress-strain characteristics of silicon‑ brass No. 1 strip (0.040 in. thick) having a ready-to-finish grain size of 0.090 mm.; 5,000-lb. capacity hydraulic testing machine and Templin automatic extensometer accurate to 0.00001 in. were used (77.74% copper, 1.30% silicon, balance zinc).

Source: R. A. Wilkins and E. S. Bunn, Copper and Copper Base Alloys, McGraw-Hill Book Co., New York, 1943, p 181

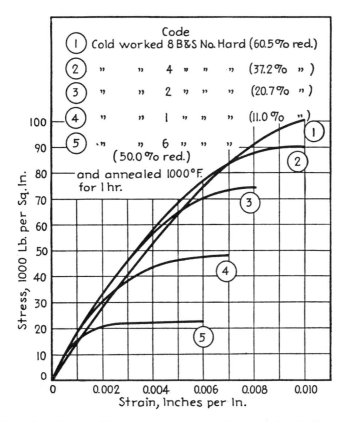

The effect of cold rolling on the stress-strain characteristics of silicon-brass No. 2 strip (0.040 in. thick) having a ready-to-finish grain size of 0.015 mm.; 5,000-lb. capacity hydraulic testing machine and Templin automatic extensometer accurate to 0.00001 in. were used (72.36% copper, 0.47% silicon, balance zinc).

Source: R. A. Wilkins and E. S. Bunn, Copper and Copper Base Alloys, McGraw-Hill Book Co., New York, 1943, p 185

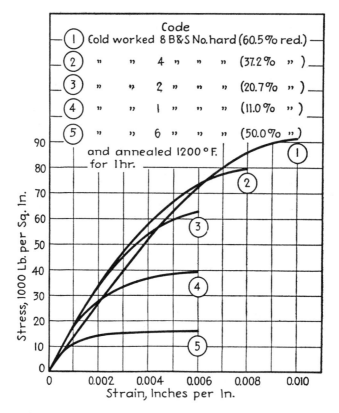

The effect of cold rolling on the stress-strain characteristics of silicon-brass No. 2 strip (0.040 in. thick) having a ready-to-finish grain size of 0.080 mm.; 5,000-lb. capacity hydraulic testing machine and Templin automatic extensometer accurate to 0.00001 in. were used (72.36% copper, 0.47% silicon, balance zinc).

Source: R. A. Wilkins and E. S. Bunn, Copper and Copper Base Alloys, McGraw-Hill Book Co., New York, 1943, p 185

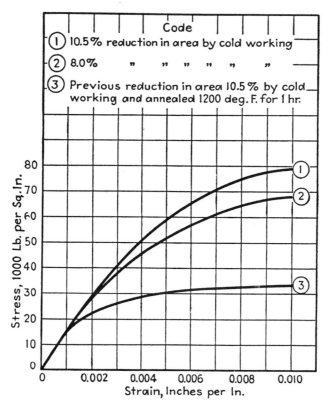

The effect of cold working and annealing on the stress-strain characteristics of silicon-aluminum-bronze rod, previously extruded (rod under 1
in. in diameter); 100,000-lb. capacity hydraulic testing machine and
Templin automatic extensometer accurate to 0.00001 in. were used (7.01%
aluminum, 1.98% silicon, balance copper).

Source: R. A. Wilkins and E. S. Bunn, Copper and Copper Base Alloys, McGraw-Hill Book Co., New York, 1943,
p 265

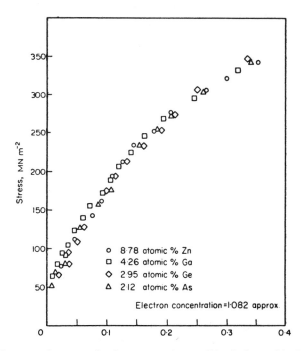

Stress-strain curves for four copper-base solid solutions with the same e/a ratio. (After Allen, Schofield and Tate.)

Source: R. W. K. Honeycombe, The Plastic Deformation of Metals, Second Edition, American Society for Metals, Metals Park OH, 1984, p 249

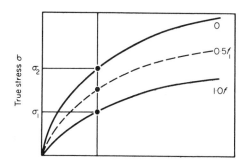

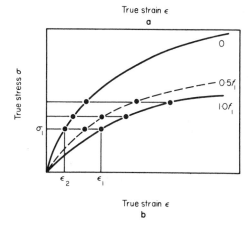

Deformation of two-phase alloys. (a) Same strain in each phase. (b) Same stress in each phase. (After Unkel.)

Ductile-Brittle Phases. It is perhaps difficult to define a brittle phase, for many phases apparently brittle can be heavily deformed in favorable conditions (for example, cementite in steels). The mode of distribution of the brittle phase is of great importance, for if it forms a continuous film around the grain boundaries disastrous brittleness occurs (for example, bismuth in copper). On the other hand, if the brittle phase spheroidizes as a result of a high interfacial energy, the alloy is not only ductile but also can have improved strength.

Source: R. W. K. Honeycombe, The Plastic Deformation of Metals, Second Edition, American Society for Metals, Metals Park OH, 1984, p 254

16-97. Copper-Nickel-Aluminum Alloy (Sand Cast)

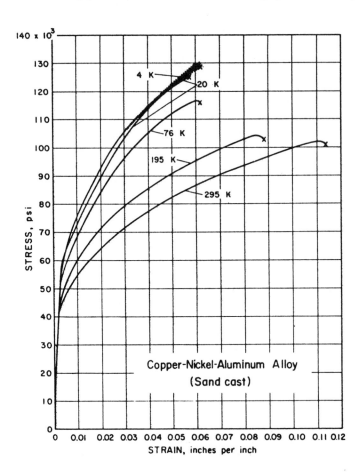

Source: Source Book on Copper and Copper Alloys, American Society for Metals, Metals Park OH, 1979, p 37

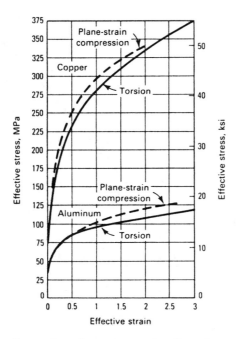

Comparison of room-temperature flow curves from torsion and plane-strain compression tests on copper and aluminum.

Source: Metals Handbook, Ninth Edition, Volume 8, Mechanical Testing, American Society for Metals, Metals Park OH, 1985, p 164

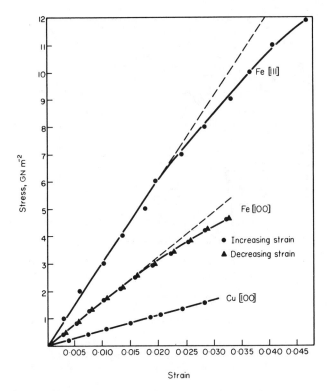

Stress-strain curves of copper and iron whiskers. (After Brenner.)

Properties of Metal Whiskers. A large number of whiskers of different metals, including iron, copper, silver, zinc, and also Al_2O_3 and SiO_2, have been deformed in tension and have exhibited elastic strains between 2 and 5%, i.e. greatly in excess of normal materials in single and polycrystalline form. Typical stress-strain curves for copper and iron whiskers are shown in the figure above, in which deviations from elastic behavior occur above 2% elongation.

Source: R. W. K. Honeycombe, The Plastic Deformation of Metals, Second Edition, American Society for Metals, Metals Park OH, 1984, p 257

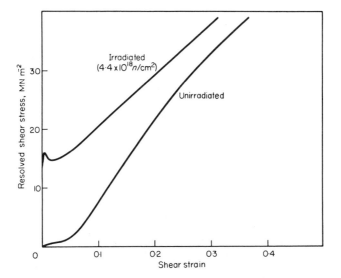

The effect of neutron irradiation (4.4×10^{18} neutron/cm^2) on the stress-strain curve of a copper crystal. (After Makin.)

Effect of Irradiation on Mechanical Properties: Single Crystals. Neutron irradiation has a marked effect on the mechanical properties of crystals. For example, the critical shear stress of copper crystals is raised from 0.5 MN m^{-2} up to 15 MN m^{-2} by a dose of 4.4×10^{18} neutrons/cm^2. Moreover, a yield point is developed and the whole level of the stress-strain curve is substantially raised (see figure above). The slip lines undergo the same change as in the quenched crystals, becoming coarse and widely spaced.

Source: R. W. K. Honeycombe, The Plastic Deformation of Metals, Second Edition, American Society for Metals, Metals Park OH, 1984, p 281

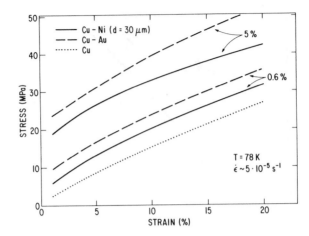

Tensile stress-strain curves for copper-nickel and copper-gold alloys of two concentrations. (After den Otter and van den Beukel.)

Source: Deformation, Processing, and Structure, George Krauss, Ed., papers presented at the ASM Materials Science Seminar, 23 October 1982, St. Louis MO, sponsored by the Seminar Committee of the Materials Science Division of the American Society for Metals, Metals Park OH, 1984, p 94

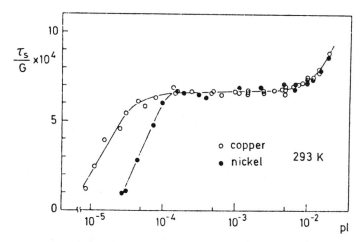

Cyclic stress-strain curves for copper and nickel single crystals in which the stress has been normalized against the shear modulus. The logarithmic scale on the abscissa has been chosen for convenient presentation only.

Except for the shear modulus, most of these parameters do not vary much among different metals having a similar crystal structure. Consequently, it is no surprise to find, as shown by Mughrabi *et al.*, that the cyclic stress-strain curves for monocrystalline copper and nickel effectively overlie one another when the flow stress is normalized against the shear modulus (see figure above). Minor differences among metals can, however, be expected, insofar as stacking-fault energy influences the cyclic-deformation processes (see, for example, the behavior of silver). Much more work is necessary to elucidate these differences.

The question of the relation between monocrystalline- and polycrystalline-fatigue results has been a recurring one. Kettunen showed an interesting correlation between the Wohler curves (plots of applied stress versus life) of polycrystals and single crystals.

Source: Fatigue and Microstructure, papers presented at the 1978 Materials Science Seminar, 14–15 October 1978, St. Louis MO, sponsored by the Materials Science Division of the American Society for Metals, Metals Park OH, 1979, p 171

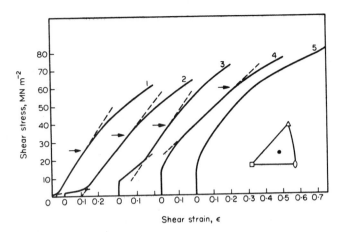

Shear stress–shear strain curves of copper crystals with different volume fractions of SiO₂ particles, particle diam. about 900 Å. (1) Pure Cu; (2) Cu, oxidized then deoxidized, $f = 0$; (3) $f = 0.33 \times 10^{-2}$; (4) $f = 0.66 \times 10^{-2}$; (5) $f = 1 \times 10^{-2}$. (After Ebeling and Ashby.)

Source: R. W. K. Honeycombe, The Plastic Deformation of Metals, Second Edition, American Society for Metals, Metals Park, OH, 1984, p 191

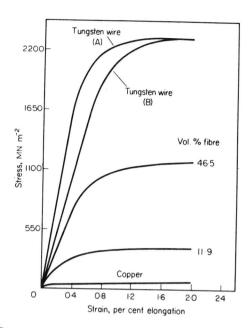

Stress-strain curves of composites of copper with tungsten wires. (After McDanels *et al.*)

Source: R. W. K. Honeycombe, The Plastic Deformation of Metals, Second Edition, American Society for Metals, Metals Park OH, 1984, p 260

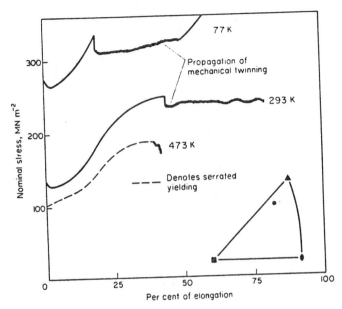

Stress-strain curves of copper-10 at.% indium crystals showing the onset of twinning (Corderoy).

Twin formation has also been found in copper-base solid solutions above room temperature, if the solute atom concentration is high, thus raising the value of the shear stress that the alloy can reach by cold work. For example, copper-10 at.% indium crystals twin at 473 K while 5% indium crystals twin at 293 K and 1% indium crystals must be deformed at 77 K before they twin. In all cases the shear stress at which twinning commences is approximately 100 MN m^{-2} (see figure above).

Source: R. W. K. Honeycombe, The Plastic Deformation of Metals, Second Edition, American Society for Metals, Metals Park OH, 1984, p 214

16-106. Copper-Zinc Crystals

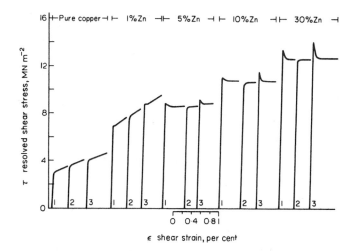

Stress-strain curves of copper and copper-zinc crystals at room temperature. Curve 1: first loading; curve 2: immediate reloading; curve 3: after 2 h at 473 K. (After Ardley and Cottrell.)

Source: R. W. K. Honeycombe, The Plastic Deformation of Metals, Second Edition, American Society for Metals, Metals Park OH, 1984, p 164

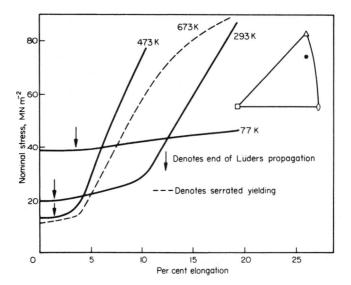

Stress-strain curves of copper-5 at.% zinc crystals at several temperatures. (Brindley *et al.*)

Source: R. W. K. Honeycombe, The Plastic Deformation of Metals, Second Edition, American Society for Metals, Metals Park OH, 1984, p 169

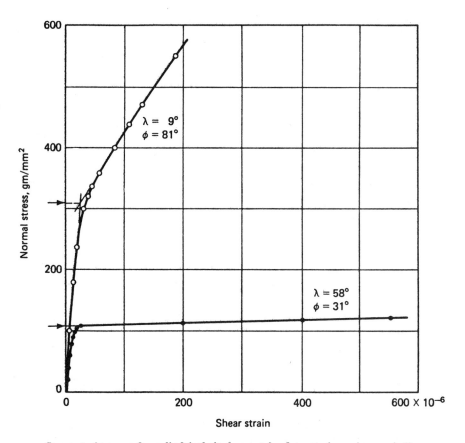

Stress-strain curve for cylindrical single crystals of magnesium. Arrows indicate yield strengh. (Adapted from E. C. Burke and W. R. Hibbard.)

Source: Charlie R. Brooks, Heat Treatment, Structure and Properties of Nonferrous Alloys, American Society for Metals, Metals Park OH, 1982, p 6

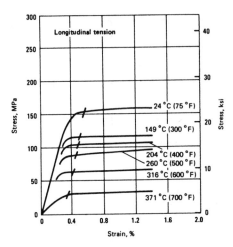

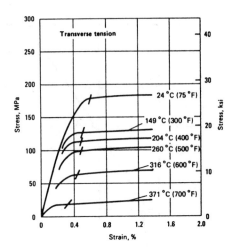

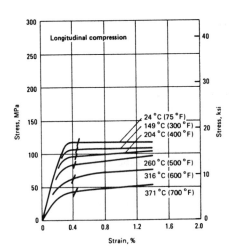

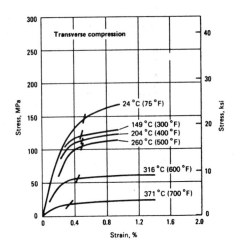

Typical stress-strain curves for HK31A-0 sheet 1.63 mm (0.064 in.) thick.

Source: Metals Handbook, Ninth Edition, Volume 2, Properties and Selection: Nonferrous Alloys and Pure Metals, American Society for Metals, Metals Park OH, 1979, p 559

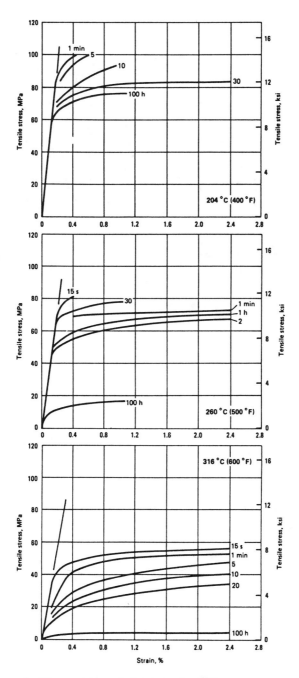

Isochronous stress-strain curves for HK31A-0 sheet 1.63 mm (0.064 in.) thick. Specimens exposed at testing temperature for 3 h before loading.

Source: Metals Handbook, Ninth Edition, Volume 2, Properties and Selection: Nonferrous Alloys and Pure Metals, American Society for Metals, Metals Park OH, 1979, p 561

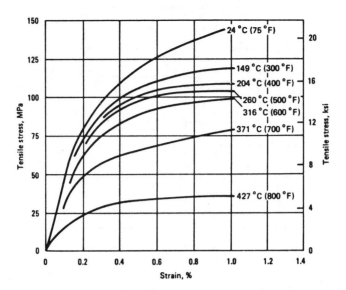

Typical stress-strain curves for HK31A separately cast test bars.

Source: Metals Handbook, Ninth Edition, Volume 2, Properties and Selection: Nonferrous Alloys and Pure Metals, American Society for Metals, Metals Park OH, 1979, p 583

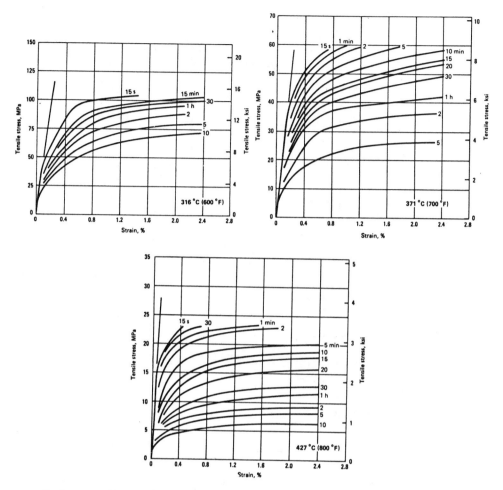

Isochronous stress-strain curves for HK31A-T6 separately cast test bars. Specimens exposed at testing temperature for 3 h before loading.

Source: Metals Handbook, Ninth Edition, Volume 2, Properties and Selection: Nonferrous Alloys and Pure Metals, American Society for Metals, Metals Park OH, 1979, p 583–584

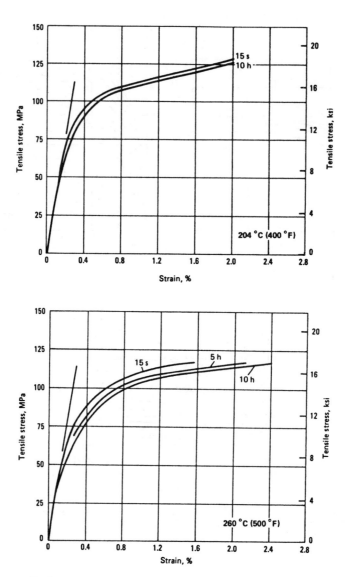

Isochronous stress-strain curves for HK31A-T6 separately cast test bars.

Source: Metals Handbook, Ninth Edition, Volume 2, Properties and Selection: Nonferrous Alloys and Pure Metals, American Society for Metals, Metals Park OH, 1979, p 583

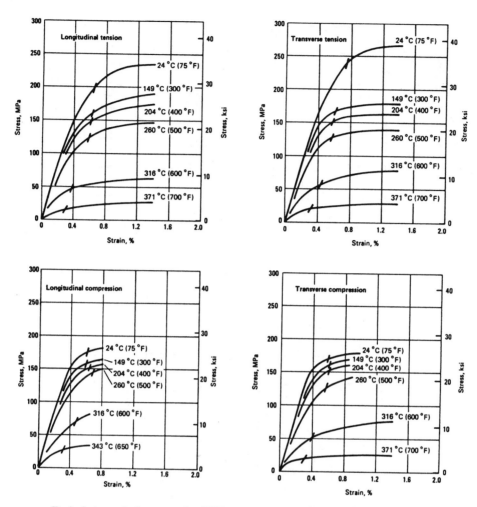

Typical stress-strain curves for HK31A-H24 sheet 1.63 mm (0.064 in.) thick.

Source: Metals Handbook, Ninth Edition, Volume 2, Properties and Selection: Nonferrous Alloys and Pure Metals, American Society for Metals, Metals Park OH, 1979, p 558

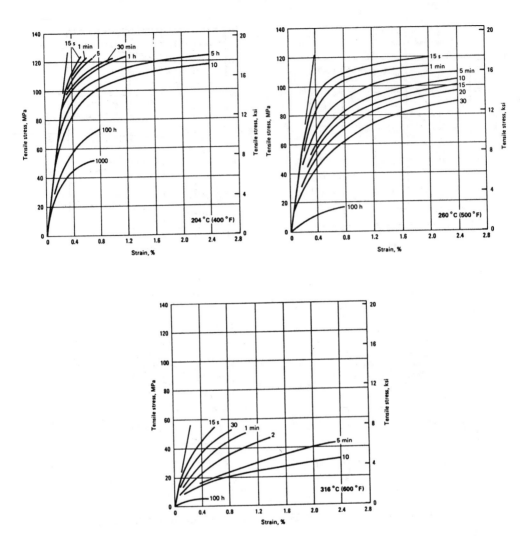

Isochronous stress-strain curves for HK31A-H24 sheet 1.63 mm (0.064 in.) thick. Specimens exposed at testing temperatures for 3 h before loading.

Source: Metals Handbook, Ninth Edition, Volume 2, Properties and Selection: Nonferrous Alloys and Pure Metals, American Society for Metals, Metals Park OH, 1979, p 560

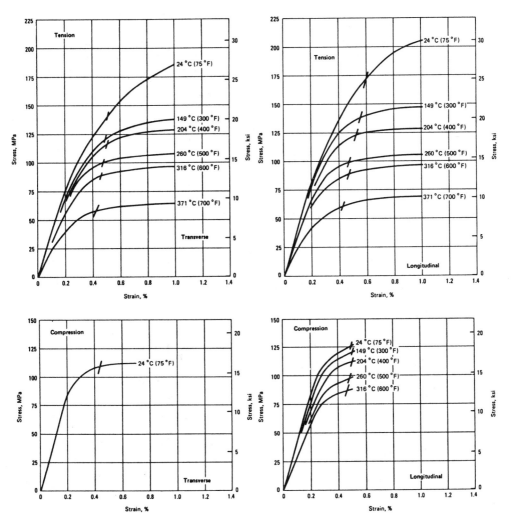

Typical stress-strain curves for HM21A-T8 sheet. Specimens held at test temperature 3 h before testing.

Source: Metals Handbook, Ninth Edition, Volume 2, Properties and Selection: Nonferrous Alloys and Pure Metals, American Society for Metals, Metals Park OH, 1979, p 562

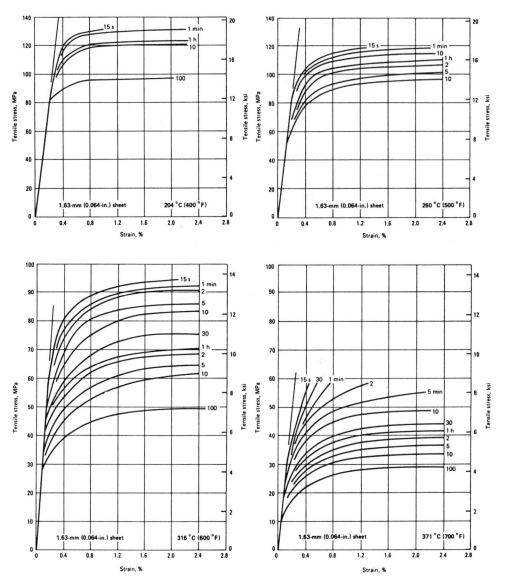

Isochronous stress-strain curves for HM21A-T8 sheet. Specimens held at test temperature 3 h before testing.

Source: Metals Handbook, Ninth Edition, Volume 2, Properties and Selection: Nonferrous Alloys and Pure Metals, American Society for Metals, Metals Park OH, 1979, p 563

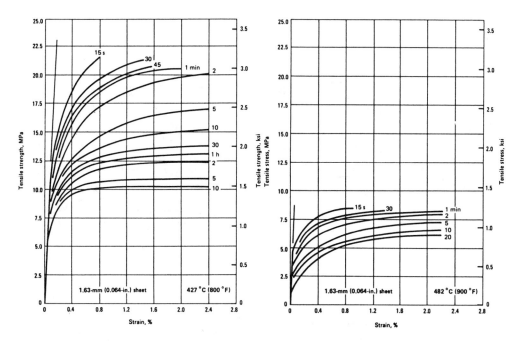

Isochronous stress-strain curves for HM21A-T8 sheet.

Source: Metals Handbook, Ninth Edition, Volume 2, Properties and Selection: Nonferrous Alloys and Pure Metals, American Society for Metals, Metals Park OH, 1979, p 564

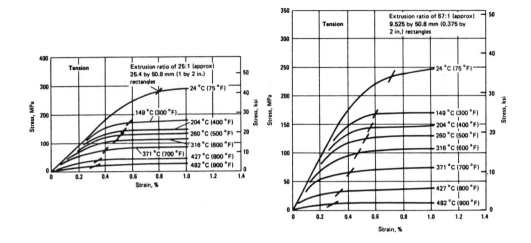

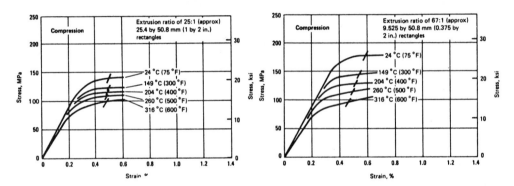

Typical stress-strain curves for HM31A extrusions. Tested in the longitudinal direction.

Source: Metals Handbook, Ninth Edition, Volume 2, Properties and Selection: Nonferrous Alloys and Pure Metals, American Society for Metals, Metals Park OH, 1979, p 566

17-13. Magnesium Alloy HZ32A-T5

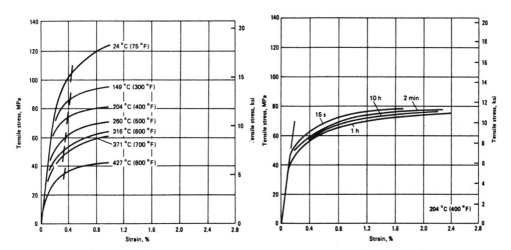

Typical stress-strain curves for HZ32A-T5 separately sand cast test bars.

Isochronous stress-strain curves for HZ32A-T5 separately sand cast test bars. Specimens exposed at testing temperatures for 3 h before loading.

Source: Metals Handbook, Ninth Edition, Volume 2, Properties and Selection: Nonferrous Alloys and Pure Metals, American Society for Metals, Metals Park OH, 1979, p 585

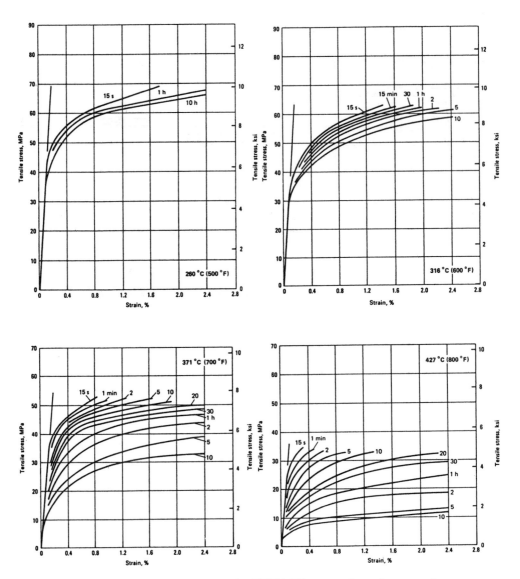

Isochronous stress-strain curves for HZ32A-T5, separately sand cast test bars.

Source: Metals Handbook, Ninth Edition, Volume 2, Properties and Selection: Nonferrous Alloys and Pure Metals, American Society for Metals, Metals Park OH, 1979, p 586

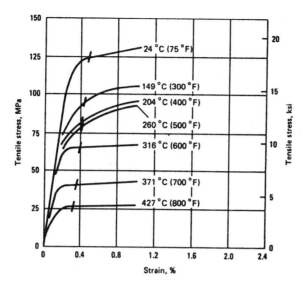

Typical stress-strain curves for EZ33A-T5 separately sand cast test bars.

Source: Metals Handbook, Ninth Edition, Volume 2, Properties and Selection: Nonferrous Alloys and Pure Metals, American Society for Metals, Metals Park OH, 1979, p 579

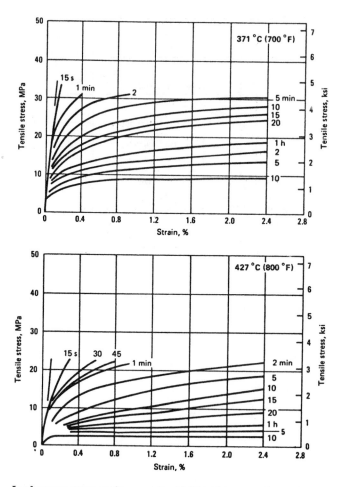

Isochronous stress-strain curves for EZ33A-T5 separately sand cast test bars.

Source: Metals Handbook, Ninth Edition, Volume 2, Properties and Selection: Nonferrous Alloys and Pure Metals, American Society for Metals, Metals Park OH, 1979, p 581

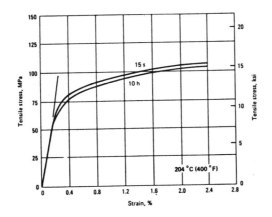

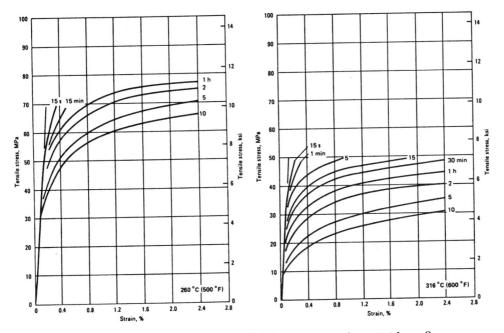

Isochronous stress-strain curves for EZ33A-T5 separately sand cast test bars. Specimens exposed at testing temperatures for 3 h before loading.

Source: Metals Handbook, Ninth Edition, Volume 2, Properties and Selection: Nonferrous Alloys and Pure Metals, American Society for Metals, Metals Park OH, 1979, p 580

17-18. Magnesium Alloy ZE41A-T5

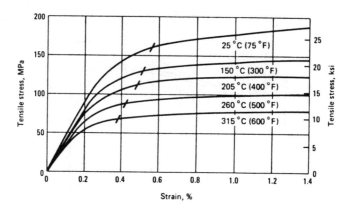

Typical stress-strain curves for ZE41A-T5 separately sand cast test bars.

Source: Metals Handbook, Ninth Edition, Volume 2, Properties and Selection: Nonferrous Alloys and Pure Metals, American Society for Metals, Metals Park OH, 1979, p 591

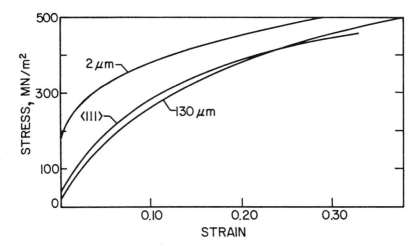

Stress-strain curves for pure nickel, as 2-μm and 130-μm grain-sizę polycrystalline and as ⟨111⟩-oriented single crystal.

Source: Marc André Meyers and Krishan Kumar Meyers, Mechanical Metallurgy: Principles and Applications, Prentice-Hall, Inc., Englewood NJ, 1984, p 345

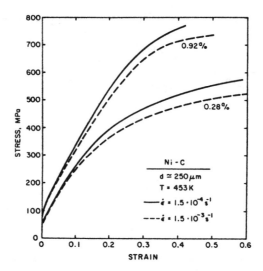

True compressive stress-strain curves for two nickel-carbon alloys. The higher strain rate (dashed) produces less strain hardening, indicating an influence of dynamic strain-aging on dynamic recovery.

The graph above shows the case of an especially strong influence of solute concentration on stress-strain behavior in Ni-C. Note, however, that the curves diverge much more strongly at large strains than at small. In first approximation one can say that the predominant effect of solutes is here to decrease Θ_r.

These stress strain curves were taken in the regime where jerky flow (the Portevin-LeChatelier effect) is observed. It appears, then, that dynamic strain-aging affects not only the mobility of the mobile dislocations (and thus the character of flow), but also the ease of rearrangement of the previously stored dislocations (and thus dynamic recovery). Further evidence for this interpretation is that the rate sensitivity *of strain hardening* is negative in the dynamic-recovery regime, not only the rate sensitivity of the flow stress (note above).

Source: Deformation, Processing, and Structure, George Krauss, Ed., papers presented at the ASM Materials Science Seminar, 23 October 1982, St. Louis MO, sponsored by the Seminar Committee of the Materials Science Division of the American Society for Metals, Metals Park OH, 1984, p 96

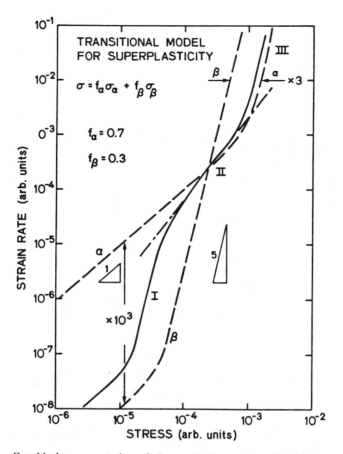

Graphical representation of the predictions of the Ghosh-Raj transitional model for superplasticity. The dashed curves represent the flow behavior of the individual phases; the solid curve is the resulting flow behavior of the two-phase alloy.

Source: Superplastic Forming, Suphal P. Agrawal, Ed., proceedings of a symposium, 22 March 1984, sponsored by the Los Angeles chapter of the American Society for Metals, Metals Park OH, 1985, p 11

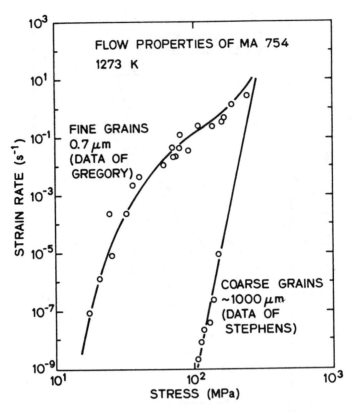

Steady-state flow properties of MA 754 in both fine-grained and coarse-grained conditions. A transition in flow behavior is evident in the fine-grained data.

Steady-state flow properties of an oxide-dispersion-strengthened alloy, MA 754 (essentially Ni-20% Cr containing a fine dispersion of Y_2O_3), are shown for two widely different grain sizes. Because of the presence of the oxide dispersion, the stress exponent for flow in the coarse-grained material is very high, about 11. Extrapolation of this data to high strain rates appears to coincide with the high-strain-rate data for the fine-grained material. This suggests a sigmoidal relationship for flow of the fine-grained material, with a maximum strain-rate sensitivity of about 0.30. Like classical superplastic materials, the fine-grained MA 754 shows a maximum elongation when the strain-rate sensitivity reaches its maximum value.

Source: Superplastic Forming, Suphal P. Agrawal, Ed., proceedings of a symposium, 22 March 1984, sponsored by the Los Angeles chapter of the American Society for Metals, Metals Park OH, 1985, p 10

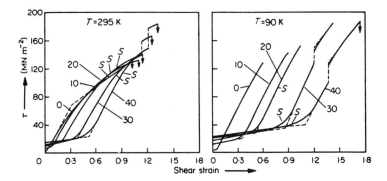

Shear stress–shear strain curves of nickel-cobalt single crystals at 295 K and 90 K. S indicates the stress at which the crystal crossed the symmetry line. Numbers represent the cobalt concentration. (Breaks in the curves arise from the use of a different criterion to calculate τ from the point when duplex slip commences.) (After Pfaff.)

Stage 1, as in pure metals, corresponds to deformation on the primary slip plane with only very limited amounts of slip on other systems, and as a consequence, the rate of work hardening is low. Stage 1 in alloy crystals is, however, more extensive and increases with increasing solute concentration. As much as 60% shear strain has been measured at room temperature for copper -0.50 at.% silver crystals with $\tau \simeq 6$ MN m^{-2}. A pure copper crystal of similar orientation had an easy glide range of about 6% and $\tau_0 \simeq 0\,6$ MN m^{-2}, so the easy glide is increased in the same ratio as the critical shear stress for glide. A similar trend over a much wider compositional range is shown by the nickel-cobalt single crystals; at 90 K, nickel crystals have only a small stage 1 strain, whereas Ni-20 at.% Co crystals exhibit 40% shear strain in this stage, and crystals with 40 at.% Co deform to about 80% shear strain before entering stage 2 of the stress-strain curve. The extent of stage 1 at two temperatures as a function of cobalt content is shown in the figure above; the length reaches a maximum at 60 at.% cobalt.

Source: R. W. K. Honeycombe, The Plastic Deformation of Metals, Second Edition, American Society for Metals, Metals Park OH, 1984, p 166

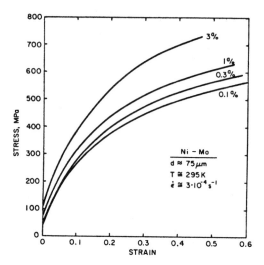

True compressive stress-strain curves for nickel-molybdenum alloys. Curves diverge monotonically.

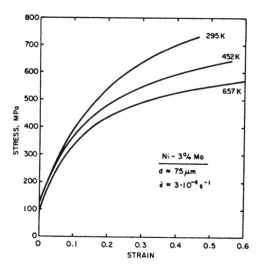

True compressive stress-strain curves for one Ni-Mo alloy at three temperatures. Curves coincide at low strains, but diverge in the dynamic-recovery regime.

Source: Deformation, Processing, and Structure, George Krauss, Ed., papers presented at the ASM Materials Science Seminar, 23 October 1982, St. Louis MO, sponsored by the Seminar Committee of the Materials Science Division of the American Society for Metals, Metals Park OH, 1984, p 100–101

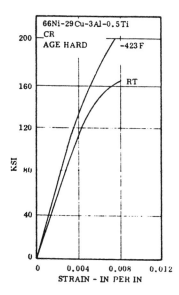

Stress-strain curves at low and room temperature for cold rolled and age-hardened alloy.

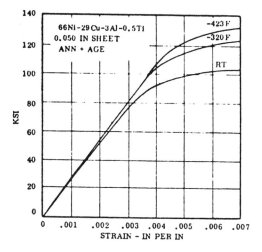

Stress-strain curves at room and low temperatures for annealed and aged sheet.

Source: Aerospace Structural Metals Handbook, Volume 4, Mechanical Properties Data Center, Battelle Columbus Laboratories, Columbus OH, 1981, p 17

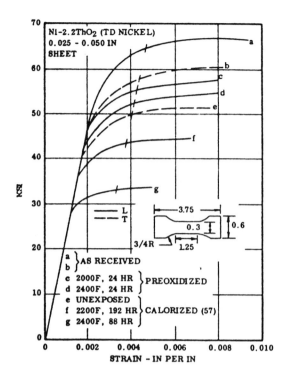

Stress-strain curves at room temperature for sheet.

Source: Aerospace Structural Metals Handbook, Volume 4, Mechanical Properties Data Center, Battelle Columbus Laboratories, Columbus OH, 1981, p 9

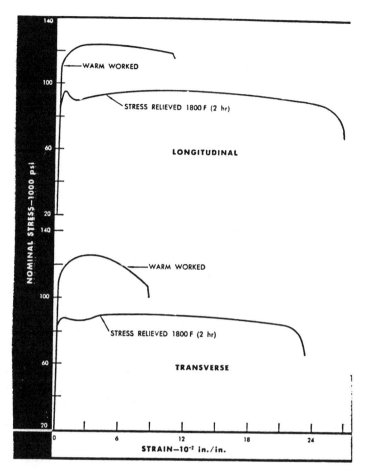

Effect of stress relieving on the stress-strain curves of unalloyed arc-cast molybdenum sheet with nominal thickness of 0.030 to 0.040 in., strain rate 0.025 in./in./min.

Source: Molybdenum Metal, Climax Molybdenum Company, 1960, p 29

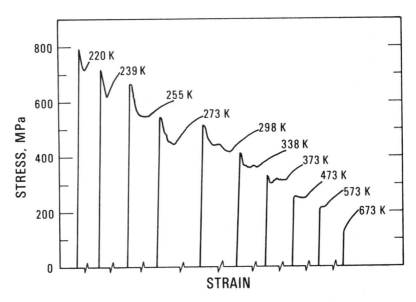

Effect of temperature on the yield point of molybdenum (After Hahn.)

Low temperatures and high impurity concentrations promote the appearance of a yield point in stress-strain curves; the temperature effect is shown in the figure above. In metals that have a ductile-brittle transition, lowering the temperature below the transition usually produces brittle fracture immediately after yield so that the lower yield point is not observed. Yield points are observed in materials of each of the three main metal and alloy crystal structures and are generally believed to be caused by the sudden unlocking of dislocations from impurity atoms. The yield-point phenomenon has been studied intensively in steels, often in connection with the process of strain aging, which does not occur at cryogenic temperatures in ordinary materials.

Source: Materials at Low Temperatures, Richard P. Reed and Alan F. Clark, Eds., American Society for Metals, Metals Park OH, 1983, p 251

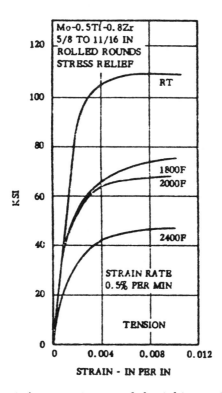

Stress-strain curves at room and elevated temperatures.

Source: Aerospace Structural Metals Handbook, Volume 5, Mechanical Properties Data Center, Battelle Columbus Laboratories, Columbus OH, 1978, p 6

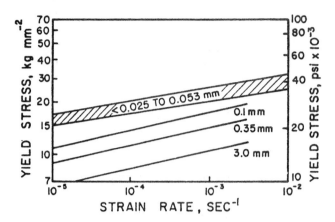

Typical effects of grain diameter and strain rate on the yield stress of niobium. Absolute values may be doubled by increasing the content of interstitials. These data show that, for any strain rate, room-temperature strength increases with smaller grain size.

Source: Walter D. Wilkinson, Properties of Refractory Metals, Gordon and Breach Science Publishers, New York, 1969, p 211

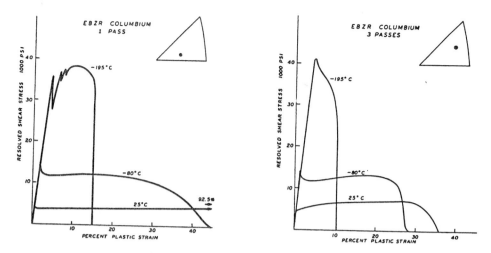

Resolved shear stress versus strain of columbium (niobium) after (left) one pass and (right) three passes.

The resolved shear stress as a function of engineering strain for the one- and three-pass electron-beam zone-refined columbium is graphed in the figures above. Their orientations are shown in the unit triangles with each curve.

Source: High Temperature Refractory Metals, R. W. Fountain, Joseph Malt and L. S. Richardson, Eds., based on a symposium, 16–20 February 1964, sponsored by the High Temperature Metals Committee (Extractive Metallurgy Division) and the Refractory Metals Committee (Institute of Metals Division) of the Metallurgical Society of the American Institute of Mining, Metallurgical, and Petroleum Engineers, Gordon and Breach Science Publishers, New York, 1966, p 460

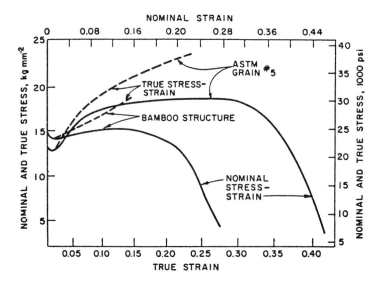

Effect of grain structure on the stress-strain behavior of niobium.

The room-temperature tensile-flow properties of refractory metals depend on grain structure as shown in the figure above. For normal fine-grain niobium: $a = K\varepsilon^n$, where a is the flow stress, K is the strength coefficient, ε is the strain, and the strain-hardening coefficient n is equal to 0.24. Actually, the flow stress may vary with the square root up to the 0.7 root of the dislocation density of niobium, depending on purity, grain size, strain rate, and annealing history. The strain rate governs not only the occurrence of a yield point in testing niobium, but also the strength properties (particularly the yield strength, which rises with increasing strain rate).

Source: Walter D. Wilkinson, Properties of Refractory Metals, Gordon and Breach Science Publishers, New York, 1969, p 211

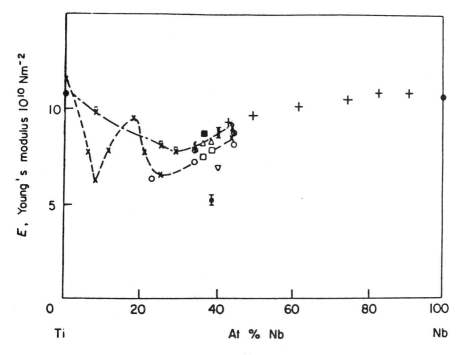

Young's modulus of Nb-Ti alloys.

Source: Niobium, Harry Stuart, Ed., proceedings of the International Symposium Niobium '81, 8–11 November 1981, San Francisco CA, AIME, New York, 1984, p 480

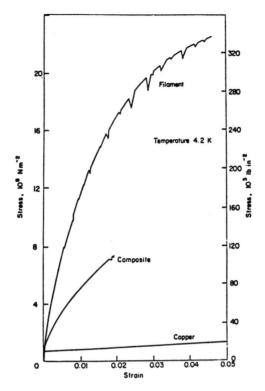

Stress-strain curves for matrix filament, and composite (Cu/Nb − Ti = 2.8/1).

Deformation of a composite such as high-modulus filaments dispersed in a ductile matrix is generally characterized by four stages on a stress-strain plot:

1. Both filaments and matrix deform elastically and the overall modulus is the weighted mean of the individual moduli.

2. The matrix yields while the filaments remain elastic and the overall modulus is now the weighted mean of the filament elastic modulus and the plastic work-hardening modulus of the matrix.

3. Both filament and matrix deform plastically.

4. The composite fails, failure being initiated by filament fracture.

The figure above represents the typical stress-strain curves of the components of a composite. A serrated yielding of Nb-Ti filaments is observed in the figure.

Source: Niobium, Harry Stuart, Ed., proceedings of the International Symposium Niobium '81, 8–11 November 1981, San Francisco CA, AIME, New York, 1984, p 481

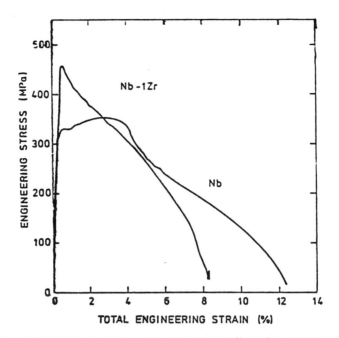

Tensile curves for irradiated Nb-1Zr (3.7×10^{26} n/m^2 at 450 °C) and Nb (3.0×10^{26} n/m^2 at 460 °C) tested at 650 at 0.02 min^{-1}.

Void Lattice. In ion-irradiated niobium and Nb-1Zr ordered arrays of voids oriented parallel to the host metal lattice (e.g. bcc) have been observed by both Brimhall and Kulcinski and Loomis and co-workers. Brimhall and Kulcinski found that the void lattice was not as well defined at 900 °C as at 800 °C and that the ordering of voids first occurred after 60 dpa and was fairly well developed by 90 dpa. Loomis *et al.* established the important result that ordered void arrays were only formed in niobium and Nb-1Zr if the oxygen impurity concentration exceeded a threshold level. The threshold concentration was between 60 and 400 at. ppm for niobium and between 400 and 2700 at. ppm for Nb-1Zr. The ordering of the voids occurred at a low dose (<5 dpa) and the void lattice parameters were again very sensitive to irradiation temperature with the greatest degree of perfection occurring at 780 °C and 800 °C.

Source: Niobium, Harry Stuart, Ed., proceedings of the International Symposium Niobium '81, 8–11 November 1981, San Francisco CA, AIME, New York, 1984, p 308

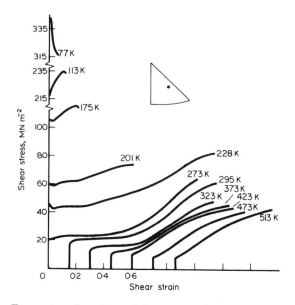

Temperature dependence of the stress-strain curves of zone-refined niobium (strain rate = 4.5 × 10⁻⁵ s⁻¹). (After Mitchell *et al.*)

Source: R. W. K. Honeycombe, The Plastic Deformation of Metals, Second Edition, American Society for Metals, Metals Park OH, 1984, p 111

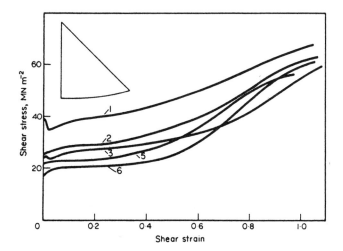

Stress-strain curves of niobium crystals of increasing purity at 295 K. (Figures refer to the number of zone passes.) (After Mitchell *et al.*)

Stress-Strain Curves of Body-Centered Cubic Crystals. The early part of the stress-strain curve of a body-centered cubic crystal is frequently interrupted by the presence of a sharp yield point due to interstitial impurities such as carbon, nitrogen or oxygen. The subsequent form of the stress-strain curve is also sensitive to such impurities as shown by work on niobium crystals of identical orientation, but subject to different numbers of zone-purifying passes during their preparation (see figure above). The purest crystal shows no yield point and has the lowest stress-strain curve.

Source: R. W. K. Honeycombe, The Plastic Deformation of Metals, Second Edition, American Society for Metals, Metals Park OH, 1984, p 110

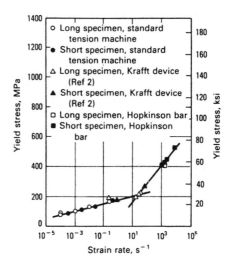

Variation of yield stress versus the logarithm of the strain rate for niobium single crystals.

The strength results are summarized in the figure above. At low deformation rates, relatively large rate changes are required to effect noticeable changes in the yield point. For example, at the lowest rates about a 1000-fold increase in strain rate is required to double the yield strength. Above some critical rate, however, strength changes become much more sensitive to the rate.

Source: Metals Handbook, Ninth Edition, Volume 8, Mechanical Testing, American Society for Metals, Metals Park OH, 1985, p 39

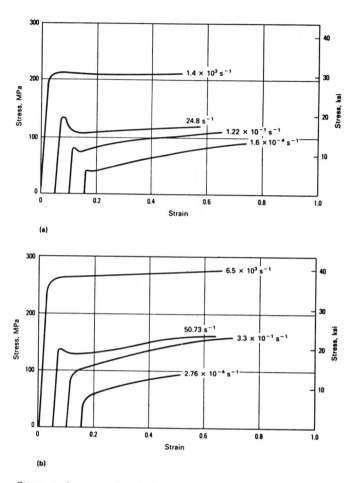

Stress-strain curves for single crystals of niobium. (a) For long specimens, 6.4 mm (0.25 in.) in length and 4.8 mm (0.19 in.) in diameter. (b) For short specimens, 2 mm (0.08 in.) in length and 4.3 mm (0.17 in.) in diameter.

The figure above shows stress-strain curves obtained from niobium monocrystals deformed at various rates. Clearly, the yield strength of the material increases as the deformation rate increases. This is the most consistently observed effect of strain rate. The figure also shows a trend for strain hardening to decrease (a decrease in the slope of the curve) with increasing strain rate.

Source: Metals Handbook, Ninth Edition, Volume 8, Mechanical Testing, American Society for Metals, Metals Park OH, 1985, p 39

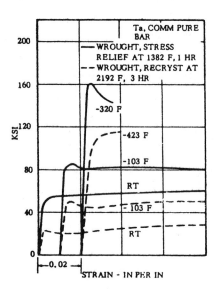

Stress-strain curves at various temperatures for stress relieved and recrystallized wrought bar.

Source: Aerospace Structural Metals Handbook, Volume 5, Mechanical Properties Data Center, Battelle Columbus Laboratories, Columbus OH, 1978, p 4

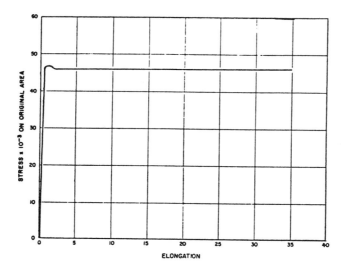

Stress-strain curve for recrystallized tantalum.

Dislocation Morphology. The stress-strain tensile curve of tantalum is similar to other bcc metals, showing the typical yield point (see graph above). A yield drop was observed in all specimens, the average yield drop being 3,000 psi. Dislocation substructures were examined in specimens after 1, 2, 4, 8, 15, 25, 30, and 34% elongation. Microhardness measurements taken from inside and outside Lüders bands are given in the table below.

Microhardness Data (Rockwell B)

Deformation, %	Inside Lüders Band*	Outside Lüders Band*
1	77.5	63.0
2	75.5	61.0
4	77.0	55.0
8	73.5	57.5
15	75.5	
25	79.5	
30	80.0	
34	80.0	

*Average of eight readings converted from Tukon Measurements.

Source: High Temperature Refractory Metals, R. W. Fountain, Joseph Malt and L. S. Richardson, Eds., based on a symposium, 16–20 February 1964, sponsored by the High Temperature Metals Committee (Extractive Metallurgy Division) and the Refractory Metals Committee (Institute of Metals Division) of the Metallurgical Society of the American Institute of Mining, Metallurgical, and Petroleum Engineers, Gordon and Breach Science Publishers, New York, 1966, p 163

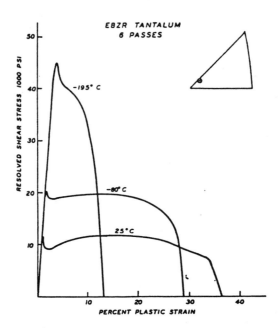

The graph above shows the resolved shear stress as a function of engineering strain for 6-pass electron-beam zone-refined tantalum. Orientation is shown in the insert triangle.

Source: High Temperature Refractory Metals, R. W. Fountain, Joseph Malt and L. S. Richardson, Eds., based on a symposium, 16–20 February 1964, sponsored by the High Temperature Metals Committee (Extractive Metallurgy Division) and the Refractory Metals Committee (Institute of Metals Division) of the Metallurgical Society of the American Institute of Mining, Metallurgical, and Petroleum Engineers, Gordon and Breach Science Publishers, New York, 1966, p 461

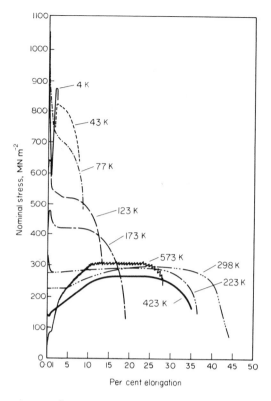

Stress-strain curves of tantalum.

The above figure gives stress-strain curves of tantalum over the temperature range 4 K to 573 K in which is shown the typical development of a yield point followed at low temperatures by localized flow that leads to a falling stress and to fracture. In the intermediate temperature range, a yield elongation zone is formed, while at the highest temperatures, serrations or localized yielding occur throughout the stress-strain curve.

Source: R. W. K. Honeycombe, The Plastic Deformation of Metals, Second Edition, American Society for Metals, Metals Park OH, 1984, p 244

19-18. Tantalum-10% Tungsten

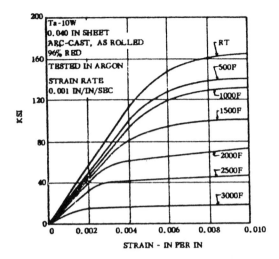

Stress-strain curves at various temperatures for arc-cast sheet.

Source: Aerospace Structural Metals Handbook, Volume 5, Mechanical Properties Data Center, Battelle Columbus Laboratories, Columbus OH, 1978, p 6

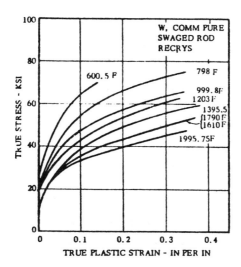

Stress-strain curves for recrystallized swaged rod.

Source: Aerospace Structural Metals Handbook, Volume 5, Mechanical Properties Data Center, Battelle Columbus Laboratories, Columbus OH, 1978, p 5

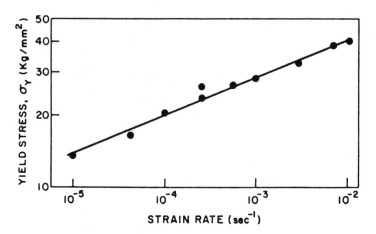

Effect of strain rate on yield stress for tungsten at 525 K (0.14 T_m) (After Bechtold.)

Source: Stephen W. H. Yih and Chun T. Wang, Tungsten Sources: Metallurgy, Properties, and Applications, Plenum Press, 1979, p 290

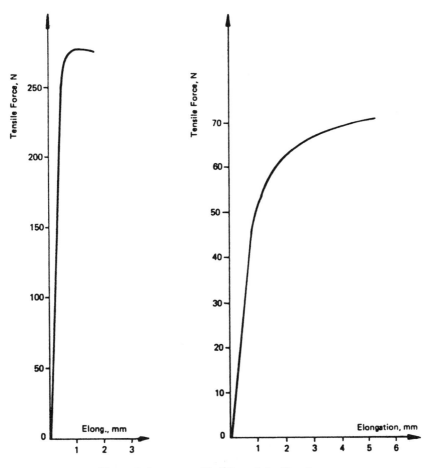

Stress-strain curves of brittle and ductile wires.

It is typical of a brittle material, that there is no significant difference between yield strength and tensile strength or, in other words, the stress-strain curve has no work-hardening section. This can be illustrated by comparing the stress-strain curve of a brittle material with that of a ductile one (see above).

Source: Tungsten 1985, proceedings of the Third International Tungsten Symposium, 13–17 May 1985, Madrid, MRP Publishing Services Ltd., 1985, p 89

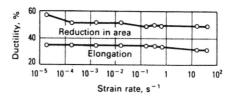

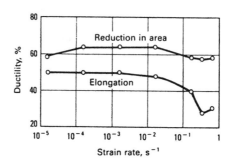

Effect of strain rate on the ductility of commercially pure titanium. (a) At room temperature, ductility is not affected. (b) At −195 °C (−319 °F), the ductility decreases at higher strain rates due to a change in deformation mechanism (slip versus twinning).

The effect of strain rate on the ductility of metals is relatively slight. For moderate strain rates and temperatures, the ductility of titanium does not vary substantially, as shown in (a) above. For conditions of low temperature and high extension rates, the ductility decreases somewhat, as shown in (b) above.

Source: Metals Handbook, Ninth Edition, Volume 8, Mechanical Testing, American Society for Metals, Metals Park OH, 1985, p 42

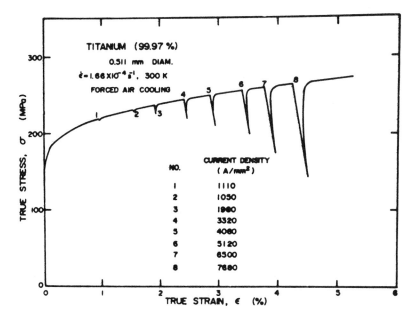

Electroplastic effect in titanium: effect of current pulses on the true stress–true strain curve.

The above figure shows the drops in stress observed in a tensile curve of titanium upon applying current densities of increasing amplitude. The drops in stress are the result of almost instantaneous increases in length of the specimen. These stress drops are dependent on the machine stiffness: the stiffer the machine, the larger the stress drop. This stress drop can be converted into an electroplastic strain that represents the effect in a normalized way (independent of machine characteristics and specimen dimensions).

Source: Marc André Meyers and Krishan Kumar Chawla, Mechanical Metallurgy: Principles and Applications, Prentice-Hall, Inc., Englewood Cliffs NJ, 1984, p 594

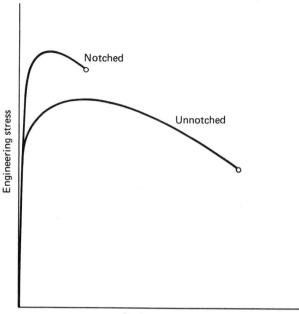

Schematic stress-strain curves for an unnotched and a notched tensile specimen. In the notched specimen, the yield and tensile strengths are higher and the elongation at fracture is less.

Another important consideration in using high-strength alloys is their notch sensitivity. This is a measure of the ability of the material to plastically deform locally at regions of high stress, relative to the average stress, instead of forming a crack and fracturing. The notch sensitivity is frequently measured by conducting a tensile test on a pre-notched specimen, and comparing the tensile strength (or yield strength or fracture stress) to that obtained using an unnotched specimen. The figure above shows schematically the stress-strain curves for both conditions. Here the stress is engineering stress, obtained by dividing the load by the original area. In the notched specimen, this is the area at the root of the notch. Note that for the notched specimen the yield and tensile strength are higher, and the elongation at fracture is less.

The behavior in these curves can be explained by the following qualitative description. At a distance removed from the notch, the axial load is evenly distributed across the cross section. Thus the stress is uniform. At the notch level, the notch area supports the load, and the free surface outside the notch supports no axial load.

Source: Charlie R. Brooks, Heat Treatment, Structure and Properties of Nonferrous Alloys, American Society for Metals, Metals Park OH, 1982, p 345

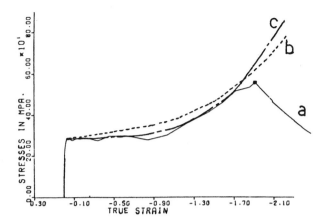

Stress-strain curves for a compression test. Curve a: Experimental results. Curves b and c: Upper bound solutions.

The graph above is a plot of the average axial stress (pressure at interface) versus average logarithmic axial strain (curve a). The period of creeping that starts in the region indicated by the square is seen to involve a considerable strain, although the specimen does not decrease in height greatly. This is because the specimen is now a thin disc with a height-to-diameter ratio of less than 1 to 20. This curve is better understood (at least up to the region of the square) when it is plotted in conjunction with an upper bound solution for axisymmetric compression. There is an upper bound solution which assumes no barreling and friction governed by a shear coefficient at the boundary.

Source: Production to Near Net Shape Source Book, C. J. Van Tyne and B. Avitzur, American Society for Metals, Metals Park OH, 1983, p 171

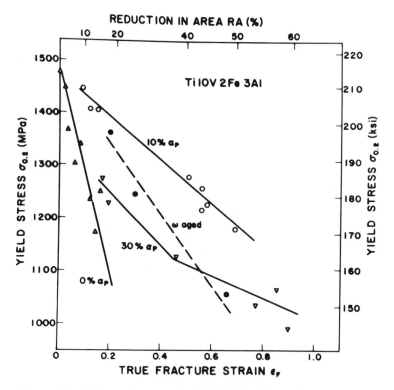

Strength-ductility trend curves for Ti-10V-2Fe-3Al containing various amounts of primary α.

The effects of TMP on the strength and ductility of β-phase alloys tend to follow trends similar to those shown earlier for α + β alloys except that the strength levels are typically considerably higher. Three major factors that are affected by TMP appear to be important: primary α morphology, primary α volume fraction, and grain-boundary α. The strengths of these alloys are controlled by the scale of the α-phase precipitates in the matrix, and this can be controlled independent of the primary α morphology and volume fraction to a significant extent. As a result, data on yield strength versus tensile fracture strain can be plotted for each of several primary α volume fractions, as shown for alloy Ti-10V-2Fe-3Al in the graph above. These data show that the alloy in the most ductile condition at any of the strength levels studied is that which contains a small (~0.1) volume fraction of primary α. This condition represents a compromise in the sense that alloys containing no primary α unavoidably have GBα, whereas at higher volume fractions of primary α, strain localization tends to occur between the primary α particles. Both GBα and strain localization lead to premature fracture initiation, and thus the alloy that does not exhibit either of these has better ductility.

Source: Deformation, Processing, and Structure, George Krauss, Ed., papers presented at the ASM Materials Science Seminar, 23 October 1982, St. Louis MO, sponsored by the Seminar Committee of the Materials Science Division of the American Society for Metals, Metals Park OH, 1984, p 323

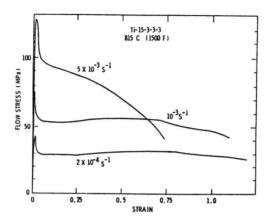

Flow stress versus strain at constant strain rate for Ti-15V-3Cr-3Sn-3Al alloy at 815 °C.

It was found that this alloy exhibits deformation behavior quite different from that of the two-phase alloys in that it tends to flow soften with about 2 to 4% strain, as shown in the graph above.

Source: Superplastic Forming, Suphal P. Agrawal, Ed., proceedings of a symposium, 22 March 1984, Los Angeles CA, sponsored by the Los Angeles chapter of the American Society for Metals, Metals Park OH, 1985, p 19

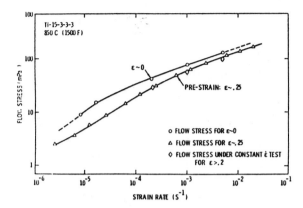

Flow stress vs strain rate for Ti-15V-3Cr-3Sn-3Al at 850 °C showing the effect of a small prestrain.

The softening is reflected in the stress versus strain-rate characteristics as shown in the graph above, where the pre-strain can be seen to reduce the flow stresses and increase the strain-rate sensitivity observed over a wide range of strain rates. The microstructural changes that occur during the softening include the development of a subgrain structure.

Source: Superplastic Forming, Suphal P. Agrawal, Ed., proceedings of a symposium, 22 March 1984, Los Angeles CA, sponsored by the Los Angeles chapter of the American Society for Metals, Metals Park OH, 1985, p 19

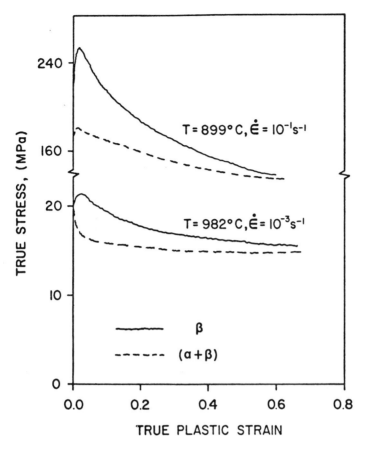

Flow stress data for Ti-6242 at various test conditions.

The flow stress of the alpha-beta titanium alloy of interest, Ti-6242Si, is depicted above. Although this was obtained for both an equiaxed and an acicular microstructure (see figure above), only that for the acicular microstructure was used.

The critical strains and temperatures for which the acicular alpha microstructure transformed to an equiaxed microstructure are shown above. It is found that deformation to strains of the order of 1.0 at a temperature of 1650 °F (900 °C), followed by heat treatment at 1750 °F (955 °C), produced the desired transformation.

Source: Forging Handbook, T. G. Byrer, S. L. Semiatin and Donald C. Vollmer, Eds., Forging Industry Association of America, Cleveland OH, 1985, p 116

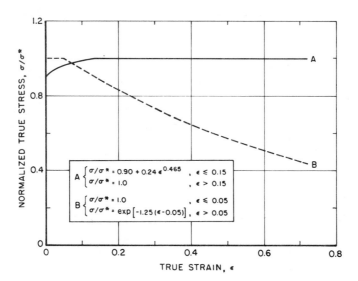

Hypothetical flow curves for metals deformed in the hot working regime. Curves show extremes of no flow softening (curve A) and a large amount of flow softening (curve B).

At a constant strain rate, the flow-stress dependence on strain, $f(\varepsilon)$, of most metals at hot working temperatures lies between two extremes (above figure). (Note in this figure that the normalization factor σ^* includes a constant term, as well as the strain-rate dependence $g(\dot{\varepsilon})$ of the flow stress.) One of the extremes of $f(\varepsilon)$ is a flow curve that shows an initial work-hardening interval followed by a flow-stress plateau (curve A above). This curve is typical of much of the α + β microstructure data. The other extreme is a curve showing little or no initial work hardening, followed by a large amount of flow softening (curve B above); the latter is typical of much of the β microstructure flow-stress data. The large strain behavior of these two flow curves is characterized by approximately constant values of γ'. The rate sensitivity of the flow stress of the hypothetical materials, $g(\dot{\varepsilon})$, was assumed to follow a power-law relation, namely, $g(\dot{\varepsilon}) \sim \dot{\varepsilon}^m$, and the rate sensitivity exponent m was assumed to have values of either 0.0, 0.125, or 0.30. These values span those typically measured in compression tests. With these assumptions, the flow stress is completely described by $\sigma = $ (const.) $\dot{\varepsilon}^m f(\varepsilon)$. As before, the temperature dependence of the flow stress is implicitly included in $f(\varepsilon)$. With these properties, α's between 0 and ∞ were obtained, suggesting a wide range of tendencies to form shear bands.

Source: S. L. Semiatin and J. J. Jonas, Formability and Workability of Metals, American Society for Metals, Metals Park OH, 1984, p 80

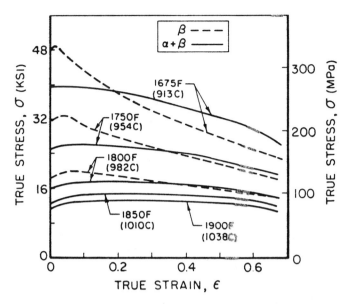

Compression flow curves for Ti-6242Si with either an equiaxed alpha (α + β) or Widmanstätten alpha (β) starting microstructure, determined as a function of temperature for a strain rate of 10 s⁻¹.

Source: S. L. Semiatin and J. J. Jonas, Formability and Workability of Metals, American Society for Metals, Metals Park OH, 1984, p 58

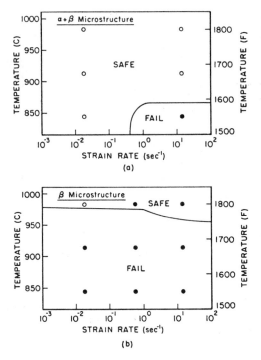

Workability maps for occurrence of shear bands in isothermal sidepressing of Ti-6242Si with (a) an α + β microstructure and (b) a β microstructure. Workability predictions (—) and forging conditions in which shear bands were (●) and were not (○) observed are noted.

The comparison of α parameter predictions and experimental observations is best done on temperature–strain rate maps. Such maps or workability diagrams are shown in the figure above for Ti-6242Si having the two different preform microstructures. It can be seen that, with the exception of two points, the loci corresponding to $\alpha_{max} \geq 5$ separate regimes in which shear bands are and are not observed for this alloy. The small disagreement between the predicted locus and the β microstructure data may be due to the choice of an α value of 5 rather than 4 for the occurrence of noticeable flow localization. Also, experimental errors in the flow-stress data affect the value of γ' and can change the position of the α = 5 loci in this way. Thus, the α parameter provides an insight into the tendency to form shear bands as well as the likely degree of localization or severity of shear banding. Although the α = 5 criterion is principally a rule of thumb, process modeling using finite element methods has confirmed the usefulness of this parameter.

Source: S. L. Semiatin and J. J. Jonas, Formability and Workability of Metals, American Society for Metals, Metals Park OH, 1984, p 74

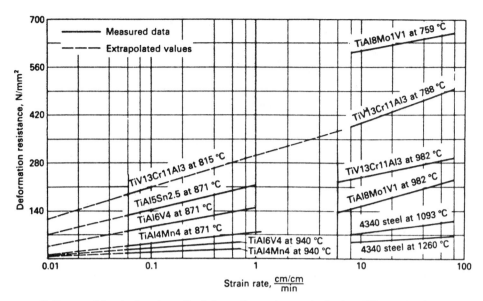

Influence of the strain rate on the deformation resistance in forging different titanium alloys and AISI 4340 steel at different deformation temperatures (deformation resistance determined at 10% upset reduction).

Source: Kurt Laue and Helmut Stenger, Extrusion Processes: Machinery Tooling, trans. A. F. Castle and Gernot Lang, American Society for Metals, Metals Park OH, 1976, p 183

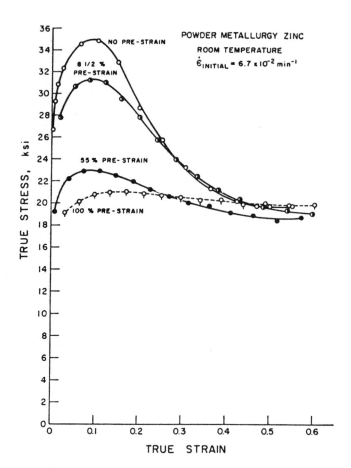

The effect of various amounts of compressive prestrain at 240 °C on the mechanical behavior of powder metallurgy zinc rod compressed longitudinally at room temperature.

It is shown above that longitudinal specimens of powder metallurgy zinc which had been prestrained 55% or more at 240 °C no longer strain-softened appreciably, and were considerably weaker than material that contained the much larger, elongated grains.

Source: Glen R. Edwards, John C. Payne and Oleg D. Sherby, Strain Softening in Powder Metallurgy Zinc, Met. Trans. A, October 1971, American Society for Metals, Metals Park OH, p 2956

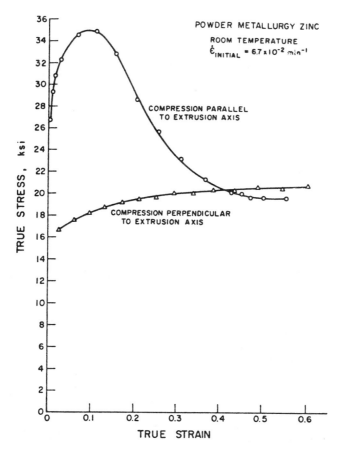

A comparison of longitudinal and transverse mechanical behavior for powder metallurgy zinc rod compressed at room temperature.

Source: Glen R. Edwards, John C. Payne and Oleg D. Sherby, Strain Softening in Powder Metallurgy Zinc, Met. Trans. A, October 1971, American Society for Metals, Metals Park OH, p 2957

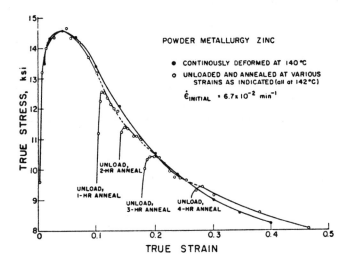

The effect of strain-aging at 0.6 T_m on the compressive mechanical behavior of powder metallurgy zinc rod.

These curves compare true stress–true strain curves for a continuously deformed sample and for a sample that was unloaded and annealed at several points in strain. Both samples were compressed parallel to the extrusion axis at 140 °C (0.6 T_m) and at an initial strain rate of 6.7×10^{-2} min^{-1}. No drop in flow stress was ever observed when the interrupted text was continued, even after a 4-h anneal at 0.6 T_m on a sample deformed to 25% true strain. The effects of strain rate and temperature on the degree of strain softening in powder metallurgy zinc were also inconsistent with dynamic recovery. Strain softening was enhanced by high strain rate and low temperature, being most prominent at -76 °C and 1.7 min^{-1}.

Source: Glen R. Edwards, John C. Payne and Oleg D. Sherby, Strain Softening in Powder Metallurgy Zinc, Met. Trans. A, October 1971, American Society for Metals, Metals Park OH, p 2956

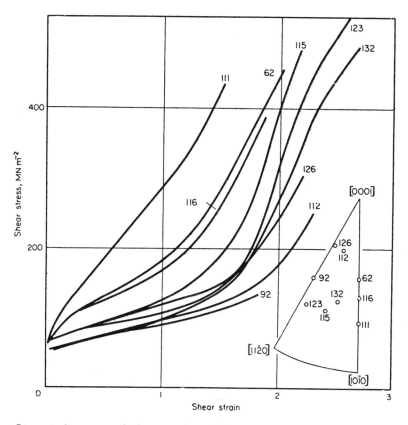

Stress-strain curves of zinc crystals at 294 K. (After Lücke, Masing and Schröder.)

Source: R. W. K. Honeycombe, The Plastic Deformation of Metals, Second Edition, American Society for Metals, Metals Park OH, 1984, p 121

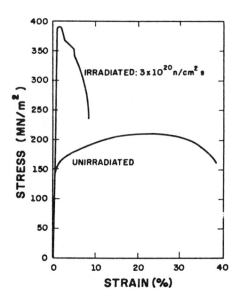

Stress-strain curves for irradiated and unirradiated Zircaloy.

The exact nature of defects introduced by radiation that are responsible for these changes in Zircaloy are not yet very well characterized. There is a considerable variation in the observed microstructures. One of the few observations about which there exists general agreement is the absence of radiation-induced vacancies in Zircaloy, which is a significant difference when compared with, say, the behavior of stainless steels. Stainless steels show swelling due to neutron irradiation.

Source: Marc André Meyers and Krishan Kumar Chawla, Mechanical Metallurgy: Principles and Applications, Prentice-Hall, Inc., Englewood Cliffs NJ, 1984, p 533

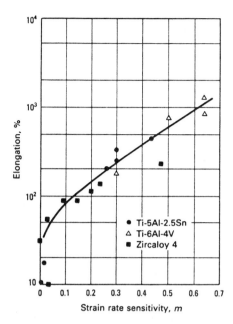

Percent elongation versus strain-rate sensitivity parameter, m, for two titanium alloys and Zircaloy 4. Note that materials with a high m value possess high ductility. The phenomenon of superplasticity occurs in materials with high m values.

The ductility of two titanium alloys and Zircaloy has been shown to increase with the strain-rate sensitivity parameter, m, as shown in the figure above. This can be rationalized by the reduced tendency to form a neck as m approaches 1.

Source: Metals Handbook, Ninth Edition, Volume 8, Mechanical Testing, American Society for Metals, Metals Park OH, 1985, p 42

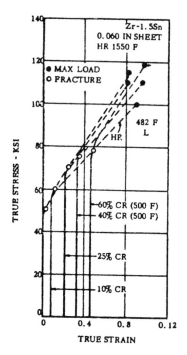

True strain curves at 482 °F for cold rolled sheet.

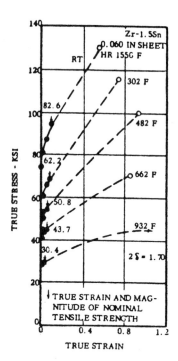

Effect of the true strain on nominal tensile strength of sheet.

Source: Aerospace Structural Metals Handbook, Volume 5, Mechanical Properties Data Center, Battelle Columbus Laboratories, Columbus OH, 1978, p 5

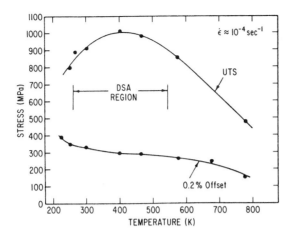

Yield stress and tensile strength as functions of temperature for Hadfield steel, showing the pronounced influence of dynamic strain aging on the UTS. (After Dastur and Leslie.)

It has, in fact, long been known that the "humps" in diagrams of flow stress versus temperature in the dynamic strainaging regime are more pronounced for ultimate tensile strength than for yield strength (e.g., in Fe-N and type 316 stainless steel). The curves above show this feature for Hadfield steel.

Source: Deformation, Processing, and Structure, George Krauss, Ed., papers presented at the ASM Materials Science Seminar, 23 October 1982, St. Louis MO, sponsored by the Seminar Committee of the Materials Science Division of the American Society for Metals, Metals Park OH, 1984, p 96

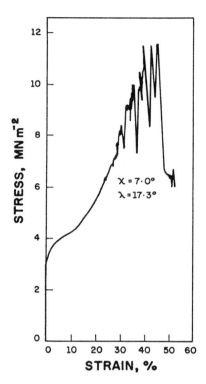

**Serrated stress-strain curve due to twinning in
a Cd single crystal.**

The "cry of tin" heard when a polycrystalline sample of tin is bent plastically is caused by the sudden formation of deformation twins. The bursting
of twins during straining can lead to a serrated form of stress-strain curve
(above). In many HCP metals, the slip is restricted to basal planes. Thus,
twinning can contribute to plastic deformation by the shear that it produces,
but this is generally small. More important, however, the twinning process
serves to reorient the crystal lattice to favor further basal slip. In HCP
metals, the common twinning elements are $(10\bar{1}2)$ plane and $[10\bar{1}1]$ direction. Twinning results in a compression or elongation along the c axis depending on the c/a ratio. For $c/a > \sqrt{3}$ (the case of Zn and Cd), twinning
occurs on $(10\bar{1}2)[10\bar{1}1]$ when compressed along the c axis. When $c/a =
\sqrt{3}$ the twinning shear is zero. For $c/a < \sqrt{3}$ (the case of Mg and Be),
twinning occurs under tension along the c axis.

Source: Marc André Meyers and Krishan Kumar Chawla, Mechanical Metallurgy: Principles and Applications, Prentice-Hall, Inc., Englewood Cliffs NJ, 1984, p 286

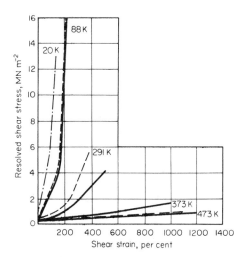

Stress-strain curves of cadmium crystals. The dashed curves were obtained at strain rates 100 times greater than those for the full curves. (After Schmid and Boas.)

The figure above gives data for cadmium that illustrate three points:

1. Large glide strains (for the appropriate orientations).

2. Low linear rates of hardening (at room temperature and above) over a large part of the stress-strain curve.

3. A marked temperature dependence of the stress-strain curves.

The pronounced effect of the deformation temperature should be particularly noted.

Source: R. W. K. Honeycombe, The Plastic Deformation of Metals, Second Edition, American Society for Metals, Metals Park OH, 1984, p 27

23-4. 12% Cobalt-Cemented Tungsten Carbide

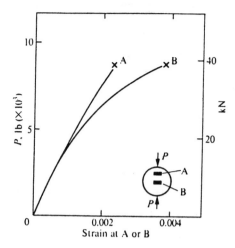

Load (P) strain curves for 12 wt.% Co cemented tungsten carbide with strain gage mounted transversely at center of disc (A) and approximately at midpoint between point A and the load (B). The material at B goes plastic before fracture occurs at A while fracture at A occurs very near the yield point. (After Takagi and Shaw.)

Source: Milton C. Shaw, Metal Cutting Principles, Clarendon Press, Oxford, 1984, p 128

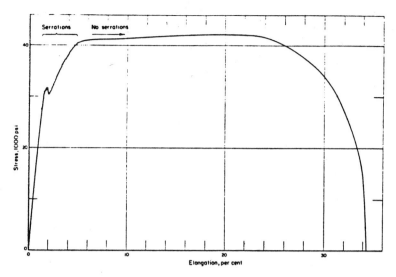

Stress-strain curve for quenched medium-grain-size chromium. Test temperature 340 °C.

The chromium displayed an average rate of work hardening of 4000 psi/ percent strain between 0 and 3% strain, compared with a rate of 500 psi between 3 and 20% strain (see above).

Source: High Temperature Refractory Metals, R. W. Fountain, Joseph Malt and L. S. Richardson, Eds., based on a symposium, 16–20 February 1964, sponsored by the High Temperature Metals Committee (Extractive Metallurgy Division) and the Refractory Metals Committee (Institute of Metals Division) of the Metallurgical Society of the American Institute of Mining, Metallurgical, and Petroleum Engineers, Gordon and Breach Science Publishers, New York, 1966, p 200

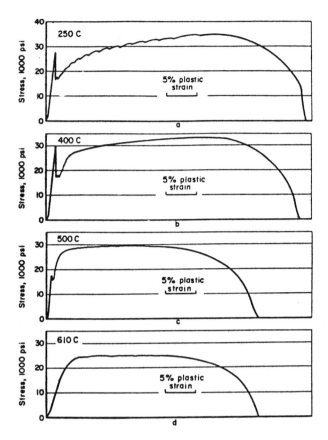

Stress-strain curves for fine-grain chromium at the temperatures indicated.

Serrated yielding occurred from 200 to 300 C, but the serrations were generally small in amplitude. In the quenched samples, serrations were only apparent on the rapid-work-hardening portion of the curve. The most pronounced example of irregular flow was in the fine-grain specimens tested at temperatures from 250 to 350 °C when a "wavy" type of flow occurred (upper graph). The lower three graphs show the change in nature of the stress-strain curve as the testing temperature is varied.

Source: High Temperature Refractory Metals, R. W. Fountain, Joseph Malt and L. S. Richardson, Eds., based on a symposium, 16–20 February 1964, sponsored by the High Temperature Metals Committee (Extractive Metallurgy Division) and the Refractory Metals Committee (Institute of Metals Division) of the Metallurgical Society of the American Institute of Mining, Metallurgical, and Petroleum Engineers, Gordon and Breach Science Publishers, New York, 1966, p 201

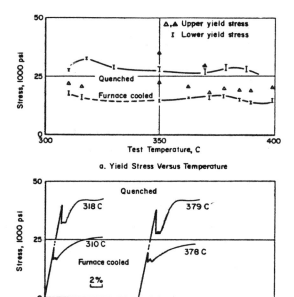

a. Yield Stress Versus Temperature

b. Effect of Cooling Rate on the Shape of the Stress-Strain Curves

Comparison of furnace-cooled and quenched samples, medium grain size chromium. (a) Yield stress versus temperature; (b) Effect of cooling rate on the shape of the stress-strain curves.

The effect of quenching on the yield properties is shown in the figure above. The quenched specimens were all strained 8% in the strain-aging range and, compared to the furnace-cooled samples, had higher upper and lower yield-stress values and markedly different stress-strain curves which showed an unusually high rate of work hardening. After about 3% strain, the rate of work hardening decreased substantially.

Source: High Temperature Refractory Metals, R. W. Fountain, Joseph Malt and L. S. Richardson, Eds., based on a symposium, 16–20 February 1964, sponsored by the High Temperature Metals Committee (Extractive Metallurgy Division) and the Refractory Metals Committee (Institute of Metals Division) of the Metallurgical Society of the American Institute of Mining, Metallurgical, and Petroleum Engineers, Gordon and Breach Science Publishers, New York, 1966, p 199

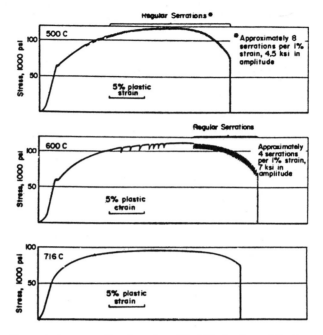

Stress-strain curves for chromium-35 at.% rhenium at the temperatures indicated.

Source: High Temperature Refractory Metals, R. W. Fountain, Joseph Malt and L. S. Richardson, Eds., based on a symposium, 16–20 February 1964, sponsored by the High Temperature Metals Committee (Extractive Metallurgy Division) and the Refractory Metals Committee (Institute of Metals Division) of the Metallurgical Society of the American Institute of Mining, Metallurgical, and Petroleum Engineers, Gordon and Breach Science Publishers, New York, 1966, p 205

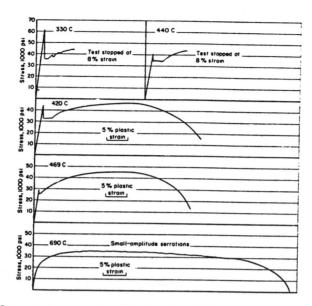

Stress-strain curves for chromium-1 at.% rhenium at the temperatures indicated.

Source: High Temperature Refractory Metals, R. W. Fountain, Joseph Malt and L. S. Richardson, Eds., based on a symposium, 16–20 February 1964, sponsored by the High Temperature Metals Committee (Extractive Metallurgy Division) and the Refractory Metals Committee (Institute of Metals Division) of the Metallurgical Society of the American Institute of Mining, Metallurgical, and Petroleum Engineers, Gordon and Breach Science Publishers, New York, 1966, p 163

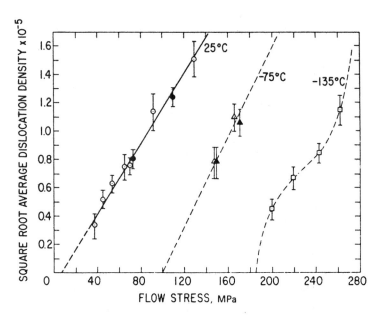

Relationship between square root of average dislocation density and flow stress of iron. ○△□, G.S. ~15 μm; ●▲, G.S. ~100 μm. (A. S. Keh and S. Weissmann.)

Source: William C. Leslie, The Physical Metallurgy of Steels, McGraw-Hill Book Co., New York, and Hemisphere Publishing Corp., Washington, D.C., 1981, p 6

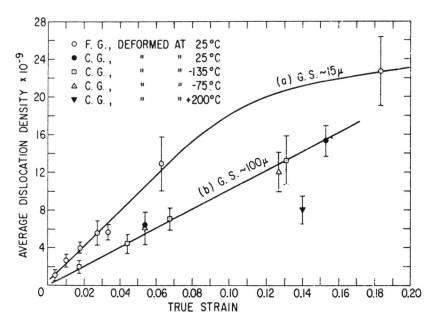

Relationship between average dislocation density and plastic strain of iron. CG = coarse grain, FG = fine grain. (A. S. Keh and S. Weissmann.)

Source: William C. Leslie, The Physical Metallurgy of Steels, McGraw-Hill Book Co., New York, and Hemisphere Publishing Corp., Washington, D.C., 1981, p 6

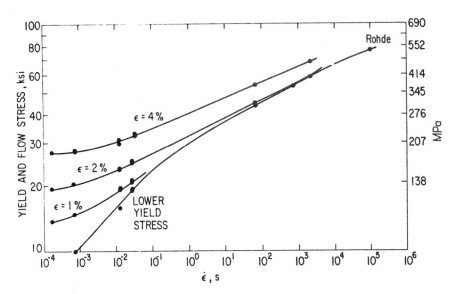

The effect of strain rate on the strength of polycrystalline bcc iron at room temperature. (W. C. Leslie, R. J. Sober, S. G. Babcock, and S. J. Green.)

If the yield stress of a metal is highly sensitive to temperature, it will also be highly sensitive to strain rate. The figure above shows the effect of strain rate, over nine orders of magnitude, on the yield and flow stress of iron. The yield strength of iron at room temperature, at a strain rate of 10^5 s^{-1}, is nearly as great as the yield strength at -196 °C at conventional slow testing-machine strain rates. In both instances, the mechanism of plastic deformation changed from slip to twinning, and thereafter the yield strength remained constant.

Source: William C. Leslie, The Physical Metallurgy of Steels, McGraw-Hill Book Co., New York, and Hemisphere Publishing Corp., Washington, D.C., 1981, p 29

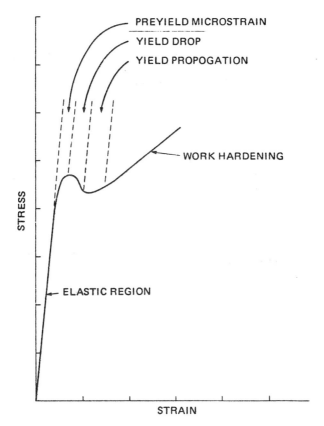

Typical flow curve of polycrystalline bcc iron in uniaxial tension.

For high-purity, polycrystalline, annealed bcc iron containing essentially no interstitial solutes, the flow curve in tension appears as in the figure above at conventional testing-machine strain rates and at temperatures slightly below ambient. As load is applied in a uniaxial test the iron first yields elastically, but the extent of this elastic deformation is severely limited. It is followed by preyield microstrain, during which dislocations move in a few grains, presumably at sites of stress concentration.

Source: William C. Leslie, The Physical Metallurgy of Steels, McGraw-Hill Book Co., New York, and Hemisphere Publishing Corp., Washington, D.C., 1981, p 3

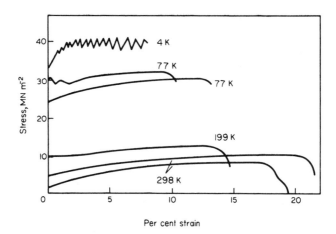

Stress-strain curves of zone-purified iron at various temperatures.
(After Smith and Rutherford.)

Source: R. W. K. Honeycombe, The Plastic Deformation of Metals, Second Edition, American Society for Metals, Metals Park OH, 1984, p 239

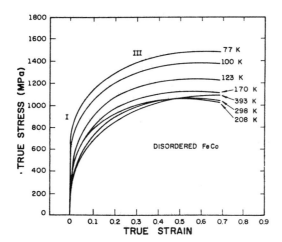

Stress-strain curves of fully disordered FeCo alloys at different temperatures.

The stress-strain curves of the FeCo alloy in the disordered state are shown in the figure above. The sharp yield point and stage II are absent in the disordered alloy. The disordered alloy goes straight into stage III after gradual yielding. Fully ordered alloys deform by means of the movement of superlattice dislocations at rather low stresses. However, the superdislocations (i.e., closely spaced pairs of unit dislocations bound together by an antiphase boundary) must move as a group in order to maintain the ordered arrangement of atoms. This makes cross-slip difficult. Long-range order thus leads to high strain-hardening rates and frequently to brittle fracture.

Source: Marc André Meyers and Krishan Kumar Chawla, Mechanical Metallurgy: Principles and Applications, Prentice-Hall, Inc., Englewood Cliffs NJ, 1984, p 541

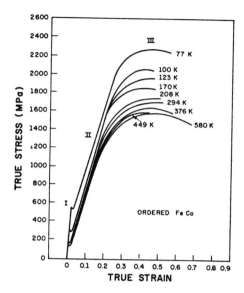

Stress-strain curves of ordered FeCo alloys at different temperatures.

The presence of atomic order leads to a marked change in the flow curve of the alloy. The figure above shows the flow curves of a fully ordered FeCo alloy at low temperatures where the order is not affected. Stage I in the figure is associated with a well-defined yield point. This is followed by a high linear work-hardening stage, II. Finally there occurs stage III, with nearly zero work hardening.

Source: Marc André Meyers and Krishan Kumar Chawla, Mechanical Metallurgy: Principles and Applications, Prentice-Hall, Inc., Englewood Cliffs NJ, 1984, p 541

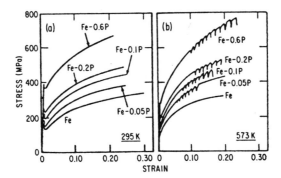

True tensile stress-strain curves for iron-phosphorus alloys tested (a) below and (b) within the jerky-flow regime. The approximate phosphorus content is given in wt %. (After Spitzig.)

The figure above shows how the influence of phosphorus on strain hardening in iron is pronounced (and similar) both in the region of jerky flow and at lower temperatures. This leads to the conclusion either that phenomena akin to dynamic strain-aging occur far beyond the temperature range where jerky flow is observed and exert their influence on dynamic recovery over this wider range, or that an influence of solutes on dynamic recovery is more general than by way of dynamic strain aging (or, most probably, both).

Source: Deformation, Processing, and Structure, George Krauss, Ed., papers presented at the ASM Materials Science Seminar, 23 October 1982, St. Louis MO, sponsored by the Seminar Committee of the Materials Science Division of the American Society for Metals, Metals Park OH, 1984, p 99

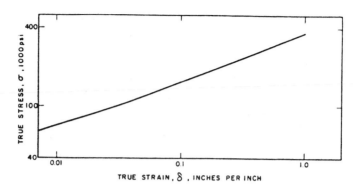

Average true stress-strain diagram for rhenium.

Room-temperature tensile properties of fabricated rhenium forms

Form	Sheet (0.01 in.)		Wire (0.05 in.)		Rod (0.125 in.)
Condition	Annealed	Reduced 30.7%	Annealed	Reduced 15%	Annealed
0.2% yield strength (p.s.i.)	135,000	311,000	—	—	46,000
Ultimate tensile strength (p.s.i.)	168,000	322,000	170,000	337,000	164,000
Elongation (%)	28	2	10	—	24
Reduction of area (%)	30	<1	16	2	22

Source: Rhenium, B. W. Gonser, Ed., papers presented at the symposium on rhenium, 3–4 May 1960, Chicago IL, sponsored by the Electrothermics and Metallurgy Division of the Electrochemical Society, Elsevier Publishing Co., 1962, p 34

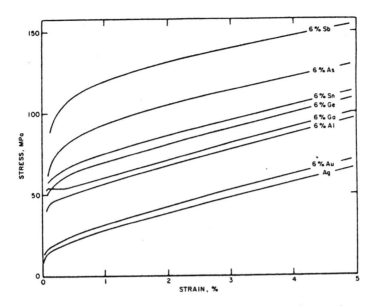

True tensile stress-strain curves for silver alloys (d ≃ 50 μm), tested at 77 K. (After Hutchison and Honeycombe.)

Source: Deformation, Processing, and Structure, George Krauss, Ed., papers presented at the ASM Materials Science Seminar, 23 October 1982, St. Louis MO, sponsored by the Seminar Committee of the Materials Science Division of the American Society for Metals, Metals Park OH, 1984, p 94

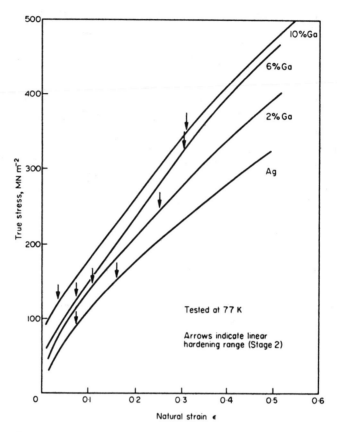

Stress-strain curves of silver and silver-gallium solid solutions at 77 K (constant grain size). (After Hutchison.)

Source: R. W. K. Honeycombe, The Plastic Deformation of Metals, Second Edition, American Society for Metals, Metals Park OH, 1984, p 235

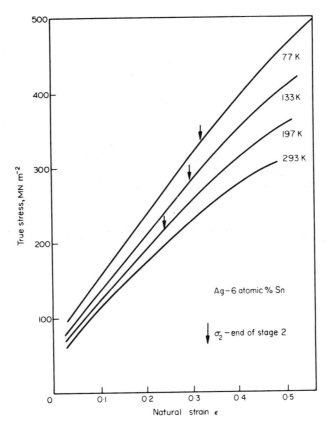

Stress-strain curves of a silver-6 at.% tin solid solution at different temperatures. (After Hutchison.)

Source: R. W. K. Honeycombe, The Plastic Deformation of Metals, Second Edition, American Society for Metals, Metals Park OH, 1984, p 233

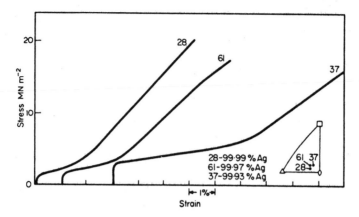

Influence of soluble impurities on stress-strain curves of silver crystals. (After Rosi.)

Source: R. W. K. Honeycombe, The Plastic Deformation of Metals, Second Edition, American Society for Metals, Metals Park OH, 1984, p 83

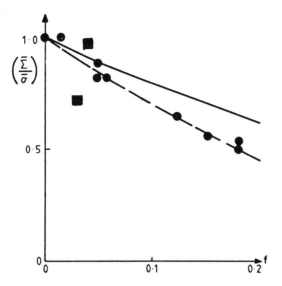

The effect of porosity on the 0.2% proof stress of a range of powder compacted steels. ●—●, Gurson yield surface; ●————●, Tvergaard yield surface; ■, extreme behavior of a large number of tests. (After Eiselstein.)

The obvious feature is the reduction in flow stress due to the porosity, and the curves above show 0.2% proof stress of a range of Eiselstein's alloys as a function of the volume fraction of voids. The square symbols represent the extreme behavior of a large number of tests in this region. The 0.2% proof stress does, however, decrease faster with porosity than would be expected simply by a loss in cross-sectional area in accord with the predictions. In fact, comparison with Gurson's yield criteria and Tvergaard's modification shows better agreement for Tvergaard's proposal.

Source: Yield, Flow and Fracture of Polycrystals, T. N. Baker, Ed., Applied Science Publishers Ltd., Essex, England, 1983, p 73

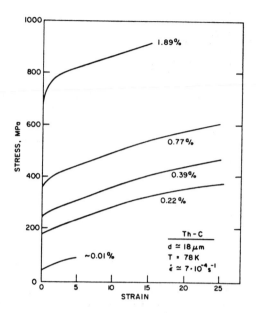

True tensile stress-strain curves for thorium-carbon alloys. (After Peterson and Skaggs.)

Source: Deformation, Processing, and Structure, George Krauss, Ed., papers presented at the ASM Materials Science Seminar, 23 October 1982, St. Louis MO, sponsored by the Seminar Committee of the Materials Science Division of the American Society for Metals, Metals Park OH, 1984, p 95

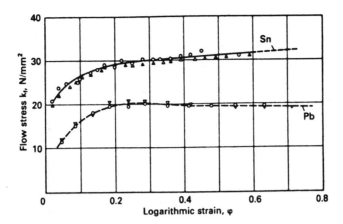

Stress-strain curves of lead and tin at room temperature.

Source: Kurt Laue and Helmut Stenger, Extrusion Processes: Machinery Tooling, trans. A. F. Castle and Gernot Lang, American Society for Metals, Metals Park OH, 1976, p 117

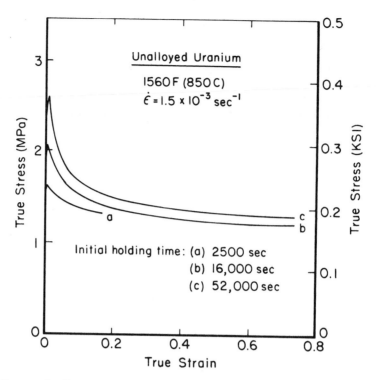

Compression flow curves for unalloyed depleted uranium tested in the gamma phase field that exhibit high levels of flow softening. Test specimens were held at 850 °C (1560 °F) in an atmosphere containing 10 ppm oxygen for the times indicated prior to testing. The external oxide layer formed on holding led to high initial flow stresses but broke up rapidly with strain beyond a few percent deformation.

Perhaps the first reported instance of flow instability under compressive loading is that of Jonas and Luton. While establishing the general sources of flow softening at elevated temperatures, they found unusual, localized bulges in specimens of oxidized uranium isothermally upset at 850 °C (1562 °F) and $\dot{\epsilon} \approx -1.5 \times 10^{-3}$ s^{-1}. For this material, the stress-strain curves (see above) showed large amounts of flow softening whose magnitude depended on the holding time prior to testing, during which oxide layers of increasing thickness developed on the specimen surface.

Source: S. L. Semiatin and J. J. Jonas, Formability and Workability of Metals, American Society for Metals, Metals Park OH, 1984, p 56

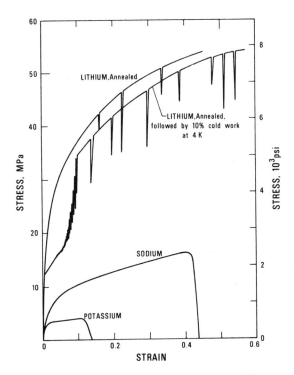

Stress-strain curves at 4 K of annealed polycrystalline sodium, potassium, lithium, and cold-worked lithium (After Reed; Hull and Rosenberg.)

Source: Materials at Low Temperatures, Richard P. Reed and Alan F. Clark, Eds., American Society for Metals, Metals Park OH, 1983, p 315

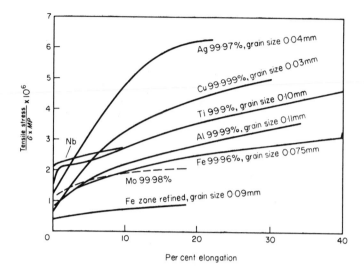

Stress-strain curves of polycrystalline metals compensated for differences in melting point and elastic modulus. (After McLean.)

Source: R. W. K. Honeycombe, The Plastic Deformation of Metals, Second Edition, American Society for Metals, Metals Park OH, 1984, p 239

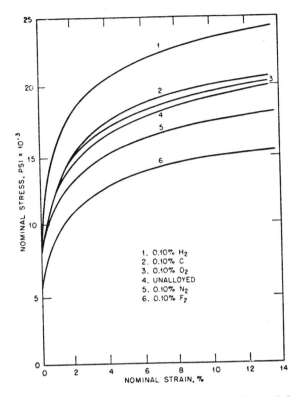

Portion of stress-strain relationships of alloys of annealed yttrium with H₂, C, O₂, N₂, and F₂.

Source: The Rare Earths, F. H. Spedding and A. H. Daane, Eds., John Wiley & Sons, Inc., London, 1961, p 447

23-30. Tensile Behavior of Polymer

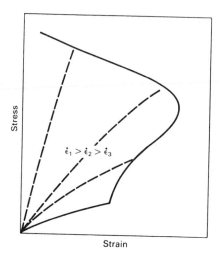

Schematic of a hypothetical failure envelope illustrating tensile behavior of a typical polymer. Solid line represents the failure envelope. The dashed lines inside the failure envelope represent stress-strain curves at different strain rates.

The effect of strain rate on the ductility of polymers is much more complex than it is for metals. The ductility is at first strongly dependent on strain rate and increases with increased strain rate. At some point, the ductility peaks and then decreases with further increases in extension rate. The tensile behavior of a representative polymer is shown in the figure above. The strain at failure reaches a maximum at some intermediate strain rate and then decreases with further increases in strain rate.

Source: Metals Handbook, Ninth Edition, Volume 8, Mechanical Testing, American Society for Metals, Metals Park OH, 1985, p 43

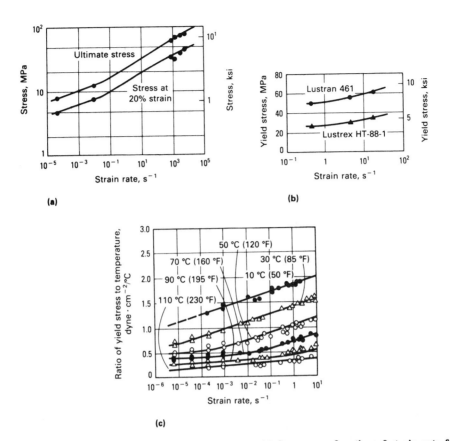

The effect of strain rate on the strength of polymers. (a) Stress as a function of strain rate for LX-04-M, a polymeric explosive material. (b) Stress as a function of strain rate for two commercial polymers. (c) Temperature-normalized stress as a function of strain rate for polypropylene at various temperatures.

The effect of strain rate on the strength of several polymeric materials is shown above. The explosive material LX-04-M in (a) has a very high strain-rate sensitivity; the stress axis is logarithmic in the figure. The effect in polypropylene (c) is shown at six different temperatures. The effect diminishes as temperature increases.

Source: Metals Handbook, Ninth Edition, Volume 8, Mechanical Testing, American Society for Metals, Metals Park OH, 1985, p 42

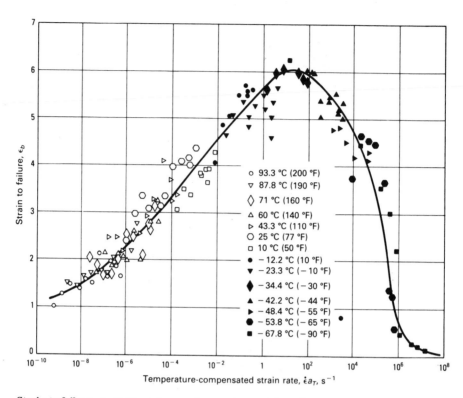

Strain to failure, ϵ_b, versus a temperature-compensated strain rate, $\dot{\epsilon}a_r$, for a cross-linked rubber. This figure depicts the peak in the ductility dependence on strain rate. The temperature compensation is possible because polymers exhibit strong thermally activated deformation dependence.

Source: Metals Handbook, Ninth Edition, Volume 8, Mechanical Testing, American Society for Metals, Metals Park OH, 1985, p 43